Lecture Notes in Computer Science 14753

Founding Editors

Gerhard Goos
Juris Hartmanis

Editorial Board Members

The series Lecture Notes in Computer Science (LNCS), including its subseries Lecture Notes in Artificial Intelligence (LNAI) and Lecture Notes in Bioinformatics (LNBI), has established itself as a medium for the publication of new developments in computer science and information technology research, teaching, and education.

LNCS enjoys close cooperation with the computer science R & D community, the series counts many renowned academics among its volume editors and paper authors, and collaborates with prestigious societies. Its mission is to serve this international community by providing an invaluable service, mainly focused on the publication of conference and workshop proceedings and postproceedings. LNCS commenced publication in 1973.

Marc Sevaux · Alexandru-Liviu Olteanu ·
Eduardo G. Pardo · Angelo Sifaleras ·
Salma Makboul

Editors

Metaheuristics

15th International Conference, MIC 2024
Lorient, France, June 4–7, 2024
Proceedings, Part I

 Springer

Editors
Marc Sevaux
Lab-STICC, UMR 6285, CNRS
Université Bretagne Sud
Lorient, France

Alexandru-Liviu Olteanu
Lab-STICC, UMR 6285, CNRS
Université Bretagne Sud
Lorient, France

Eduardo G. Pardo ⓘ
Universidad Rey Juan Carlos
Móstoles, Spain

Angelo Sifaleras ⓘ
University of Macedonia
Thessaloniki, Greece

Salma Makboul ⓘ
Université de Technologie de Troyes
Troyes Cedex, France

ISSN 0302-9743   ISSN 1611-3349 (electronic)
Lecture Notes in Computer Science
ISBN 978-3-031-62911-2   ISBN 978-3-031-62912-9 (eBook)
https://doi.org/10.1007/978-3-031-62912-9

Preface

Solving computationally hard problems is the everyday challenge of many researchers and probably all members of the metaheuristics community. Metaheuristics, as powerful solving tools, are largely used to tackle real-life hard optimization problems and are, without a doubt, held in high regard for providing good solutions within a short period of time. In their most basic implementation, metaheuristics can be viewed as simple tools to provide more quickly better solutions than simple ad-hoc heuristics. However, improvements, hybridizations, combinations with exact methods (matheuristics) and, more recently, combinations with artificial intelligence have shown that "metaheuristics" is in itself a full research discipline.

The first edition of the Metaheuristics International Conference (MIC) was held in 1995 in Breckenridge, Colorado, USA. Since then, every two years, MIC visited Sophia-Antipolis, France in 1997, Angra dos Reis, Brazil in 1999, Porto, Portugal in 2001, Kyoto, Japan in 2003, Vienna, Austria in 2005, Montréal, Canada in 2007, Hamburg, Germany in 2009, Udine, Italy in 2011, Singapore in 2013, Agadir, Tunisia in 2015, Barcelona, Spain in 2017, Cartagena, Colombia in 2019 and finally, after skipping one year due to COVID, Syracuse, Italy in 2022.

The 15th edition took place in Lorient, France June 4–7, 2024. The port-city is located amid green valleys at the mouth of the Blavet and Scorff rivers, in the Morbihan department of France. At the beginning of the 17th century, merchants who were trading with India had established warehouses in Port-Louis. They later built additional warehouses across the bay in 1628, at the location which became known as "L'Orient" (the Orient in French). Later, the French East India Company, founded in 1664 and chartered by King Louis XIV, established shipyards there, thus giving an impetus to the development of the city. In 1746 during the War of the Austrian Succession, Britain launched a raid on Lorient to destroy French shipping. In attempts to destroy German submarine pens (U-boat bases) and their supply lines, most of this city was destroyed by Allied bombing during World War II. Thus, today's Lorient reflects an architectural style of the 1950s and many architects and city planners visit Lorient for its visionary architecture and city layout. Its architectural heritage includes beautiful 18th-century mansions, the Quai des Indes dock, houses from the 1930s, the port enclosure and the Gabriel mansion, reminders of the French East India Company.

MIC 2024 was also the first time that this event was merged with the 10th edition of the International Conference on VNS (ICVNS) and the annual meeting of the EURO Working Group on Metaheuristics (EU/ME) in a unique conference. The organization of the event was supported by the Université Bretagne Sud and Lab-STICC laboratory. As for every edition, MIC focuses on the progress of the area of Metaheuristics and their applications and provides an opportunity to the international research community to discuss recent research results, to develop new ideas and collaborations, and to meet old friends and make new ones in a relaxed atmosphere. In 2024, four plenary speakers in the name of Éric Taillard, University of Applied Sciences and Arts of Western Switzerland, Belén Melián-Batista (ICVNS Plenary Speaker), University of La Laguna, Spain, Daniele Vigo, University of Bologna, Italy, and Rafael Martí, University of Valencia, Spain (EURO Plenary Speaker) contributed to the success of the conference. This year was also the first time that a series of 4 tutorials on the implementation of metaheuristics with the Julia language was launched. As demonstrated during the conference, the Julia language is particularly well adapted to rapid prototyping and also to large scale competitive programming.

A total of 100 valid submissions were received. The 70 members of the program committee made a selection of 36 regular papers, 34 short papers and 30 oral presentations. Short and regular papers are all included in this LNCS volume (in two parts). The organizing team is grateful to the PC members for their support, time and efforts to make this volume a reality. The number of participants to the MIC was 129 and they enjoyed a program over 4 days with 4 plenary talks, 4 tutorials, 25 parallel sessions, an ICVNS stream, excellent lunches with local specialties, a welcome reception on the sea shore, a boat tour and visit of the "Cité de la Voile Éric Tabarly" which invited participants on a tour of sailing and offshore racing discovery, and an exquisite conference dinner.

This Preface cannot end without extending our heartfelt gratitude to the organizing committee and local members, together with the sponsors of the Metaheuristics International Conference (with a special mention to our Platinum Sponsors, Hexaly and Entanglement Inc). Their dedication made this event an exceptional platform for knowledge exchange and collaboration in the field of metaheuristics. We are truly appreciative of the meticulous planning and seamless execution that went into every aspect of the conference. Special thanks to the sponsors whose generous support ensured the success and impact of this gathering. Their contributions enabled researchers and practitioners from around the globe to come together and explore cutting-edge developments in all aspects of metaheuristic methodologies. The diverse range of sessions, workshops, and discussions offered valuable insights and fostered meaningful connections among attendees. It's through their commitment to advancing the field that such gatherings can continue to push the boundaries of knowledge and innovation. The exchange of ideas and experiences facilitated by this conference will undoubtedly inspire future breakthroughs and collaborations. Their support not only enriches the academic community but also contributes to the advancement of science and technology worldwide. Once again, thank you for your unwavering support and dedication to promoting excellence in metaheuristic research.

April 2024

Marc Sevaux
Alexandru-Liviu Olteanu
Eduardo G. Pardo
Angelo Sifaleras
Salma Makboul

Organization

General Chairs

Alexandru-Liviu Olteanu	Université Bretagne Sud, France
Marc Sevaux	Université Bretagne Sud, France

Organizing Committee

Romain Billot	IMT Atlantique, France
Salma Makboul	Université de Technologie de Troyes, France
Patrick Meyer	IMT Atlantique, France
Alexandru-Liviu Olteanu	Université Bretagne Sud, France
Eduardo G. Pardo	Universidad Rey Juan Carlos, Spain
Quentin Perrachon	Université Bretagne Sud, France
Marc Sevaux	Université Bretagne Sud, France
Angelo Sifaleras	University of Macedonia, Greece
Owein Thuillier	Université Bretagne Sud, France
Essognim Richard Wilouwou	Université Bretagne Sud, France

Program Committee Chairs

Marc Sevaux	Université Bretagne Sud, France
Alexandru-Liviu Olteanu	Université Bretagne Sud, France
Eduardo G. Pardo	Universidad Rey Juan Carlos, Spain
Angelo Sifaleras	University of Macedonia, Greece
Salma Makboul	Université de Technologie de Troyes, France

MIC Steering Committee

Fred Glover	Entanglement, Inc., USA
Belén Melián-Batista	University of La Laguna, Spain
Celso Ribeiro	Universidade Federal Fluminense, Brazil
Éric Taillard	University of Applied Sciences of Western Switzerland, Switzerland
Stefan Voss	University of Hamburg, Germany

Program Committee

Mohammadmohsen Aghelinejad	Université de Technologie de Troyes, France
David Alvarez Martinez	Los Andes University, Colombia
Claudia Archetti	ESSEC Business School, France
Ghita Bencheikh	LINEACT Cesi Engineering School, France
Romain Billot	IMT Atlantique, France
Christian Blum	Spanish National Research Council, Spain
Eric Bourreau	Université de Montpelllier, France
Marco Caserta	University of Hamburg, Germany
Sara Ceschia	University of Udine, Italy
Marco Chiarandini	University of Southern Denmark, Denmark
Jean-Francois Cordeau	HEC Montréal, Canada
Samuel Deleplanque	JUNIA, France
Xavier Delorme	ENSM-SE, France
Bernabe Dorronsoro	University of Cadiz, Spain
Javier Faulin	Universidad Pública de Navarra, Spain
Andreas Fink	Helmut-Schmidt-University Hamburg, Germany
Frédéric Gardi	LocalSolver, France
Michel Gendreau	École Polytechnique de Montréal, Canada
Fred Glover	Entanglement, USA
Bruce Golden	University of Maryland, USA
Peter Greistorfer	Karl-Franzens-Universität Graz, Austria
Christelle guéret	Université d'Angers, France
Said Hanafi	University of Valenciennes, France
Jin-Kao Hao	Université d'Angers, France
Richard F. Hartl	University of Vienna, Austria
Colin Johnson	University of Nottingham, UK
Laetitia Jourdan	Université de Lille, France
Philippe Lacomme	Université Clermont Auvergne, France
Fabien Lehuédé	IMT Atlantique, France
Rodrigo Linfati	Universidad del Bío-Bío, Chile
Manuel López-Ibáñez	University of Manchester, UK
Salma Makboul	Université de Technologie de Troyes, France
Vittorio Maniezzo	University of Bologna, Italy
Rafael Marti	University of Valencia, Spain
Antonio Mauttone	Universidad de la República, Uruguay
Patrick Meyer	IMT Atlantique, France
Jairo R. Montoya-Torres	Universidad de La Sabana, Colombia
Alexandru-Liviu Olteanu	Université Bretagne Sud, France
Dimitri Papadimitriou	University of Antwerp, Belgium
Eduardo G. Pardo	Universidad Rey Juan Carlos, Spain

Sophie N. Parragh	Johannes Kepler University Linz, Austria
Quentin Perrachon	Université Bretagne Sud, France
Erwin Pesch	University of Siegen, Germany
Luciana Pessoa	PUC-Rio, Brazil
Jean-Yves Potvin	University of Montreal, Canada
Caroline Prodhon	Université de Technologie de Troyes, France
Jakob Puchinger	EM Normandie Business School, France
Ellaia Rachid	EMI, Morocco
Günther Raidl	Vienna University of Technology, Austria
Celso Ribeiro	Universidade Federal Fluminense, Brazil
Roger Z. Rios	Universidad Autónoma de Nuevo León, Mexico
Andrea Schaerf	University of Udine, Italy
Marc Sevaux	Université de Bretagne Sud, France
Patrick Siarry	Université de Paris 12, France
Angelo Sifaleras	University of Macedonia, Greece
Christine Solnon	INSA Lyon, France
Kenneth Sörensen	University of Antwerp, Belgium
Thomas Stützle	Université Libre de Bruxelles, Belgium
Anand Subramanian	Universidade Federal da Paraíba, Brazil
Muhammad Sulaiman	Abdul Wali Khan University, Pakistan
Owein Thuillier	Université Bretagne Sud, France
Paolo Toth	University of Bologna, Italy
Michael Trick	Carnegie Mellon University, USA
Pascal Van Hentenryck	Georgia Tech, USA
Daniel Vert	Systematic Paris-Region, France
Stefan Voss	University of Hamburg, Germany
Essognim Richard Wilouwou	Université Bretagne Sud, France
Mutsunori Yagiura	Nagoya University, Japan
Xin-She Yang	Middlesex University, UK
Nicolas Zufferey	University of Geneva, Switzerland

Keynote Speakers

Éric Taillard	University of Applied Sciences and Arts of Western Switzerland
Belén Melián-Batista	University of La Laguna, Spain
Daniele Vigo	University of Bologna, Italy
Rafael Martí	University of Valencia, Spain

Tutorial Speakers

Xavier Gandibleux   Université de Nantes, France
Jesús-Adolfo Mejia-De Dios   Autonomous University of Coahuila, Mexico
Antonio J. Nebro   University of Málaga, Spain
Alexandru-Liviu Olteanu   Université Bretagne Sud, France
Marc Sevaux   Université Bretagne Sud, France

Additional Reviewers

Murat Afsar   Yannick Kergosien
Riad Aggoune   Yury Kochetov
Rachid Benmasour   Damien Lamy
Sergio Cavero Díaz   Mariana Londe
José Manuel Colmenar   Flavien Lucas
Sergio Consoli   Raúl Martín Santamaría
Tatjana Davidović   Sergio Pérez-Peló
Amélia Durbec   Florian Rascoussier
Samia Dziri   Marcos Robles
Abdelhak El Idrissi   Nicolás Rodríguez
Lina Fahed   Jesús Sánchez-Oro Calvo
Ke Feng   Raca Todosijević
Paolo Gianessi   Dragan Urosević
Sergio Gil-Borrás   Daniel Vert
Bachtiar Herdianto   Margarita Veshchezerova
Alberto Herrán González   Bogdan Vulpescu
Panagiotis Kalatzantonakis   Essognim Wilouwou
Panagiotis Karakostas   Javier Yuste

Contents – Part I

Optimization for Forecasting

Quantum Meta-Heuristic for Operations Research

International Conference on Variable Neighborhood Search (ICVNS)

Contents – Part II

Advances in Combinatorial
Optimization

Breakout Local Search for Heaviest Subgraph Problem

He Zheng and Jin-Kao Hao$^{(\boxtimes)}$

LERIA, Université d'Angers, 2 Boulevard Lavoisier, 49045 Angers, France
`jin-kao.hao@univ-angers.fr`

Abstract. This paper presents a breakout local search (BLS) heuristic algorithm for solving the heaviest k-subgraph problem - a combinatorial optimization graph problem with various practical applications. BLS explores the search space by alternating iteratively between local search phase and dedicated perturbation strategies. Focusing on the perturbation phase, the algorithm determines its jump magnitude and perturbation type according to the search history to obtain the most appropriate degree of diversification. Computational experiments are performed on a number of large random graphs. The experimental evaluations show that the results obtained by BLS are comparable to, and in most cases superior to, those of the current state-of-the-art approaches.

Keywords: Iterated local search · heaviest subgraph problem · adaptive perturbation

1 Introduction

Given an edge-weighted undirected graph $G(V, E)$, where V is a set of vertices with $|V| = n$ and E is a set of edges, the Heaviest k-Subgraph Problem (HSP) is to determine a subset U of k vertices (k is given) such that the total edge weight of the subgraph induced by U is maximized. The NP-hard Densest k-Subgraph Problem (DSP), also known as the k-Cluster Problem [4], is a special case of HSP when the edge weight equals one. HSP is a relevant model for many important applications in areas such as social networks, protein interaction graphs, and the world wide web, etc. However, solving the problem is computationally challenging since it generalizes the NP-hard DSP.

Several exact approaches have been proposed to solve the problem [5], but they can only deal with small and sparse graphs with a small range of k. To solve large instances, heuristics and metaheuristics have been used to find approximate solutions in a reasonable time. Macambira proposed a tabu-based heuristic for solving HSP [6], which is based on three construction strategies and a neighborhood search strategy. Brimberg et al. presented a basic variable neighborhood search (BVNS) and some variants of the heuristic for the problem, using the swap

We would like to thank the reviewers for their insightful comments. The first author is supported by a CSC scholarship (No. 202306290083).

Algorithm 1. The BLS algorithm for HSP

 1: **Input:** Edge-weighted graph $G(V, E)$, integer k
 2: **Output:** The best solution S^* found
 3: $S \leftarrow Initializing(G)$
 4: $S^* \leftarrow S$ and $f(S^*) \leftarrow f(S)$
 5: $L \leftarrow L_{min}$ /* Set initial jump magnitude */
 6: $\omega \leftarrow 0$ /* Set the number for consecutive non-improving local optima */
 7: $prev \leftarrow f(S)$ /* Set the best objective value of the last descent phase */
 8: **while** stopping condition is not met **do**
 9:     $S \leftarrow LocalSearch(S, H)$ /* H is a vector with historical search information */
10:     **if** $f(S) > f(S^*)$ **then**
11:         $S^* \leftarrow S$ and $f(S^*) = f(S)$
12:         $\omega \leftarrow 0$
13:     **else if** $f(S) \neq prev$ **then**
14:         $\omega \leftarrow \omega + 1$
15:     **end if**
16:     $L \leftarrow DetermineJumpMagnitude(L, S, \omega, prev)$
17:     $T \leftarrow DeterminePerturbationType(S, \omega)$
18:     $prev \leftarrow f(S)$
19:     $S \leftarrow Perturb(S, L, T, \omega, H)$
20: **end while**
21: **return** S^*

---

neighborhood [3]. Saarinen et al. introduced an opportunistic version of the VNS heuristic (OVNS) [7], which exploits the characteristics of the problem instance during the search process. These algorithms have improved the state of the art in solving HSP. However, their performance often depends on the instances studied. The VNS algorithms lack stability in their results. BVNS mainly performs well on sparse graphs, while OVNS is more suitable for dense graphs. In this work, we present an effective algorithm for HSP based on breakout local search (BLS).

2   Breakout Local Search for HSP

2.1   General Framework

Breakout local search [1,2] follows the basic scheme of the iterated local search (ILS) approach. In general, BLS repeats a descent-based local search phase to perform an intensive search in a given region, and an adaptive perturbation phase to discover new promising regions. Special attention is paid to the design of the perturbation, which aims to introduce an appropriate degree of diversification according to the search stage. This is achieved by dynamically and adaptively determining the number of perturbation moves (the jump magnitude) and the type of the perturbation moves based on the search information.

The BLS algorithm for HSP (Algorithm 1) starts from an initial solution S given by the *Initializing* procedure and then uses the best-improvement descent *LocalSearch* procedure to attain a local optimum. At this point, BLS tries to escape from the current optimum by setting the jump magnitude L to an appropriate value and choosing a suitable perturbation type T of a certain intensity, where L and T are determined by $DetermineJumpMagnitude$ and $DeterminePerturbationType$, respectively, based on the search history. The perturbed solution becomes the new starting point for the next search round of the algorithm. This process is repeated until the stopping condition (e.g., time limit, maximum number of iterations...) is met.

2.2   Initial Solution

For a given graph $G(V, E)$ and an integer k, a candidate solution S is represented by a vector of length n, $S = \{x_1, x_2, ..., x_n\}$, where $x_v = 1$ ($1 \leq v \leq n$) if vertex v is among the k selected vertices in the current solution; $x_v = 0$ otherwise.

For each vertex v, the total weight of the edges from v to all the selected vertices is recorded in $\alpha_v = \sum_{u \in \{i \in V | x_i = 1\}} w_{vu}$, where w_{vu} denotes the weight of edge joining vertices v and u. The vector α is created when constructing an initial solution and is updated each time a move is performed. We get an initial solution of reasonable quality using the drop operator. We start by setting $x_v = 1$ for $v = 1, 2, ..., n$ and compute the α_v value for each vertex v. Then we iteratively drop $(n - k)$ vertices to obtain a solution with k selected vertices, each drop involving the vertex with the smallest α value. We then update the α vector in constant time.

2.3   Local Search

To move from one solution S to another in the search space, BLS uses the popular move operator $Swap(v, u)$, which exchanges a selected vertex v in the solution ($x_v = 1$) against a non-selected vertex u ($x_u = 0$). Let $S \oplus Swap(v, u)$ denote the neighboring solution obtained by applying $Swap(v, u)$ to solution S, then the corresponding neighborhood can be defined as $N(S) = \{S \oplus Swap(v, u) : x_v = 1, x_u = 0, 1 \leq v, u \leq n\}$. We use $\delta_{vu} = \alpha_u - \alpha_v - w_{vu}$ to compute the move gain, i.e., the change in the objective function value if vertex v is replaced by u in the solution. Each step of the local search with the best improvement strategy selects, among all neighboring solutions in $N(S)$, the one with the best (largest) move gain δ_{vu}.

2.4   Adaptive Perturbation

Jump Magnitude. The basic idea of BLS adaptive perturbation is to increase the number of perturbation moves (jump magnitude L) to redirect the search to a new and sufficiently distant area when the search seems to have stalled, as shown in Algorithm 2. The number of perturbation moves is usually set to a small value L_{min} at the beginning of the algorithm or when a new local optimum is found. If L is not large enough to escape the basin of attraction of the current local optimum, L is increased. Otherwise, it is reduced to its initial value L_{min}. If the best solution is not improved for MI successive search rounds, the jump magnitude is set to a large number L_{max} to allow for strong perturbations.

Algorithm 2. DetermineJumpMagnitude($L, S, \omega, prev$)

1: **Input:** Current jump magnitude L, local optimum S, history information ω, $prev$
2: **Output:** The jump magnitude L for the next perturbation
3: **if** $\omega > MI$ **then**
4:      $L \leftarrow L_{max}$
5:      $\omega \leftarrow 0$
6: **else if** $f(S) = prev$ **then**
7:      $L \leftarrow L + 1$
8: **else**
9:      $L \leftarrow L_{min}$
10: **end if**
11: **return** L

Algorithm 3. DeterminePerturbationType(S, ω)

1: **Input:** Current local optimum S, constant in $(0,1)$ Q, counter of successive non-improving search rounds ω
2: **Output:** The perturbation type T
3: Determine probability P of directed perturbation considering ω
4: With probability P, $T \leftarrow DirectedPerturbation$
5: With probability $(1 - P) \cdot Q$, $T \leftarrow RecencyBasedPerturbation$
6: With probability $(1 - P) \cdot (1 - Q)$, $T \leftarrow RandomPerturbation$
7: **return** T

Three Types of Perturbation Moves. To introduce different perturbation intensities, the BLS algorithm adopts three types of perturbations.

– *Directed Perturbation.* The directed perturbation applies a selection rule similar to tabu search, which favors swap moves that cause the least decrease in the objective value, with the constraint that the moves are not forbidden at the current search stage. A forbidden move involves a vertex v such that v has been removed from the solution during the last γ iterations (tabu tenure) (γ takes a random value from a given range related to k).
– *Recency Perturbation.* It relies only on the historical information stored in a vector H that counts the number of times each vertex has been moved during the search. The recency perturbation focuses on the least recently moved vertices, regardless of the objective degradation of the perturbation moves performed.
– *Random Perturbation.* The random perturbation introduces the greatest degree of diversification. It selects the two vertices to be swapped uniformly at random regardless of the objective degradation of the perturbation move.

These three types of perturbations are selected with different probabilities depending on the stage of the search (as shown in Algorithm 3). The number ω of successive non-improving search rounds is used to determine the current search state, which is reset to zero each time the best solution is improved or when ω reaches the maximum bound. Precisely, when ω is small, the search can go back to the basin of attraction of the current local optimum solution. To avoid this, the directed perturbation is applied with a higher probability. If an increasing ω fails to help the algorithm to escape from the current search region, BLS applies the Recency-based perturbation or the Random perturbation to introduce a strong degree of diversification. According to [1], the probability P of applying the directed perturbation is determined by $P = e^{-\omega/MI}$. Given P, the probability of applying the recency-based perturbation and the random perturbation is $(1 - P) \cdot Q$ and $(1 - P) \cdot (1 - Q)$ with the constant Q in $(0, 1)$.

3 Experimental Results

3.1 Test Instances

We used two sets of 129 instances generated from 43 random graphs according to [3] with integer edge weights uniformly taken in the range [100...1000].

- SET I (81 instances). This set contains 81 instances generated from 27 graphs with $|V| = 1000$ vertices, including 16 sparse graphs with an average vertex degree of 10 to 40, incremented by 2, and 11 dense graphs with an average vertex degree of 200 to 400, incremented by 20. k is set to $300, 400, 500$, giving 81 instances (27 graphs $\times$ 3 k values).
- SET II (48 instances). This set has 16 random sparse graphs with $|V| = 3000$ vertices and an average vertex degree of 10 to 40. For each graph, k is set to $900, 1200, 1500$, giving a total of 16 $\times$ 3 = 48 instances.

3.2 Results

We ran our algorithm and the two best-performing algorithms BVNS [3] and OVNS [7] to solve each instance 5 times (3600 s per run). We also ran the Branch and Bound (BB) algorithm of the CPLEX solver once on each instance with a cut-off time of 3600 s. Table 1 summarizes the comparison results between BLS and the reference algorithms. #win, #ties, and #losses respectively denote the number of instances where our BLS algorithm achieves better, equal and worse values compared to the reference algorithms in terms of the best objective values. We also give the deviation dev of each algorithm's average objective value f^{avg} from the best objective value f^* found by all algorithms, defined as %$dev = (f^* - f^{avg})/f^*$. The results show that BLS always achieves equal or better results compared to BVNS and OVNS, with a clear dominance on sparse graphs. BLS outperforms BVNS on 102 out of the 129 instances, and its results are better than those of OVNS on 87 instances, resulting in the smallest average deviation of 0.02% over all instances against 12.44% for BB, 0.16% for BVNS and 0.11% for OVNS. To check whether the proposed algorithm is statistically better than the reference algorithms, we applied the Wilcoxon signed-rank test with a significance level of 0.05 to the best results of the compared algorithms. The small p-value ($\ll$ 0.05) confirms that the difference between the results of BLS and those of each reference algorithm is statistically significant.

Table 1. Comparison between BLS and the reference algorithms BB, BVNS and OVNS

Type	k	BLS vs BVNS [3]				BLS vs OVNS [7]				%dev			
		#win	#ties	#losses	p-value	#win	#ties	#losses	p-value	BB	BVNS	OVNS	BLS
SET I - sparse	300	14	2	0	–	11	5	0	–	6.27	0.24	0.12	0.01
	400	15	1	0	–	13	3	0	–	3.15	0.13	0.08	0.01
	500	15	1	0	–	13	3	0	–	1.65	0.08	0.05	0.00
SET I - dense	300	2	9	0	–	0	11	0	–	18.17	0.05	0.01	0.00
	400	8	3	0	–	3	8	0	–	13.56	0.02	0.01	0.01
	500	0	11	0	–	1	10	0	–	10.18	0.01	0.00	0.00
SET II - sparse	900	16	0	0	–	14	2	0	–	26.71	0.53	0.34	0.07
	1200	16	0	0	–	16	0	0	–	18.95	0.22	0.26	0.04
	1500	16	0	0	–	16	0	0	–	13.29	0.12	0.16	0.02
Total		102	27	0	1.85e−18	87	42	0	2.48e−15				
Average										12.44	0.16	0.11	0.02

4   Conclusion

A heuristic approach based on breakout local search is developed to solve the NP-hard heaviest k-subgraph problem. The proposed method is characterized by its informed perturbation mechanism, which adaptively chooses between directed, recency-based, and random perturbations to introduce an appropriate degree of diversification at different search stages. Computational results on different types of instances demonstrate the effectiveness of the proposed algorithm. However, in some cases it is still time consuming for the algorithm to obtain high quality solutions. Additional strategies in combination with learning techniques can be explored to improve the method.

References

1. Benlic, U., Hao, J.K.: Breakout local search for the quadratic assignment problem. Appl. Math. Comput. **219**(9), 4800–4815 (2013)
2. Benlic, U., Hao, J.K.: Breakout local search for the vertex separator problem. In: Proceedings of the Twenty-Third International Joint Conference on Artificial Intelligence, pp. 461–467. AAAI Press (2013)
3. Brimberg, J., Mladenović, N., Urošević, D., Ngai, E.: Variable neighborhood search for the heaviest k-subgraph. Comput. Oper. Res. **36**(11), 2885–2891 (2009)
4. Corneil, D.G., Perl, Y.: Clustering and domination in perfect graphs. Discret. Appl. Math. **9**(1), 27–39 (1984)
5. Letsios, M., Balalau, O.D., Danisch, M., Orsini, E., Sozio, M.: Finding heaviest k-subgraphs and events in social media. In: 2016 IEEE 16th International Conference on Data Mining Workshops, pp. 113–120. IEEE (2016)
6. Macambira, E.M.: An application of tabu search heuristic for the maximum edge-weighted subgraph problem. Ann. Oper. Res. **117**, 175–190 (2002)
7. Saarinen, V.P., Chen, T.H.Y., Kivelä, M.: OVNS: opportunistic variable neighborhood search for heaviest subgraph problem in social networks. arXiv preprint arXiv:2305.19729 (2023)

A Biased Random Key Genetic Algorithm for Solving the α-Neighbor p-Center Problem

Sergio Pérez-Peló$^{(\boxtimes)}$ (ID), Jesús Sánchez-Oro (ID), and Abraham Duarte (ID)

Universidad Rey Juan Carlos, 28933 Móstoles, Madrid, Spain
{sergio.perez.pelo,jesus.sanchezoro,abraham.duarte}@urjc.es
https://grafo.etsii.urjc.es/en/

Abstract. In this paper, a Biased Random Key Genetic Algorithm is proposed to solve the α-neighbor p-center problem. A decoder and a local search procedure are developed obtaining competitive solutions for the problem. The objective of the ANPC is to locate p facilities serving demand points and assign a number α of facilities to each demand point. The objective function is evaluated as the maximum distance to the farthest facility assigned to each client, and the goal is to minimize this maximum distance. The proposed algorithm is compared with the best method found in the literature. The performance of the algorithm is evaluated over a large set of instances showing the robustness of the proposal.

Keywords: BRKGA · Metaheuristics · Facility Location

1 Introduction

Facility Location Problems (FLP) [1] are a set of problems in which it is required to locate a set of facilities in order to serve a set of demand points. This family of problems has application in different real-life scenarios, such as humanitarian emergencies, business decisions, engineering... Depending on the objective functions defined and the constraints considered there are different variants of the problem: capacitated [5] and uncapacitated [2], etc.

In this work, we focus on solving the α-neighbor p-center problem (ANPC). In this variant, unlike the classical approach, a client is not assigned to a single facility but to α facilities. This approach focuses on providing a robust failover solution so that if the nearest facility assigned to a customer is not available to meet its demand, the customer can obtain the same service from another facility that is available. The availability of the facilities is assumed to be information known *a priori* to the client, so the objective is to minimize the maximum

This work has been partially supported by the "Ministerio de Ciencia e Innovación" under grant ref. PID2021-125709OA-C22 and PID2021-126605NB-I00.

distance a given customer must travel in the worst-case scenario, i.e. when only one facility is available.

In this work, the problem is tackled from a discrete point of view. It can be defined as follows: let K be a set of points in a plane with $|K| = n$; $d(a, b)$ the distance between points a and b, where $a, b \in K$; and p the number of facilities that must be opened, with $1 \leq p \leq |K|$. It is important to note that, in this variant, all points in the set K can be both facilities and demand points. The goal of the ANPC is to select a set $Z \subseteq K$ with exactly p points representing the open facilities and assign α facilities to each client, i.e. those points in K that have not been selected as facilities $(K \setminus Z)$, minimizing the maximum distance between each client and its assigned αth facility. In mathematical terms, the value of the objective function assigned to a client can be defined as:

$$\alpha\text{-distance}(a, Z) = min_{S \in Z}\{max_{b \in S} d(a, b)\} \tag{1}$$

where S represents any subset of Z of size α. Therefore, the ANPC objective function is defined as:

$$OF(Z, K) = max_{a \in K \setminus Z} \alpha\text{-distance}(a, Z) \tag{2}$$

The objective in ANPC is to minimize the defined objective function. More formally:

$$min_{\substack{Z \subset K \\ |Z|=p}} OF(Z, K) \tag{3}$$

Figure 1 represents two different feasible solutions Z for the ANPC. In this case, there are 6 points and the values $p = 3$ and $\alpha = 2$ are considered.

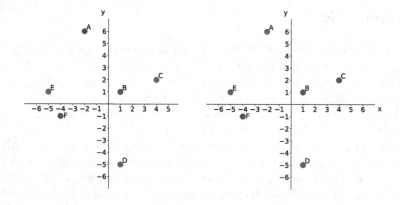

(a) Solution $Z_1 = \{C, D, E\}$ (b) Solution $Z_2 = \{A, D, E\}$

Fig. 1. Two example solutions for an instance with $|K| = 6$, $p = 3$ and $\alpha = 2$.

In these figures, the selected facilities are represented by a red circle and the clients by a green circle. In Fig. 1a the solution $Z_1 = \{C, D, E\}$ is represented, obtaining a value of the objective function $OF(Z_1, K) = 7.21$. This value is given by the distance between facility A and facility C, since node A has been assigned to facilities C and E and C is the second closest facility to node A. The same reasoning is followed to calculate the OF value of Z_2, depicted in Fig. 1b: the value obtained is $OF(Z_2, K) = 7.62$, and it is extracted from the distance between the client B and its second closest assigned facility: D. Therefore, Z_1 is better than Z_2 in terms of the objective function, since $OF(Z_1, K) < OF(Z_2, K)$. To the best of our knowledge the best method found in the literature to solve the ANPC is a metaheuristic that combines Greedy Randomized Adaptive Search Procedure with Strategic Oscillation [7].

2   Biased Random Key Genetic Algorithm

In this work, a Biased Random Key Genetic Algorithm (BRKGA) [3] is proposed to solve the ANPC. BRKGA is a metaheuristic algorithm that is based on the behavior of classical genetic algorithms: the chromosomes representing the solutions are encoded in some way; these chromosomes compose a population and are then combined in a process called crossover, generating now members in the population. In the case of BRKGA, chromosomes are encoded by vectors of real numbers in the interval $[0, 1]$. These numbers are commonly referred to as alleles, and represent the *random keys* present in name of the metaheuristic. In order to compose a solution to the problem being solved, the chromosomes must be translated into the context of the solution in a decoding phase that translates an allele into an element of the solution under construction.

The BRKGA starts from a population of size t. This initial population is composed by t chromosomes and a small group of t_e of them are selected as *elite* individuals, which are the best solutions of the population. The evolving population is produced using an elitist strategy. This means that the t_e elements selected as the elite set remain unchanged in the next generation. To avoid getting stuck at local optima, genetic algorithms introduce new elements in the population, known as mutants. In BRKGA, these mutants are introduced into the population in the same way that the initial members of the population were generated. A number of t_m mutants is introduced in the population. Finally, to reach the t elements that must compose the population, $t - t_e - t_m$ elements are generated by the crossover technique. This mechanism consist of selecting elements from one of the elite solutions and elements from one of the non-elite ones. To decide which element of the solution will be introduced in the new chromosome, a random choice is made. If the random value generated is greater than or equal to set value known as crossover or inheritance probability, then the allele from the elite solution is selected to be part of the chromosome under construction. Otherwise the allele from the non-elite chromosome is selected. These steps are repeated until the number of generations is reached, returning the best solution found during the execution of the algorithm.

In this work, each position in the vector that makes up a chromosome represents the node with the id equal to this position. This codification allows the simplification of the decoding phase, which is described in Sect. 2.1.

Finally, in order to intensify the search, the solutions belonging to the elite set at the end of all the generations established are subjected to a local improvement procedure, consisting of a local search, returning the best solution found after executing this procedure to all the solutions on which it was applied. This local search procedure is defined in Sect. 2.2

As can be seen, different parameters must be set here: initial population size, percentage of elements t_e, percentage of elements t_m and the crossover probability ρ_e. In Sect. 3 it is explained how all these parameters have been adjusted.

2.1   Decoder

To translate the information encoded in the chromosomes, it is necessary to define a method to convert the allele values into elements of the solution of the problem being solved. It is important to design a computationally efficient decoding phase, since it is one of the most repeated processes during the execution of the algorithm. With this goal in mind, the decoder proposed in this work is straightforward: the alleles of the chromosome being decoded are sorted in a descending order with respect to their associated random value and the first p elements of the vector are selected as open facilities in the solution Z being constructed. By following this decoding strategy, not only a fast decoding process is obtained, but also a highly diverse population, which will potentially lead to better results, since that the exploration of the search space is broader.

2.2   Local Improvement

Once a diverse population is obtained, a local improvement phase is executed with the objective of finding local optima in the explored neighborhoods. In this work, a local search procedure is proposed. Let us first define the movement that generates the explored neighborhood as:

$$Swap(Z, a, b) \leftarrow Z \cup \{b\} \setminus \{a\} \tag{4}$$

where $a \in Z$ and $b \notin Z$, it is, a represents an open facility while b represents a demand point.

This movement is traditionally known as *swap* movement. It consists of removing a facility from one point and placing it at another point that does not hold a facility in the current solution. Given this definition, the explored neighborhood for a given solution Z is that made up of all the solutions that can be reached by performing a *swap* movement. It is clear that the exploration of this neighborhood is quite time-consuming. In order to reduce the computational cost associated with it, in this work we reduce the number of facilities and clients that are swapped. To perform this reduction in an intelligent way, the facilities that are closed are only those that represent the αth closest facility to some client. For each facility, only those clients whose distance to the facility is less than that of its αth closest client are candidates to participate in the movement. These decisions make it possible to considerably reduce the explored region of the search space. Finally, this neighborhood is explored following a classical first improvement strategy.

3 Experiments and Results

The experiments performed in this work are devoted to test the actual state of the proposal in terms of solution quality and computing time. To perform the experimentation, a subset of 39 instances derived from the TSPLIB [6] has been used. More specifically, the set of instances used is composed of a subset of the instances used in [7]. The experiments have been executed on a machine with an AMD Ryzen 5 3600 (2.2 GHz) CPU with 16 GB RAM. All algorithms have been implemented using Java 17. The α-values considered are 1, 2 and 3 for the same subset of instances. The parameters of the algorithms have been set automatically using the irace package [4]. In this case, the population size has a fixed value of 100 elements, while the percentages t_e and t_m as well as the probability ρ_e are adjusted by irace. The values tested during the configuration are those in the ranges proposed by the authors of the metaheuristic in [3]. Table 1 summarizes the obtained results with the following information: α indicates the α-value; Algorithm includes the name of the algorithm under comparison; Avg. contains the average of the objective function value obtained by each algorithm for the evaluated instance set; Time (s) shows the average computing time, in seconds, required by each algorithm to find a solution; # Best reports the number of times each algorithm has found the best value obtained in the experiment; finally, Dev (%) reports the percentage deviation obtained by an algorithm when it is not able to obtain the best value. All the algorithms executed in the experiment have been time-limited to 1800 s, as it has been done in [7]. After running irace, the best algorithm settings provided are $t_e = 0.17$, $t_m = 0.11$ $\rho_e = 0.79$.

Table 1. Table results for the considered α-values.

α	Algorithm	Avg	Time (s)	# Best	Dev (%)
1	BRKGA	698.52	720.92	4	9.34%
	SO	604.93	301.29	13	0.00%
2	BRKGA	1043.71	615.60	5	7.07%
	SO	995.42	382.08	10	3.85%
3	BRKGA	1314.27	530.41	6	5.10%
	SO	1211.17	414.33	7	2.80%

As it can be derived from the table, the BRKGA algorithm shows a good performance, but still needs to be improved. In particular, the proposal obtains results close to the best method found in the literature, but it is still far from emerging as the state-of-the-art method.

It interesting to note that, the larger the α value, the better the performance of the algorithm both in terms of objective function value and computational time. This suggests that new decoders focused on meeting the requirements of lower alpha values could perform better, reducing the number of iterations required by the local search procedure to reach a local optimum.

4   Conclusions and Future Work

In this work, a novel approach based on BRKGA is proposed for solving the α-neighbor p-Center problem. This approach proposes a decoder and a local search method. The experimental phase shows that the algorithm provides promising quality and computational time performance. However, it still needs to be improved to reach the best method found in the literature results. As future work, new decoders and improvement methods can be proposed. In addition a parallel version of the algorithm could provide better results in less computing time, taking advantage of the nowadays processors multi-thread features and performance.

References

1. Celik Turkoglu, D., Erol Genevois, M.: A comparative survey of service facility location problems. Ann. Oper. Res. **292**, 399–468 (2020)
2. Gendron, B., Khuong, P.-V., Semet, F.: Models and methods for two-level uncapacitated facility location problems. In: Contributions to Combinatorial Optimization and Applications (2023)
3. Gonçalves, J.F., Resende, M.G.C.: Biased random-key genetic algorithms for combinatorial optimization. J. Heuristics **17**(5), 487–525 (2011)
4. López-Ibáñez, M., et al.: The irace package: iterated racing for automatic algorithm configuration. Oper. Res. Perspect. **3**, 43–58 (2016)
5. Maia, M.R.H., et al.: Metaheuristic techniques for the capacitated facility location problem with customer incompatibilities. Soft. Comput. **27**(8), 4685–4698 (2023)
6. Reinhelt, G.: TSPLIB: a library of sample instances for the TSP (and related problems) from various sources and of various types (2014). http://comopt.ifi.uniheidelberg.de/software/TSPLIB95
7. Sáchez-Oro, J., et al.: RASP with strategic oscillation for the α-neighbor p-center problem. Eur. J. Oper. Res. **303**(1), 143–58 (2022)

A Continuous-GRASP Random-Key Optimizer

Antonio A. Chaves[1] , Mauricio G. C. Resende[2(✉)] ,
and Ricardo M. A. Silva[3]

[1] Departamento de Ciência e Tecnologia, Universidade Federal de São Paulo,
São José dos Campos, SP, Brazil
chaves@unifesp.br
[2] Industrial and Systems Engineering, University of Washington, Seattle, WA, USA
mgcr@uw.edu, mgcr@berkeley.edu
[3] Centro de Informática, Universidade Federal de Pernambuco, Recife, PE, Brazil
rmas@cin.ufpe.br
http://mauricio.resende.info

Abstract. This paper introduces a problem-independent GRASP metaheuristic for combinatorial optimization implemented as a random-key
optimizer (RKO). CGRASP, or continuous GRASP, is an extension of the
GRASP metaheuristic for optimization of a general objective function in
the continuous unit hypercube. The novel approach extends CGRASP
using random keys for encoding solutions of the optimization problem in
the unit hypercube and a decoder for evaluating encoded solutions. This
random-key GRASP combines a universal optimizer component with a
specific decoder for each problem. As a demonstration, it was tested on
five NP-hard problems: Traveling salesman problem (TSP); Tree hub
location problem in graphs (THLP); Steiner triple set covering problem
(STCP); Node capacitated graph partitioning problem (NCGPP); and
Job sequencing and tool switching problem (SSP).

Keywords: Continuous-GRASP · Random-Key Optimizer ·
Combinatorial Optimization

1 Introduction

In this paper, a problem-independent GRASP metaheuristic is proposed using
the random-key optimizer (RKO) paradigm. GRASP [3,4] (greedy randomized
adaptive search procedure) is a metaheuristic for combinatorial optimization
which repeatedly applies a semi-greedy construction procedure followed by a
local search procedure. The best solution found over all iterations is returned as
the solution of the GRASP. Continuous GRASP (CGRASP) [6] is an extension
of GRASP for continuous optimization in the unit hypercube. A random-key
optimizer (RKO) [12] makes use of a vector of random keys to encode a solution to a combinatorial optimization problem. It uses a decoder to evaluate

M. Sevaux et al. (Eds.): MIC 2024, LNCS 14753, pp. 15–20, 2024.
https://doi.org/10.1007/978-3-031-62912-9_3

a solution encoded by the vector of random keys. A random-key GRASP is a CGRASP where points in the unit hypercube are evaluated by means of a decoder. We describe a random key GRASP made up of a problem-independent component and a problem-dependent decoder. As a proof of concept, three variants of the random-key GRASP are tested on five NP-hard combinatorial optimization problems: traveling salesman problem, tree of hubs location problem, Steiner triple covering problem, node capacitated graph partitioning problem, and job sequencing and tool switching problem.

2   Random-Key Optimizer

Random-key genetic algorithms (RKGA) were first introduced by Bean [2]. In a RKGA, solutions are encoded as vectors of random keys, i.e., randomly-generated real numbers in the interval $(0, 1]$. A population of p random-key vectors is evolved over a number of generations. The initial population consists of randomly generated n-vectors. At each iteration, each random-key vector is decoded and evaluated. The *decoder* is a deterministic algorithm, usually a heuristic, that takes as input a vector of random keys and returns a feasible solution of the problem being solved along the cost of the solution. Biased random-key genetic algorithms [5] are an extension of RKGA where a bias is applied in the selection of one of the parents as well as in the mating process.

Both RKGA and BRKGA are problem-independent algorithms in the sense that there is a clear separation between the solver and the problem being solved where the connection of the solver with the problem is done by way of a decoder. For each type of problem to be solved, a new decoder is implemented to serve as the link between the solver and the problem. These algorithms are example of *Random-Key Optimizers* (RKO) [12]. An RKO is an optimization heuristic algorithm that solves a discrete optimization problem indirectly in the continuous unit n-dimensional hypercube $\mathcal{H}_n$. For each solution $x \in \mathcal{H}_n$ a decoder $\mathcal{D}$ maps x to the solution $\mathcal{D}(x)$ in the solution space of the discrete optimization problem. With such a separation of solver and decoder, one needs only to implement the solver once and then it can be reused to solve a number of different problems by simply devising a decoder for that problem. Examples of APIs for BRKGA are Toso and Resende [14], Andrade et al. [1], and Oliveira et al. [10].

3   Random-Key GRASP

We now introduce Random-Key GRASP, or RK-GRASP, a problem-independent GRASP which solves discrete optimization problems through continuous optimization and a decoder. An advantage of using this GRASP is that the user only needs to implement a decoder since the algorithm-specific components of GRASP are implemented as an Application Programming Interface or API. When using a standard GRASP the user needs to tailor both the semi-greedy construction and the local search for the problem on hand.

3.1   GRASP

The metaheuristic *Greedy Randomized Adaptive Search Procedure* (GRASP) [11] was introduced by Feo and Resende [3,4]. A GRASP is a multi-start procedure in which at each iteration a semi-greedy solution is constructed and local search is applied to this solution. The best locally optimal solution visited over all iterations is returned as the GRASP solution.

3.2   Continuous GRASP

Continuous GRASP, or simply C-GRASP, is an extension of GRASP for solving continuous optimization problems subject to box or simple bounding constraints [6,7],

$$\min_{x \in \mathbb{R}^n} \{f(x) \mid L_n \leq x \leq U_n\},$$

where L_n and U_n are vectors of lower and upper bounds on x, i.e. $L_n(i) \leq x_i \leq U_n(i)$, for $i = 1, \ldots, n$ and $f(x)$ is the cost of solution x. Cost $f(x)$ can be evaluated in a multitude of ways, e.g. analytically, through simulation, via a mathematical program, or with a decoder. The objective is to find a global optimum. Like GRASP, C-GRASP is a multi-start procedure in which each iteration consists of a construction, or diversification, phase followed by a local search, or intensification, phase. C-GRASP evaluates points on a dynamic grid, with grid size initially set to $h = h_0$. Each construction phase starts at the current solution x (initially a random point $x \in \mathbb{R}^n \mid L_n \leq x \leq U_n$). To build an RCL C-GRASP performs a line search on $f(x)$ in each direction $e_i = (0, 0, \ldots, 0, 1, 0, \ldots, 0, 0)$, for $i = 1, 2, \ldots, n$, where the only nonzero is a 1 in position i. For $i = 1, 2, \ldots, n$, the line search is limited to evaluating $f(x + e_i \cdot h \cdot k)$ for all values of $k \in \{0, 1, -1, 2, -2, \ldots\}$ such that $0 \leq x_i + e_i \cdot h \cdot k \leq 1$. The result of line search i is z_i with cost $g_i = f(z_i)$. Let $g_m = \min\{g_i \mid i = 1, \ldots, n\}$ and $g_M = \max\{g_i \mid i = 1, \ldots, n\}$. The best line search solutions, i.e. those with $g_i \leq (1 - \alpha) \cdot g_m + \alpha \cdot g_M$ for some $\alpha \in [0, 1]$ are placed in a RCL and one is selected at random to be to be fixed in x with $x_i = z_i$. Direction e_i is flagged to no longer be explored in this construction iteration. This is repeated until a solution is constructed.

Once a semi-greedy solution x^* has been constructed, a local search or intensification phase, is applied around x^*. Several implementations of local search have been described. Suppose the current semi-greedy solution is $\bar{x}$. In the first paper, Hirsch et al. [6] suggest examining a given maximum numbers of points MaxDirToTry of the form $\bar{x} + h \cdot \{-1, 0, 1\}^n$. This local search examines only grid points. Hirsch, Pardalos and Resende [7] sample a user-defined number of feasible grid points and projects each one of them onto the surface of the hyper-sphere of radius h, centered at $\bar{x}$. Each point is evaluated as it is projected and a first-improvement policy is applied. If an improving point is found, then $\bar{x}$ is set to this improving solution. If no improving solution is found, then the grid-size is reduced by a specified factor and the local search continues. See [6,7,13] for details.

3.3   Random-Key-GRASP

We consider the Random-Key Optimization (RKO) Problem of the type

$$\min\{f(x) \mid x \in \mathcal{H}_n\},$$

where $\mathcal{H}_n$ is the unit hypercube in $\mathbb{R}^n$ and $f(x)$ is the cost value returned by a decoder when given as input $x \in \mathcal{H}_n$. Random-Key-GRASP, or RK-GRASP applies the CGRASP algorithm of Sect. 3.2 to solve the above-defined RKO Problem. The first paper to extend the concept of RKO to algorithmic frameworks other than RKGA and BRKGA was Schuetz et al. [12] where in addition to a BRKGA for robot motion planning the authors propose a RKO using dual annealing, an extension of generalized simulated annealing [15,16]. Recently, RKO has been implemented in simulated annealing, iterated local search, and variable neighborhood search for the tree hub location problem [9].

The main contribution of this paper is a problem-independent GRASP for combinatorial optimization. In the classical GRASP, a specially tailored semi-greedy algorithm as well as a custom-made local search were needed for its implementation. For any new problem both algorithmic components had to be developed and tested in the implementation phase of GRASP. CGRASP was limited to continuous optimization on the unite hypercube and therefore could not be applied directly to combinatorial optimization problems.

With RK-GRASP, on the other hand, new construction and local search algorithms are not needed. Rather, the practitioner only needs to develop a decoder for the problem. Decoders are often simple to create, as the vast literature [8] of Biased Random-Key Genetic Algorithms (BRKGAs) can attest to.

4   Experimental Results

To test the idea presented in this paper we propose three implementations of RK-GRASP, each using the same greedy randomized construction procedure but a different local improvement procedure. Pseudo-codes of the algorithms implemented in this paper as well as all instances and experimental results are given at https://github.com/antoniochaves19/RK-GRASP. The three local improvement procedures implemented are Grid search, Nelder-Mead search, and Random variable neighborhood descent.

These three variants of RK-GRASP are applied to five combinatorial optimization problems: Traveling salesman problem (TSP); Tree hub location problem in graphs (THLP); Steiner triple set covering problem (STCP); Node capacitated graph partitioning problem (NCGPP); and Job sequencing and tool switching problem (SSP).

The RK-GRASP was coded in C++ and compiled with GCC. The computer used in all experiments was a Dual Xenon Silver 4114 20c/40t 2.2 Ghz processor with 96 GB of DDR4 RAM and running CentOS 8.0 x64. We tested 20 instances of the TSP, 36 instances of the THLP, five instances of the STCP, 20 instances of the NCGPP, and 40 instances of the SSP. The RK-GRASP was run 5 times for each instance. A time limit as a stopping criterion was selected proportional to the instance size.

The three versions of the RK-GRASP generated effective solutions for the problems on hand. Notably, no local improvement technique was dominant regarding solution quality. The RK-GRASP variant with Grid Search emerged as the most effective, delivering superior solutions for the STCP, SSP, and large-scale instances of the TSP. Meanwhile, RK-GRASP with Nelder Mead search found good solutions for TSP, STCP, and NCGPP. However, it exhibited results with a poor average percentage deviation in the case of THLP. In contrast, the RK-GRASP version employing RVND demonstrated noteworthy results for THLP, NCGPP, and MTSP, outperforming state-of-the-art methods, particularly in the case of THLP. In summary, all versions of RK-GRASP demonstrated efficiency in both solution quality and computational time, showing their effectiveness across diverse problem instances.

References

1. Andrade, C.E., Toso, R.F., Gonçalves, J.F., Resende, M.G.: The multi-parent biased random-key genetic algorithm with implicit path-relinking and its real-world applications. Eur. J. Oper. Res. **289**(1), 17–30 (2021)
2. Bean, J.C.: Genetic algorithms and random keys for sequencing and optimization. ORSA J. Comput. **6**(2), 154–160 (1994)
3. Feo, T., Resende, M.: A probabilistic heuristic for a computationally difficult set covering problem. Oper. Res. Lett. **8**, 67–71 (1989)
4. Feo, T., Resende, M.: Greedy randomized adaptive search procedures. J. Global Optim. **6**(1), 109–133 (1995)
5. Gonçalves, J.F., Resende, M.G.C.: Biased random-key genetic algorithms for combinatorial optimization. J. Heurist. **17**(1), 487–525 (2011)
6. Hirsch, M., Meneses, C., Pardalos, P., Resende, M.: Global optimization by continuous GRASP. Optim. Lett. **1**, 201–212 (2007)
7. Hirsch, M., Pardalos, P., Resende, M.: Speeding up continuous GRASP. Eur. J. Oper. Res. **205**, 507–521 (2010)
8. Londe, M.A., Pessoa, L.S., Andrade, C.E., Resende, M.G.C.: Biased random-key genetic algorithms: a review. Technical report 2312.00961, arXiv (2023)
9. Mangussi, A.D., et al.: Meta-heurísticas via chaves aleatórias aplicadas ao problema de localização de hubs em árvore. In: ANAIS DO SIMPÓSIO BRASILEIRO DE PESQUISA OPERACIONAL. Galoá, São José dos Campos (2023). https://proceedings.science/sbpo-2023/trabalhos/meta-heuristicas-via-chaves-aleatorias-aplicadas-ao-problema-de-localizacao-de-h?lang=pt-br
10. Oliveira, B.B., Carravilla, M.A., Oliveira, J.F., Resende, M.G.C.: A C++ application programming interface for co-evolutionary biased random-key genetic algorithms for solution and scenario generation. Optim. Methods Softw. **37**(3), 1065–1086 (2022)
11. Resende, M.G.C., Ribeiro, C.C.: Optimization by GRASP: Greedy Randomized Adaptive Search Procedures. Springer, New York (2016). https://doi.org/10.1007/978-1-4939-6530-4
12. Schuetz, M., et al.: Optimization of robot trajectory planning with nature-inspired and hybrid quantum algorithms. Phys. Rev. Appl. **18**(5) (2022)

13. Silva, R.M.A., Resende, M.G.C., Pardalos, P.M., Hirsch, M.J.: A Python/C library for bound-constrained global optimization with continuous GRASP. Optim. Lett. **7**, 967–984 (2013)
14. Toso, R., Resende, M.: A C++ application programming interface for biased random-key genetic algorithms. Optim. Methods Softw. **30**(1), 81–93 (2015)
15. Tsallis, C., Stariolo, D.A.: Generalized simulated annealing. Phys. A **233**(1), 395–406 (1996)
16. Xiang, Y., Sun, D., Fan, W., Gong, X.: Generalized simulated annealing algorithm and its application to the Thomson model. Phys. Lett. A **233**(3), 216–220 (1997)

Adaptive Ant Colony Optimization Using Node Clustering with Simulated Annealing

Nozomi Kotake[✉] ⓘ, Rikuto Shibutani ⓘ, Kazuma Nakajima ⓘ,
Takafumi Matsuura ⓘ, and Takayuki Kimura ⓘ

Nippon Institute of Technology, 4–1 Gakuendai, Miyashiro,
Minami-Saitama, Saitama 345-8501, Japan
nozomi52317@gmail.com

Abstract. Multiple fields, including transport, and engineering, require finding the shortest route. This problem is known as the traveling salesman problems (TSP), which is an $\mathcal{NP}$-hard combinatorial optimization problems. As the number of cities increases, finding the shortest path in TSP, considering all combinations, becomes challenging. The ant colony optimization (ACO) has been proposed as a solution to TSP. However, the performance of the ACO heavily depends on its parameters. In addition, finding appropriate settings of parameters for each problem is time-consuming. To address this, an adaptive ant colony optimization with node clustering (AACO-NC) has been proposed. AACO-NC uses both node clustering and effective pheromone evaporation to find the shorter route. Despite its strength, our preliminary experiments suggest that the pheromone updates restrict the solution search range. Consequently, finding a good solution becomes increasingly difficult as the search progresses. Therefore, we implement the simulated annealing (SA) method to expand the solution search space and to escape from the local optimum solution. Numerical experiments demonstrate that the proposed method outperforms the conventional method on a variety of benchmark problems, yielding a smaller error rate between the optimal solution and the mean value.

Keywords: Combinatorial optimization · Ant colony optimization · Simulated annealing

1 Introduction

To address the traveling salesman problems (TSP) [1], an ant colony optimization (ACO) was introduced [2]. The performance of ACO heavily relies on evaporation parameter setting for pheromone. However, findings appropriate settings of evaporation parameters for each problem can be time-consuming. To tackle these issues, an adaptive ant colony optimization with node clustering (AACO-NC) was proposed [3]. AACO-NC uses a node clustering technique, which effectively limits the search range of solution. By combining both node clustering and pheromone trail, this method finds shorter routes for numerous TSP benchmark problems [3]. However, our preliminary experiments revealed that the rules for updating the pheromone trail can limit problem-solving performance. This is because the added pheromone amount depends on the improvement of solutions, leading to on a local minimum trap if the local search method significantly improves the solution. To overcome this, we introduce the simulated annealing

© The Author(s), under exclusive license to Springer Nature Switzerland AG 2024
M. Sevaux et al. (Eds.): MIC 2024, LNCS 14753, pp. 21–27, 2024.
https://doi.org/10.1007/978-3-031-62912-9_4

(SA) [4], which allows pheromone updates stochastically based on solution quality, thereby expanding the solutions space. Additionally, while AACO-NC uses the modified k-Opt method [5] as a local search method, we use the 2-Opt method [6] to reduce the computational time. Numeral experiments indicate that our proposed method has lower error rates between the average value and optimal solution in various TSP benchmark problems [7] compared to the conventional AACO-NC method.

2   Adaptive Ant Colony Optimization

Ant colony optimization (ACO) is a technique inspired by natural behaviour of ants [2]. When ants find food in nature, they leave pheromones along their path. They then construct the shortest route based on this pheromone and distance information. ACO has a high dependence on parameter, leading to proposal of adaptive ant colony optimization with node clustering (AACO-NC) [3]. The flowchart of AACO-NC is provided below (Fig. 1).

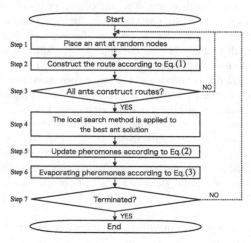

Fig. 1. The flowchart of the AACO-NC

AACO-NC uses distance and pheromone information to probabilistically determine the next visit node by using the following equation:

$$p(k \rightarrow j) = \frac{\eta_{kj}^{\alpha} \tau_{kj}^{\beta}}{\sum_{i \in V_{\text{unvisited}}} \eta_{ki}^{\alpha} \tau_{ki}^{\beta}} \quad \forall j \in V_{\text{unvisited}}, \tag{1}$$

where $p(k \rightarrow j)$ is a probability of an ant at node k selecting node j as the next visit node, $V_{\text{unvisited}}$ is a set of nodes that have not been visited, η_{kj} is an inverse distance between node k and j, and τ_{kj} is a pheromone trail between node k and j. After all ants have generated a route, local search methods are applied to the best ant solution. AACO-NC uses the modified k-Opt method ($O(kn^2)$) as a local search method, where k is the number of consecutively selected nodes and n is the number of nodes in the problems, but this study uses the 2-Opt method ($O(n^2)$) to reduce the computational time. After the solution is updated by the local search method, the pheromone is updated for each route that each ant takes. The updating equation of pheromone is defined as follows:

$$\tau_{ij} = \tau_{ij} + \delta x_{ij} \frac{C_{best}}{C_{ants}} \quad \forall i, j \in V, \tag{2}$$

where τ_{ij} is a pheromone trail between node i and node j, x_{ij} is a decision variable representing a visit between node i and node j in the best solution of the ant set, δ is a pheromone update coefficient, C_{ants} is the best solution value before applying the local search method in one iteration, C_{best} is the global best solution value after applying the local search method, and V is a set of all nodes. After the pheromone is renewed, the pheromone is evaporated for all edges. The equation of pheromone evaporation is defined as follows:

$$\tau_{ij} = (1 - \rho)\tau_{ij} \quad \forall i, j \in V, \tag{3}$$

where ρ is a value of the evaporation coefficient. In ACO, setting evaporation coefficient, ρ, to a constant low value can yield a shorter route. However, this extends the search time. On the other hand, if ρ is a constant high value, a solution can be found quickly but it may not be the short route. AACO-NC involves increasing ρ at the start of the solution search and decreasing it nearing a local solution [3]. This strategy, which adjusts ρ based on solution quality, allows for a short route and faster solution time. To achieve this, pheromone coefficients for AACO-NC is defined as follows:

$$\rho = \rho_{min} + \frac{(\rho_{max} - \rho_{min})(H - H_{min})}{H_{max} - H_{min}}, \tag{4}$$

where ρ_{min} and ρ_{max} are the lower and upper limits of the evaporation coefficients, H is the information entropy, defined as follows:

$$H = -\sum_{i=2}^{n} \sum_{j=1}^{i-1} p_{ij} \log_2 p_{ij}, \tag{5}$$

$$p_{ij} = \frac{E_{ij}}{nm_{ants}}, \tag{6}$$

where p_{ij} is a proportion of ants that transversed the edge between node i and j, E_{ij} is the number of times ants have transversed the edges between nodes i and j, n is the total number of nodes, and m_{ants} is the total number of ants. In ACO, the pheromone converges to a specific solution as the search progresses, reducing diversity of ants. If the termination condition of ACO is set solely to only the maximum number of iterations, finding a good solution may become challenging. Therefore, a method has been proposed to reduce the search time by terminating the search when the ant diversity is low. This scheme is defined by the following equation:

$$H \leq H_{min}(1 + \omega), \tag{7}$$

where ω is a coefficient of tolerance relative to the lower limit of information entropy H_{min}. The termination condition for AACO-NC in this study is either when the maximum number of iterations reaches 10000 or when the Eq. (7) is satisfied. In addition, we define one iteration as the journey from step 1 to step 7.

3 Simulated Annealing

Our preliminary experiments indicate that AACO-NC combined with 2-Opt method converges solutions rapidly but tend to fall into local solutions just as quickly. As a result, it loses the diversity, which is the key advantage of ant colony optimization, and struggles to search for optimal solution. We believe this issue arises because pheromone updating is only executed on worthwhile solutions. We suggest resolving this issue by incorporating simulated annealing (SA) [4] that allows pheromone updates even for stochastically inferior solutions. Permitting pheromone updates for both superior and inferior solutions broadens the search space for solutions, thereby enhancing the diversity of ACO. Our SA technique for pheromone update is defined as follows:

$$\tau_{ij} = \tau_{ij} + \delta x_{ij} \frac{C_{\text{select}}}{C_{\text{ants}}}, \tag{8}$$

$$C_{\text{select}} = \begin{cases} S_{\text{best}} & (\text{if } C_{\text{best}} \geq S_{\text{best}} \text{ or } \exp\left(\frac{\Delta H \Delta S}{\Delta T}\right) > \text{rand}[0,1]), \\ C_{\text{best}} & (\text{otherwise}), \end{cases} \tag{9}$$

$$\Delta H = \frac{H_{\text{old}} - H_{\text{new}}}{H_{\text{new}} - H_{\text{min}}}, \tag{10}$$

$$\Delta S = \frac{(C_{\text{best}} - C_{\text{ants}})}{C_{\text{best}}}, \tag{11}$$

$$\Delta T = \frac{t}{t_{\text{max}}}, \tag{12}$$

where C_{select} is a solution value for pheromone update, S_{best} is the best solution value after applying local search method in one iteration, rand[0, 1] is a uniformly distributed random number from 0 to 1, H_{old} and H_{new} are the information entropy in the previous and current searches, t is the count of unsuccessful attempts to update the best solution, t_{max} is maximum number of iterations. Our method updates the pheromone for the best solution in one iteration when $C_{\text{select}} = S_{\text{best}}$. It also updates the pheromone if the current solution value after applying local search method slightly worsens the best solution value.

4 Numerical Experiment

In this study, we used six benchmark problems from TSPLIB [7], eil51, kroA100, d198, kroA200, rd400, and pcb442. The parameters were set to $\alpha = 1, \beta = 1, \delta = 1, m_{\text{ants}} = 96, \omega = 0.002$. In addition, $\rho_{\text{min}} = 0.001$ and $\rho_{\text{max}} = 0.1$ when n is smaller than 400, $\rho_{\text{min}} = 0.02$ and $\rho_{\text{max}} = 0.5$ when n is greater than or equals to 400. We conducted 30 experiments on a personal computer equipped with Apple M1 Ultra, 3.2 GHz CPU, and 64GB RAM. Each method was run on a single thread, and calculated the best solution. We also calculated the error rate between the average and the best known solution, and the average real time from the start of the calculation to the end of the calculation. There exists a discrepancy in performance between the experimental machine in Ref. [3]

Table 1. Comparison of conventional and proposed methods

instance	optimal solution	best solution		error rate [%]		time [s]	
		conventional	proposed	conventional	proposed	conventional	proposed
eil51	426	426	**426**	0.18	0.44	2.58	**1.80**
kroA100	21282	21282	**21282**	0.06	**0.01**	11.99	**4.59**
d198	15780	15854	**15796**	1.00	**0.28**	200.79	222.87
kroA200	29368	29382	**29368**	0.83	**0.25**	98.93	**37.38**
rd400	15281	15458	**15353**	2.72	**1.11**	256.40	**166.69**
pcb442	50778	51306	**51126**	1.99	**0.83**	656.11	**284.25**

and the one utilized in this study. In this study, we realized a conventional method parallel with the proposed method. This approach facilitated comprehensive numerical experiment, allowing for more robust comparison and analysis.

Table 1 illustrates a comparison between two methods: the "conventional" method using AACO-NC and "proposed" method employing AACO-NC-SA. The conventional method applied the modified k-Opt method, a local search method, once every ten iterations. Pheromone updates take place on the best solution prior to executing this local search method. On the other hand, the proposed method utilizes the 2-Opt method, a local search method used in every iteration. Pheromone updates were performed by using Eqs. (8) (12). The data in Table 1 shows that the proposed method yields a smaller best solution value than the conventional method for all problems, and a lower error rate for five problems. This evidence indicate that the proposed method tends to outperform the conventional methods on numerous benchmark problems. The proposed method also shows shorter computational time than the conventional method for five problems. These results validate that the superior performance of the proposed method over the conventional method in TSP. However, the results from the table alone do not confirm whether the SA is functioning correctly. Therefore, we used rd400 problem to investigate the information entropy and the solution transition diagrams for the best solution and the best solution per one iterations.

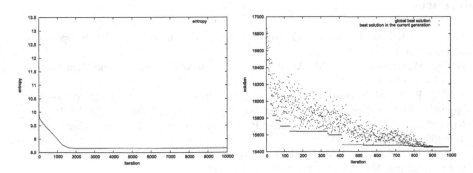

Fig. 2. Information entropy per iteration   **Fig. 3.** Solutions per iteration

Figure 2 confirms that the information entropy converges as the solution search progresses. Figure 3 verifies that the search range is broad in the initial stages and narrows down as the search progresses. This suggest that the pheromone update for both superior and inferior solutions in SA is appropriate. Consequently, the information entropy of the proposed method converges slower than the conventional method, allowing for more good solutions to be found. It was noted that the error rate of proposed method improves more significantly for larger problems compared to the conventional method. This is likely because the solution search space for larger problems is more extensive than that for smaller ones.

5   Conclusion

This study proposed ant colony optimization with node clustering with simulated annealing (AACO-NC-SA), and compared it with the conventional AACO-NC method. Numerical experiments confirm that the proposed method outperforms the conventional methods. In addition, we observed a large variation of solutions in the early stages of the solution search, confirming that the SA is moving in the right direction. These results suggest that the proposed method performs well because the SA extends the solution search range. In future work, we aim to extend the application of the proposed method to a range of problems, including the vehicle routing problem (VRP), the capacitated vehicle routing problem with time windows problems (CVRPTW), and additional benchmark problems in TSP. Furthermore, we intend to address larger instances within the TSP. These applications will be compared with the state of the art methods, such as the Lin-Kernighan heuristic [8], the artificial bee colony optimization [9], and the deer hunting optimization [10]. Additionally, a comprehensive analysis will be conducted to elucidate the reasons of underpinning the superior performance of the proposed method.

Acknowledgment. The research of T.K. and T.M. was partially supported by a Grant-in-Aid for Scientific Research (C) from JSPS (No. 23K04274, 22K04602).

References

1. Jünger, M., Reinelt, G., Rinaldi, G.: The traveling salesman problem. Handb. Oper. Res. Manage. Sci. **7**, 225–330 (1995)
2. Dorigo, M., Gambardella, L.M.: Ant colonies for the travelling salesman problem. Biosystems **43**(2), 73–81 (1997)
3. Stodola, P., Otřísal, P., Hasilová, K.: Adaptive ant colony optimization with node clustering applied to the travelling salesman problem. Swarm Evol. Comput. **70**, 101056 (2022)
4. Kirkpatrick, S., Gelatt, C.D., Jr., Vecchi, M.P.: Optimization by simulated annealing. Science **220**(4598), 671–680 (1983)
5. Stodola, P.: Hybrid ant colony optimization algorithm applied to the multi-depot vehicle routing problem. Nat. Comput. **19**(2), 463–475 (2020)
6. Englert, M., Röglin, H., Vöcking, B.: Worst case and probabilistic analysis of the 2-opt algorithm for the TSP. Algorithmica **68**(1), 190–264 (2014)
7. tsplib95. http://comopt.ifi.uni-heidelberg.de/software/TSPLIB95/. Accessed 29 Jan 2024

8. Lin, S., Kernighan, B.W.: An effective heuristic algorithm for the traveling-salesman problem. Oper. Res. **21**(2), 498–516 (1973)
9. Karaboga, D., et al.: An idea based on honey bee swarm for numerical optimization. Technical report, Technical report-tr06, Erciyes University, Engineering Faculty, Computer Engineering Department (2005)
10. Brammya, G., Praveena, S., Ninu Preetha, N., Ramya, R., Rajakumar, B., Binu, D.: Deer hunting optimization algorithm: a new nature-inspired meta-heuristic paradigm. Comput. J. bxy133 (2019)

Job-Shop Scheduling with Robot Synchronization for Transport Operations

Jean Philippe Gayon, Philippe Lacomme$^{(\boxtimes)}$, and Amine Oussama

Université Clermont-Auvergne, CNRS, Mines de Saint-Étienne, Clermont-Auvergne-INP, LIMOS, 63000 Clermont-Ferrand, France
`j-philippe.gayon@uca.fr`, `philippe.lacomme@isima.fr`,
`Amine.OUSSAMA@doctorant.uca.fr`

Abstract. We consider a Job Shop Scheduling Problem with transport (JSPT) which consists in jointly scheduling machines and robots. In contrast with the literature, we assume that a transport operation may involve several robots simultaneously, which requires resource synchronization over time. We formulate this problem as a Mixed Integer Linear Programming (MILP) formulation. Then we propose a GRASP-ELS meta-heuristic and a local search procedure where we use a Bierwith's sequence approach to evaluate a solution. In a numerical study, we have adapted instances from the literature to our problem. The meta-heuristic competes with the exact resolution providing high quality solution in reduced computation time, which lead us to consider that both the modeling and local search are accurate.

Keywords: Job shop · Synchronization · Scheduling · Transportation · Disjunctive graph · Meta-heuristic · Mixed integer linear programming

1 Introduction

The job-shop scheduling (JSP) problem has received an enormous amount of attention in the research literature (see [10] for an introduction). The job-shop scheduling problem with transport (JSPT) consists in jointly scheduling machine operations and transport operations (see [15] for a survey on this problem). This problem was first studied in FMS (Flexible Manufacturing System) literature [2, 13, 14] for the first meta-heuristic). This problem has also been studied as a flow-shop with transport in [4] and as a job-shop with transport in [5]. Two data sets have been introduced early [6, 13].

The context for this work is the increasing robotization of manufacturing and logistics processes. New-generation mobile robots can cooperate to handle or transport heavy or large objects (see e.g. the Mecabotix Company [9]). For example, small loads can be transported by a single robot, while bulky or heavy loads require the collaboration of several robots. A review on modular reconfigurable robotics system can be found in [12].

The originality of our work lies in the assumption that transport operations may require the simultaneous intervention of several robots, which means that correct synchronization between robots is necessary. Note that synchronization issues appear in other problems including the vehicle routing problem (see [3] for a survey) where the synchronization can involve, for example, vehicles, drivers or medical staff.

2 Problem Description

In this section, we describe the job-shop scheduling problem with transport and synchronization.

Jobs, Machines and Robots. We consider a set of N jobs $\mathcal{J} = \{J_1, ..J_N\}$ that have to be processed on a set of M machines $\mathcal{M} = \{M_1, ..M_M\}$. The process of transferring a job from one machine to another is referred to as a transport operation. Transport operations are performed by a set of NR robots $\mathcal{R} = \{R_1, .., R_{NR}\}$. . A transport operation can be accomplished either by using a single robot (referred to as a single-robot transport operation), or by employing multiple robots simultaneously (referred to as a multi-robot transport operation).

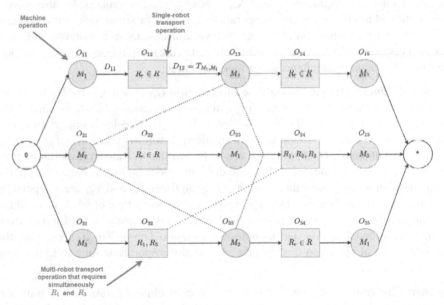

Fig. 1. Partial representation of the undirected disjunctive graph used to represent a problem of 3 jobs, 3 machines and 3 robots.

Undirected Graph. We illustrate notations in Fig. 1 for a problem with 3 jobs, 3 machines and 3 robots. The conjunctive (solid) arcs represent the ordered sequence of operations per job. Two machine operations that must be processed on the same machine are connected by disjunctive edges (dash lines in Fig. 1). To avoid overloading Fig. 1, we only show the disjunctive edges for machine M_2. Two transport operations that must be processed by the same robot cannot be executed simultaneously and are also connected by disjunctive edges. For example, transport operations O_{32} and O_{24} have R_1 and R_3 in common and are therefore linked by a disjunctive edge.

Operations. Without loss of generality, we assume that each job J_i consists of a sequence of NO operations denoted $\mathcal{O}_i = (O_{i1}, ..O_{i,NO})$ defining the precedence constraints. The j-th operation of job J_i is denoted by O_{ij}. A machine operation is followed by a transport operation, which is in turn followed by a machine operation. Hence, operation O_{ij}, when index j is odd, corresponds to a machine operation, and to a transport operation when index j is even. Each Job J_i consists of M machine operations and $M - 1$ transport operations. Hence, the total number of operations is $NO = 2M - 1$. For example, in Fig. 1, operations O_{11}, O_{13} and O_{15} correspond to the machine operations of job J_1, while O_{12} and O_{14} correspond to its transport operations.

Machine Operations. When O_{ij} is a machine operation, we denote by M_{ij} the corresponding machine. For instance, in Fig. 1, we have $M_{11} = M_1, M_{13} = M_2$.

Transport Operations. When O_{ij} is a transport operation, it transfers job J_i from machine $M_{i,j-1}$ to machine $M_{i,j+1}$. There are two types of transport operations. A transport operation O_{ij} can either require a single robot R_r chosen in the set of robots $\mathcal{R}$, or it can be multi-robot, requiring a subset $\mathcal{R}_{ij} \subset \mathcal{R}$ of at least two robots. In other words, the execution of multi-robot transport operations requires the simultaneous involvement of multiple robots (a minimum of two robots). We illustrate these notations with Fig. 1. Transport operation O_{12} requires a robot R_r chosen in the set of robots $\mathcal{R}$ while transport operation O_{32} requires simultaneously robots R_1 and R_3.

Processing Times. The processing time of operation O_{ij} is denoted by D_{ij} for either a machine or a transport operation. For a machine operation O_{ij}, D_{ij} is simply the processing time on machine M_{ij}. We also denote by T_{sd} the loaded transportation time from machine s (source) to machine d (destination). It follows that $D_{ij} = T_{M_{i,j-1},M_{i,j+1}}$ for a transport operation O_{ij}. Finally, we denote by V_{sd} the unloaded transportation time from machine s to machine d. Let illustrate these notations on Fig. 1. Note that, for the sake of simplicity, we assume that the transportation times T_{sd} and V_{sd} are independent of the involved robots. Duration D_{11} represents the processing time of job J_1 on machine M_1. Duration $D_{12} = T_{M_1,M_2}$ represents the loaded transportation time of job J_1 from machine M_1 to machine M_2. Let us note that duration $D_{32} + V_{M_2,M_1}$ represents the processing time of operation O_{32} plus the unloaded transportation time from machine M_2 to machine M_1.

Objective Function. The objective is to schedule machine operations and transport operations in order to minimize the makespan, i.e. the time at which all jobs have been processed.

Binary Variables. We end this section by introducing some binary variables that will help us formulating the problem. To differentiate between machine operations and transport operations, we introduce the binary variable L_{ij} which is equal to 0 if O_{ij} is a machine operation, and equal to 1 if it's a transport operation.

To distinguish between the two types of transport operations, we introduce the binary variable Y_{ij} which is equal to 0 if O_{ij} is a single-robot transport operation, and to 1 if it is a multi-robot transport operation.

For a multi-robot transport operation O_{ij}, we represent the robots assigned to this operation by a vector E_{ij} of NR binary variables, denoted as $\left(E_{ij}^r\right)_{r=1..NR}$, where $E_{ij}^r = 1$ if robot R_r is required for this specific transport operation and $E_{ij}^r = 0$ otherwise.

For any two distinct multi-robot transport operations O_{ij} and O_{kp}, let F_{ijkp} the binary variable equal to 1 if O_{ij} and O_{kp} share at least one robot, and equal to 0 otherwise. Note that F_{ijkp} is the scalar product of E_{ij} and E_{kp}: $F_{ijkp} = \left(E_{ij}\right)^t.E_{kp}$, i.e. $F_{ijkp} = \sum_{r=1}^{NR} E_{ij}^r \times E_{kp}^r$.

3   Linear Formulation of the Problem

3.1   Data

N:	Number of jobs to schedule.
M:	Number of machines.
NR:	Number of robots.
NO:	Total number of operations to schedule for one job ($NO = 2 \times M - 1$).
L_{ij}:	Binary variable that equals 0 if operation O_{ij} is a machine operation, and equals 1 if O_{ij} is a transport operation. This variable is defined for $i = 1..N$ and $j = 1..NO$.
M_{ij}:	Machine required by machine operation $j = 1..NO$ of job $i = 1..N$. It is defined only if $L_{ij} = 0$.
T_{sd}:	Loaded transportation time for a robot from machine s to machine v.
V_{sd}:	Unloaded transportation time for a robot from machine s to machine v.
D_{ij}:	Processing time of operation $j = 1..NO$ of job $i = 1..N$. It is the processing time on the machine if and only if O_{ij} is a machine operation, and it is the loaded transportation time from $M_{i,j-1}$ to $M_{i,j+1}$ if O_{ij} is a transport operation.
Y_{ij}:	Binary variable that equals 1 if the operation O_{ij} is a multi-robot transport operation, and equals 0 if O_{ij} is a single-robot transport operation. This variable is defined for $i = 1..N, j = 1..NO$ if $L_{ij} = 1$.
E_{ij}^r:	Binary variable, that equals 1 if robot R_r is required to achieve the multi-robot transport operation O_{ij}. This variable is defined for $i = 1..N, j = 1..NO$ only if $L_{ij} = Y_{ij} = 1$.
$E_{ij} = \left(E_{ij}^r\right)_{r=1..NR}$:	vector of binary variables E_{ij}^r.
F_{ijkp}:	Binary variable that equals 1 if the two distinct multi-robot transport operations O_{ij} and O_{kp} require at least one common robot. This variable is defined for $i = 1..N, j = 1..NO, k = 1..N, p = 1..NO$ if $L_{ij} = L_{kp} = Y_{ij} = Y_{kp} = 1$.
H:	A large positive number.

3.2 Decision Variables

st_{ij}:   Starting time of operation $j = 1..NO$ of job $i = 1..N$.

c_{max}: Finishing time of the last operation on the last machine.

b_{ijkp}: Binary variable that equals 1 if the operation O_{ij} is scheduled before the operation O_{kp}, and equals 0 otherwise. This variable is defined for $i = 1..N, j = 1..NO$, $k = 1..N, p = 1..NO$ if $i \neq k$ or $j \neq p$.

a_{ij}^r: Binary variable that equals 1 if robot R_r is allocated to the single-robot transport operation O_{ij}, and equals 0 otherwise. This variable is defined for $i = 1..N$ and $j = 1..NO$ if $L_{ij} = 1$ and $Y_{ij} = 0$.

w_{ijkp}^r: Binary variable that equals 1 if robot R_r is allocated to the two distinct single-robot transport operations O_{ij} and O_{kp}, and equals 0 otherwise. This variable is defined for $i = 1..N, j = 1..NO, k = 1..N, p = 1..NO$ and $r = 1..NR$, if $L_{ij} = L_{kp} = 1, Y_{ij} = Y_{kp} = 0$.

w_{ijkp}: Binary variable that equals 1 if the two distinct single-robot transport operations O_{ij} and O_{kp} require the same robot, and equals 0 otherwise. This variable is defined for $i = 1..N, j = 1..NO, k = 1..N$ and $p = 1..NO$ if $L_{ij} = L_{kp} = 1$ and $Y_{ij} = Y_{kp} = 0$.

z_{ijkp}^r: Binary variable that equals 1 if robot R_r is allocated to both the single-robot transport operation O_{ij} and the multi-robot transport operation O_{kp}, and equals 0 otherwise. This variable is defined for $i = 1..N, j = 1..NO, k = 1..N, p = 1..NO$ and $r = 1..NR$ if $L_{ij} = L_{kp} = Y_{kp} = 1$ and $Y_{ij} = 0$.

z_{ijkp}: Binary variable equals 1 if both the single-robot transport operation O_{ij} and the multi-robot transport operation O_{kp} require the same robot, and equals 0 otherwise. This variable is defined for $i = 1..N, j = 1..NO, k = 1..N$ and $p = 1..NO$ if $L_{ij} = L_{kp} = Y_{kp} = 1$ and $Y_{ij} = 0$.

3.3 Objective Function

$$\text{Minimize } C_{max}$$

The objective is to minimize the makespan C_{max}, i.e. the time at which all jobs have been processed.

3.4 Constraints

Makespan

For $i = 1..N$:

$$st_{i,NO} + D_{i,NO} \leq C_{max} \tag{1}$$

The makespan C_{max} must be larger than the completion time of the last operation of each job.

Sequence of Operations

For $i = 1..N, j = 1..NO$:

$$st_{ij} + D_{ij} \leq st_{i,j+1} \tag{2}$$

The starting time of the operation $O_{i,j+1}$ is greater than the finishing time of the operation O_{ij}.

Disjunction Between Machine Operations Processed on the Same Machine
For $i = 1..N - 1, j = 1..NO, k = (i + 1)..N, p = 1..NO$ such that $M_{ij} = M_{kp}$ and $L_{ij} = L_{kp} = 0$:

$$st_{ij} + D_{ij} \leq st_{kp} + H.\left(1 - b_{ijkp}\right) \tag{3}$$

$$st_{kp} + D_{kp} \leq st_{ij} + H.b_{ijkp} \tag{4}$$

Machine operations that must be processed on the same machine cannot be executed simultaneously. Let us consider two machine operations O_{ij} and O_{kp} ($L_{ij} = L_{kp} = 0$) where both operations require the same machine ($M_{ij} = M_{kp}$). A disjunction between the two operations implies that either O_{ij} is scheduled first, followed by O_{kp}, and in this case we have $st_{ij} + D_{ij} \leq st_{kp}$, or O_{kp} is scheduled first, followed by O_{ij}, and we have $st_{kp} + D_{kp} \leq st_{ij}$.

Number of Robots Allocated to a Single-Robot Transport Operation
For $i = 1..N, j = 1..NO$ such that $L_{ij} = 1$ and $Y_{ij} = 0$:

$$\sum_{r=1}^{NR} a_{ij}^r = 1 \tag{5}$$

For each single-robot transport operation O_{ij} ($L_{ij} = 1$ and $Y_{ij} = 0$), we must ensure the assignment of exactly one robot to the transport operation.

Disjunction Between Single-Robot Transport Operations that Require the Same Robot
For $i = 1..N, j = 1..NO, k = i + 1..N, p = 1..NO, r = 1..NR$ such that $L_{ij} = L_{kp} = 1$ and $Y_{ij} = Y_{kp} = 0$:

$$a_{ij}^r + a_{kp}^r \leq 1 + w_{ijkp}^r \tag{6}$$

$$w_{ijkp}^r \leq a_{ij}^r \tag{7}$$

$$w_{ijkp}^r \leq a_{kp}^r \tag{8}$$

$$\sum_{r=1}^{NR} w_{ijkp}^r = w_{ijkp} \tag{9}$$

For $i = 1..N, j = 1..NO, k = i + 1..N, p = 1..NO$ such that $L_{ij} = L_{kp} = 1$ and $Y_{ij} = Y_{kp} = 0$:

$$st_{ij} + T_{M_{i,j-1},M_{i,j+1}} + V_{M_{i,j+1},M_{k,p-1}} \leq st_{kp} + H.\left(1 - b_{ijkp}\right) + H.\left(1 - w_{ijkp}\right) \tag{10}$$

$$st_{kp} + T_{M_{k,p-1},M_{k,p+1}} + V_{M_{k,p+1},M_{i,j-1}} \le st_{ij} + H.b_{ijkp} + H.(1 - w_{ijkp}) \tag{11}$$

Two single-robot transport operations share the same robot R_r if and only if $a^r_{ij} = a^r_{kp} = 1$. This condition can be expressed by the equation $a^r_{ij} \times a^r_{kp} = 1$. Inequalities (6), (7) and (8) are used to linearly express w^r_{ijkp} as the product $a^r_{ij} \times a^r_{kp}$. Subsequently, Eq. (9) ensures that the two single-robot transport operations share at least one robot if and only if $w_{ijkp} = 1$. If that is the case, then the two operations cannot be processed simultaneously. The disjunction between the two operations is expressed through inequalities (10) and (11).

Disjunction Between Multi-robot Transport Operations that Require at Least One Common Robot

For $i = 1..N, j = 1..NO, k = 1..N, p = 1..NO$ such that $L_{ij} = L_{kp} = Y_{ij} = Y_{kp} = F_{ijkp} = 1$:

$$st_{ij} + T_{M_{i,j-1},M_{i,j+1}} + V_{M_{i,j+1},M_{k,p-1}} \le st_{kp} + H.(1 - b_{ijkp}) \tag{12}$$

$$st_{kp} + T_{M_{k,p-1},M_{k,p+1}} + V_{M_{k,p+1},M_{i,j-1}} \le st_{ij} + H.b_{ijkp} \tag{13}$$

Multi-robot transport operations that share at least one robot cannot be processed simultaneously. Let us consider two distinct multi-robot transport operations O_{ij} and $O_{kp}(L_{ij} = L_{kp} = Y_{ij} = Y_{kp} = 1)$ that require at least one common robot ($F_{ijkp} = 1$). These operations are known in advance as F_{ijkp} is part of the problem data. The disjunction between the two operations is explicitly expressed through inequalities (12) and (13).

Disjunction Between a Single-Robot Transport Operation and a Multi-Robot Transport Operation that Require the Same Robot

For $i = 1..N, j = 1..NO, k = 1..N, p = 1..NO, r = 1..NR$ such that $L_{ij} = L_{kp} = Y_{kp} = 1$ and $Y_{ij} = 0$:

$$E^r_{kp} + a^r_{ij} \le 1 + z^r_{ijkp} \tag{14}$$

$$z^r_{ijkp} \le a^r_{ij} \tag{15}$$

$$z^r_{ijkp} \le E^r_{kp} \tag{16}$$

$$\sum_{r=1}^{NR} z^r_{ijkp} = z_{ijkp} \tag{17}$$

For $i = 1..N, j = 1..NO, k = 1..N, p = 1..NO$ such that $L_{ij} = L_{kp} = Y_{kp} = 1$ and $Y_{ij} = 0$:

$$st_{ij} + T_{M_{i,j-1},m_{i,j+1}} + V_{M_{i,j+1},M_{k,p-1}} \le st_{kp} + H.(1 - b_{ijkp}) + H.(1 - z_{ijkp}) \tag{18}$$

$$st_{kp} + T_{M_{k,p-1},M_{k,p+1}} + V_{M_{k,p+1},M_{i,j-1}} \le st_{ij} + H.b_{ijkp} + H.(1 - z_{ijkp}) \tag{19}$$

A multi-robot transport operation O_{kp} ($L_{kp} = Y_{kp} = 1$) uses a robot R_r if and only if $E_{kp}^r = 1$, and this is known in advance due to the inclusion of $\left(E_{kp}^r\right)_{r=1..NR}$ in the problem data. The same robot R_r may also be assigned to a single-robot transport operation O_{ij} ($L_{ij} = 1$ and $Y_{ij} = 0$), which is equivalent to $a_{ij}^r = 1$. In this scenario, both operations O_{ij} and O_{kp} require robot R_r, and this can be expressed by the equation $E_{kp}^r \times a_{ij}^r = 1$. Inequalities (13), (14), and (15) are used to linearly represent z_{ijkp}^r as the product $E_{kp}^r \times a_{ij}^r$. Subsequently, Eq. (16) ensures that a single-robot transport operation and a multi-robot transport operation share the same robot if and only if $z_{ijkp} = 1$. If that is the case, the two operations cannot be processed simultaneously. The disjunction between the two operations is expressed through inequalities (17) and (18).

4 Metaheuristic Based Resolution

4.1 Solution Modeling Based on Disjunctive Graph

The resolution of a JSPT encompasses jointly the sequencing of machine operations, the assignment of a robot to each single-robot transport operation and the sequencing of transport operations. The sequencing of both machine operations and transport operations is accomplished by turning all edges into directed arcs, thereby defining an oriented disjunctive graph. However, only acyclic oriented disjunctive graphs describe feasible solutions.

For two consecutive transport operations O_{ij} and O_{kp} ($L_{ij} = I_{kp} = 1$) scheduling O_{ij} before O_{kp} entails that O_{kp} can only be scheduled after R_r finishes its transfer of job J_i form machine $M_{i,j-1}$ to $M_{i,j+1}$ (with a duration of $T_{M_{i,j-1},M_{i,j+1}}$), and once the robot R_r becomes available, O_{kp} has to wait for its arrival from $M_{i,j+1}$ to pick up the job J_k from machine $M_{k,p-1}$ (with a duration of $V_{M_{i,j+1},M_{k,p-1}}$).

This is represented by an arc that goes from O_{ij} to O_{kp} and gets the weight of $T_{M_{i,j-1},M_{i,j+1}} + V_{M_{i,j+1},M_{k,p-1}}$. Similarly, scheduling O_{kp} before O_{ij} is represented by an arc that goes from O_{kp} to O_{ij} and gets the weight $T_{M_{k,p-1},M_{k,p+1}} + V_{M_{k,p+1},M_{i,j-1}}$ as depicted in Fig. 2.

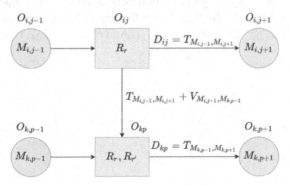

Fig. 2. Disjunction between two transport operations O_{ij} and O_{kp} that require the robot R_r. (Case when O_{ij} is scheduled before O_{kp})

It is important to note that unlike single-robot transport operations that can have at most one ongoing/outgoing disjunctive arc, a multi-robot transport operation O_{ij} that requires a subset of r robots $\mathcal{R}_{ij} = \{R_1, \ldots, R_r\}$, can have at most r outgoing/ongoing disjunctive arc (depending on its position in the vector MTS). For instance, in Fig. 3 below, multi-robot transport operation O_{12} uses robots R_1, R_2 and R_3 and is scheduled before transport operations O_{22}, O_{24} and O_{32} that require some of the robots used in O_{12}, so it has 3 outgoing disjunctive arcs, while single-robot transport operations O_{32}, O_{14} and O_{34} that use the same robot R_1 have at most only 1 ongoing/outgoing disjunctive arcs.

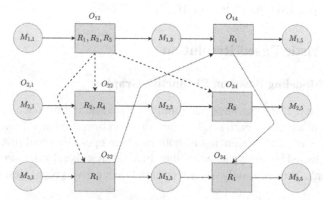

Fig. 3. Number of ongoing/outgoing of disjunctive arcs for multi-robot and single-robot transport operations

Evaluating a solution consists in constructing an acyclic directed disjunctive graph where a robot is assigned to each single-robot transport operation. Then, by performing a longest path algorithm to the graph to compute the earliest starting times of each operation and find the makespan C_{max} which is the length of the longest path. Figure 4 gives an example of an acyclic directed disjunctive graph for the example presented in Sect. 2, where we consider following order of jobs on machines: $M_1: J_1, J_2, J_3, M_2: J_2, J_1, J_3$ and $M_3: J_3, J_2, J_1$. The following assignments of robots to single-robot transport operations: R_2 for O_{12}, O_{22} and O_{34}, R_3 for O_{14}. The following order of transport operations on robots: $R_1 : O_{32}, O_{24}, R_2 : O_{12}, O_{22}, O_{34}, O_{24}$. And $R_3 : O_{32}, O_{24}, O_{14}$.

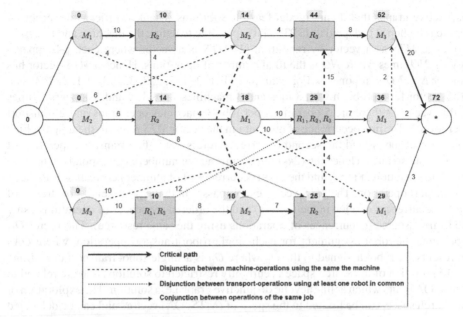

Fig. 4 Fully oriented disjunctive graph modeling a solution and the critical path

Figure 5 gives a Gantt representation of this solution.

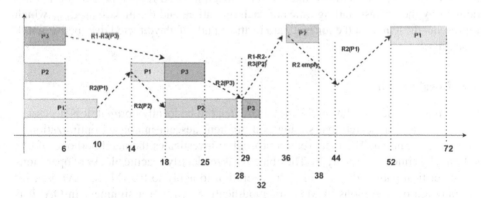

Fig. 5. Gantt representation

4.2 Indirect Representation of a Solution Using Bierwith's Vector

In this section, we show how to represent a feasible solution with a *MTS* (machine transport sequence) and a *OA* (operations assignment) vector defining the assignment of robots to single-robot transport operations. This representation is adapted from [1] and [7]. The challenging aspect of the problem consists in avoiding the generation of cyclic directed disjunctive graphs that do not model feasible solutions. Thanks to Bierwith's proposal for the job-shop [1] adapted to JSPT [7] it is possible to define the machine and transport sequence *MTS* as a vector by repetition [1] *MTS* is a vector where each job appears exactly *NO* times where *NO* is the total number of operations. Hence a *MTS* vector has exactly $N \times NO$ components. For example, (1, 1, 2, 1, 2, 2, 3, 3, 3, 3, 1, 1, 2, 2, 3) is a *MTS* vector for a problem with 3 jobs and 3 machines. Jobs 1, 2 and 3 appears 5 times in this vector as there are 5 operations per job (3 machines operations and 2 transport operations). The first occurrence of number i in the vector *MTS* refers to the first machine operation of job J_i, and the second occurrence refers to the first transport operation of the job, and so forth. Hence the $(2k$-$1)$-th occurrence of number i corresponds to the k-th machine operation of job J_i and the $2k$-th occurrence of of number i corresponds to the k-th transport operation. The *MTS* vector encompasses both the precedence constraints of a job (occurrences of a job in the vectors) and the ordered sequences of operations using the same resource (occurrences of operations using the same resource). The vector *OA* represents the robot assignment for each single-robot transport operation, where OA_{ij} represents the robot assigned to the O_{ij} where O_{ij} is a single-robot transport operation.

Exploration of the search space is currently restricted to solutions that are related to (MTS, OA) pairs which define acyclic disjunctive graph i.e. a solution. The exploration of the search space, can be limited to the space of (MTS, OA) vectors and can be delegated to any metaheuristic-based scheme, such as a memetic algorithm, as emphasized in (Lacomme et al., 2013), even in the case of the "classical" JSPT, which represents a special instance of our problem where all transport operations are single-robot. Note that, (*MTS*, *OA*) pair, can be mapped into a solution S $O(N \times M \times N$ R) where N, M and *NR* are respectively the number of jobs, machines and robots. A solution is fully defined by the earliest starting times of each operation, and the makespan C_{max} which represents the length of the longest path (critical path) in the directed disjunctive graph modeling S.

4.3 Local Search

Since the constructed solution S is not guaranteed to be locally optimal, it is necessary to perform a local search that simulates a gradient descent in convex optimization to find a local minima. The local search procedure investigates the neighborhood of this solution by critical path analysis. The objective is to identify specific blocks of operations on the critical path and to identify the operation to apply to the (MTS, OA) pair. By permutation of operations in MTS, or modification of robot assignment in OA, it is possible to define a new directed disjunctive graph modeling a neighbor solution S'. The quality of the obtained solution S' is then compared to that of S, and if it's better, the solution is updated. This process is repeated until a locally optimal solution is found. We distinguish multiple types of blocks defined in Table 1 below.

Table 1. The 5 types of blocks on the critical path

Block	Definition
MDB	A machine-disjunctive-block which is a set of at least two consecutive machine operations using the same machine on the critical path of S, following the definition of (Grabowski et al., 1986)
SRCB	A single-robot-conjunctive-block which is a set of two consecutive operations of a same job on the critical path of S where one of them is a single-robot transport operation (the other one is necessarily a machine operation)
SRDB	A single-robot-disjunctive-block which is a set of at least two consecutive single-robot transport operations using the same robot on the critical path of S
MRDB	A multi-robot-disjunctive-block which is a set of at least two consecutive multi-robot transport operations using at least one robot in common on the critical path of S
HDB	A hybrid-disjunctive-block which is a set of two consecutive transport operations on the critical path of S, one of these operations is a single-robot transport operation and the other one is a multi-robot transport operation

For instance in Fig. 4, the two machine operations O_{13} and O_{33} define a machine-disjunctive-block since they use the same operation M_2, but operation O_{21} that also uses machine M_2 doesn't belong to the block because the disjunctive arc from O_{21} to O_{13} isn't on the critical path. Furthermore, the first 2 operations of job J_1 form a single-robot-conjunctive-block because O_{11} is a machine operation and O_{12} is a single-robot transport operation, but O_{13} and O_{14} don't form a conjunctive-block because the conjunctive arc between the two operations is not in the critical path.

As mentioned previously, the concept of local search involves examining the critical path from the dummy node * to reach dummy node 0. Whenever one of these 5 blocks is identified, specific operators tailored to each block are applied to alter the critical path of the current solution, thereby producing a new directed disjunctive graph representing a neighboring solution. In Table 2, we outline the permitted operators for each type of block.

Note that the local search procedure defines operators for *MTS* and *OA* vectors, related to the solution, produce MTS and/or OA vector that define (after modification) a new solution. The definition of *MTS* and *OA* vectors guarantee that the directed disjunctive graph modeling the new solution is acyclic.

4.4 Metaheuristic

We finally use the well-known GRASP-ELS metaheuristic introduced by [11] which provides a good balance between diversification (multi-start) and intensification (further investigation of the neighborhood of a local optimum).

Table 2. Operators per block

Block	Allowed operators
MDB	Permutation of two consecutive machine operations using the same machine. This is done by swapping the corresponding value in MTS
SRCB	Assignment of a different robot for the single-robot transport operation
SRDB	1) Permutation of two consecutive single-robot transport operations O_{ij} and O_{kp} using the same robot. This is done by swapping the related values in MTS 2) Permutation of a different robot for the single-robot transport operation O_{ij}. This is achieved by updating the OA vector 3) Permutation of a different robot for the single-robot transport operation O_{kp}. This is achieved by updating the OA vector
MRDB	Permutation of two consecutive multi-robot transport operations that require at least one robot in common. This is done by swapping the two related values in MTS
HDB	1) Permutation of the two consecutive transport operations the use the same robot. This is done by swapping the related values in MTS 2) Assignment of a different robot for the single-robot transport operation. This is achieved by updating the OA vector

5 Numerical Experiments

We have tuned instances from the literature [8] by adding transport operations and robot data. The experiments were carried out using 4 instances with a number of job that vary from 3 to 10, a number of machines that varies from 3 to 5 and a number of robots that varies from 3 to 7. To favor fair future research all the instances can be downloaded at: https://perso.isima.fr/~lacomme/JSSynchro/.

Table 3. Instances characteristic

Instance	N	M	NR	Number of multi-robot transport operation	Number of binary variables	Number of constraints
1	3	3	3	2	67	141
2	4	4	6	5	485	1103
3	5	5	3	6	1186	2641
4	6	6	6	10	3323	8389
5	10	5	3	13	1574	3354
6	10	5	3	13	1574	3354
7	10	5	5	20	5185	12453
8	10	5	7	10	8809	22369

The results are reported in Tables 3 and show that the number of jobs, machines, and robots has a substantial impact on both the quantity of binary variables and constraints. Moreover, the experimental results demonstrate that the proposed GRASP-ELS metaheuristic exhibits effective performance in terms of solution quality and computational time. It consistently produces near-optimal solutions within remarkably short computational times, underscoring the effectiveness of the metaheuristic-based approach and the relevance of the disjunctive graph model (Table 4).

Table 4. Numerical experiments for 60 s

	CPLEX		MetaHeuristic (best solution found with 3 runs)	
	S_{best}	$tt(s)$	S_{best}	$t_{best}(s)$
1	48*	1	48*	<1
2	74*	1	74*	<1
3	330*	1	330*	<1
4	452*	5.81	452*	<1
5	666*	60	666*	6
6	1861	60	1695	8
7	1411	60	1279	4
8	838	60	841	21

(tt: total time, t_{best}: time to best, S_{best}: best found solution)

6 Conclusion

This paper represents a significant advancement in the generalization of the disjunctive graph model to encompass multiple robots engaged in multi-robot transport operations. The key innovation lies in introducing synchronized actions among multiple robots to accomplish transportation tasks between machines.

We propose an efficient modeling approach to generalize the job-shop problem by incorporating transport logistics, leveraging the introduction of a disjunctive graph to represent both the problem and its solutions. We derive specific properties from the longest path to generate neighborhoods, facilitating a highly targeted local search methodology tailored to this problem domain. To favor further research, we have introduced a new set of problem instances, with all relevant data available on a dedicated web page to favor fair future investigations.

References

1. Bierwirth, C.: A generalized permutation approach to job shop scheduling with genetic algorithms. OR Spektrum **17**, 87–92 (1995)

2. Bilge, U., Ulusoy, G.: A time window approach to simultaneous scheduling of machines and material handling system in an FMS. Oper. Res. **43**(6), 1058–1070 (1995)
3. Drexl, M.: Synchronization in vehicle routing - a survey of VRPs with multiple synchronization constraints. Transp. Sci. **46**(3), 297–316 (2012)
4. Hurink, J., Knust, S.: Makespan minimization for flow-shop problems with Bibliographie 214 transportation times and a single robot. Discret. Appl. Math. **112**, 199–216 (2001)
5. Hurink, J., Knust, S.: A tabu search algorithm for scheduling a single robot in a job-shop environment. Discret. Appl. Math. **119**(1–2), 181–203 (2002)
6. Hurink, J., Knust, S.: Tabu search algorithms for job-shop problems with a single transport robot. Eur. J. Oper. Res. **162**(1), 99–111 (2005)
7. Lacomme, P., Larabi, M., Tchernev, N.: Job-shop based framework for simultaneous scheduling of machines and automated guided vehicles. Int. J. Prod. Econ. **143**(1), 24–34 (2013)
8. Lawrence S.: Resource constrained project scheduling: an experimental investigation of heuristic scheduling techniques (Supplement). Graduate School of Industrial Administration, Carnegie-Mellon University, Pittsburgh, Pennsylvania (1984)
9. MecaBotiX (2023). https://www.mecabotix.com/
10. Pinedo, M.: Scheduling - Theory, Algorithms and Systems. Springer, Heidelberg (2012)
11. Prins, C.: A GRASP × evolutionary local search hybrid for the vehicle routing problem. In: Pereira, F.B., Tavares, J. (eds.) Bio-inspired Algorithms for the Vehicle Routing Problem. Studies in Computational Intelligence, vol. 161, pp. 35–53. Springer, Heidelberg (2009). https://doi.org/10.1007/978-3-540-85152-3_2
12. Seo, J., Paik, J., Yim, M.: Modular reconfigurable robotics. Annu. Rev. Control Robot. Auton. Syst. **2**, 63–88 (2019)
13. Ulusoy, G., Bilge, U.: Simultaneous scheduling of machines and material handling system in an FMS. Int. J. Prod. Res. **31**(12), 2857–2873 (1993)
14. Ulusoy, G., Sivrikaya-Serfioglu, F., Bilge, U.: A genetic algorithm approach to the simultaneous scheduling of machines and automated guided vehicles. Comput. Oper. Res. **24**(4), 335–351 (1997)
15. Yao, Y.J., Liu, Q.H., Li, X.Y., Gao, L.: A novel MILP model for job shop scheduling problem with mobile robots. Robot. Comput.-Integr. Manuf. **81** (2023)

AI and Metaheuristics for Routing

SIRO: A Deep Learning-Based Next-Generation Optimizer for Solving Global Optimization Problems

Olaide N. Oyelade[1], Absalom E. Ezugwu[2]($\boxtimes$) (iD), and Apu K. Saha[3]

[1] School of Electronics, Electrical Engineering and Computer Science, Queen's University Belfast, Belfast, UK
[2] Unit for Data Science and Computing, North-West University, 11 Hoffman Street, Potchefstroom 2520, South Africa
absalom.ezugwu@nwu.ac.za
[3] Department of Mathematics, National Institute of Technology Agartala, Agartala 799046, Tripura, India

Abstract. This paper introduces the SIR Optimizer (SIRO), a novel next-generation learned metaheuristic algorithm inspired by biological systems and deep learning techniques. The optimizer uses the susceptible-infected-removed (SIR) epidemiological model to predict the population's susceptibility, active infections, and recoveries. To enhance the search process, SIRO incorporates deep learning into its initialization and parameter tuning components, enabling intelligent and autonomous behaviour. By generating initial solutions based on neural models, the algorithm achieves efficient, effective, and robust search outcomes. To validate the effectiveness of SIRO, a set of numerical hybrid test functions from the CEC 2017 benchmark, each characterized by 30 dimensions were utilized. The experimental results were compared against various state-of-the-art algorithms, demonstrating that SIRO outperforms its competitors. Moreso, it delivers high-quality solutions while utilizing fewer control parameters. The incorporation of a learning process in SIRO leads to superior precision and computational efficiency compared to other optimization approaches in the existing literature.

Keywords: SIR-model · SIRO · optimization algorithms · bio-inspired computing · deep learning · machine learning

1 Introduction

In recent years, there has been a growing research interest in the development of efficient, effective, and robust global search techniques for solving numerical and combinatorial optimization problems [1]. However, the complexity of real-world optimization problems has significantly increased over the past decades due to advancements in industrial processes and societal evolution, posing a challenge for existing optimization techniques, particularly classical methods [2]. The literature suggests that these optimization techniques have limitations in performing intelligent and robust search operations within

M. Sevaux et al. (Eds.): MIC 2024, LNCS 14753, pp. 45–61, 2024.
https://doi.org/10.1007/978-3-031-62912-9_6

problem-solution search spaces [3]. Nevertheless, the optimization community continues to propose new design variants and implementations of optimization techniques to address these challenges [4].

Metaheuristic algorithms have gained popularity and acceptance as the preferred optimization technique, despite their inability to generate precise or exact solutions for candidate optimization problems [5]. While these algorithms provide approximate solutions, they suffer from the requirement of problem-specific information or techniques, lack of guaranteed optimality in terms of convergence, absence of a theoretical or mathematical basis, reliance on multiple search parameters, stochastic search processes resulting in different solutions for the same problem, and the need for stopping criteria declaration. However, metaheuristics offer a valuable alternative to exact or mathematical optimization methods due to their flexibility, global optimization capability, robustness to problem size and randomness, and practical applicability to challenging real-world problems. These algorithms have also found widespread applications in engineering design problems, deep learning parameter optimization, facility layout problems, medical image segmentation and classification, parallel machine scheduling, and many others [6].

Despite the existence of numerous state-of-the-art optimization algorithms in the literature, the pursuit of new optimization methods remains constant. This need for new metaheuristic algorithms is driven by the no-free lunch (NFL) theorem of optimization, as proposed by Wolpert and Macready [7]. The NFL theorem suggests that there is no single optimization algorithm that works best for all optimization problems. In other words, there is no "free lunch" or universally superior optimization strategy. Consequently, the design of a universal, general-purpose optimization strategy is deemed impossible. This realization has motivated optimization experts to develop new algorithms.

From an intelligent and learning system perspective, the focus has shifted towards the development of more intelligent metaheuristic algorithms capable of learning from historical datasets. Extensive research on classical metaheuristic algorithms has highlighted that these algorithms generate substantial volumes of datasets during the solution search process. These datasets may contain valuable knowledge regarding the properties of good and bad solutions, the performance of different operators at different stages of the search process, and the precedence of search operators [5]. Surprisingly, classical optimization algorithms have not effectively utilized the knowledge hidden within these generated datasets.

Recent studies have demonstrated that machine learning (ML) techniques can complement metaheuristics by extracting useful knowledge from the generated data throughout the search process [6]. Integrating such knowledge into the search process can guide metaheuristics to make more informed decisions, enhancing their intelligence and significantly improving solution quality, convergence rate, and robustness. Thus, the present study proposes a new metaheuristic optimization algorithm for solving both single-objective and multi-objective problems. Notably, no classical algorithm has employed ML techniques in its initial algorithmic design stage.

In this study, a novel next-generation metaheuristic optimization algorithm called SIRO is proposed. Inspired by the SIR (susceptible-infected-removed) epidemiological model, the algorithm mimics the propagation of disease to guide the search process. A machine learning component is integrated into SIRO to facilitate the intelligent automation of the algorithm's initialization and parameter configuration settings. This learning component assists the algorithm in selecting optimal initial high-quality solutions, thereby guiding an intelligent search process within the solution search spaces. The SIRO algorithm was evaluated using various numerical test functions. To validate the superior performance of SIRO, its results were compared against other state-of-the-art metaheuristic techniques.

2 Model Description

This section presents an overview of the proposed SIRO algorithm, including its inspiration and the optimization modeling design steps. Moreover, the section is divided into two main subsections for a smooth presentation of the various aspects of SIRO: the modeling source of inspiration for the SIRO from the popular SIR model is first discussed, followed by a description of the SIRO's implementation.

2.1 SIRO Algorithm Modelling

In this section, the design and model formulation describing the complete procedure for the proposed SIRO algorithm are discussed. An optimization process of SIRO is first described using a SIR-based model comprising the three compartments. Secondly, the mathematical model of the SIRO method is outlined from the concept of population initialization to the update of the compartment. Furthermore, a pseudocode describing the algorithmic representation of SIRO and an analysis of the algorithm's complexity is also presented in this section. Also, the neural learning method for improving the population and optimization of the parameter combination is described.

The SIR model applied in this study follows the classical model for modeling disease propagation so that all redundant compartments are eliminated. In Fig. 1, the combined representation of the SIR model and its optimization process are illustrated. The three compartments captured by the model include susceptible (S), infected (I), and recovered (R) individuals. The initial population of the SIR model is assigned to the S compartment since all contagious diseases target the susceptible population to generate the infected population. Considering the generic nature of the classical SIR model, the propagation pattern of each disease determines the assignment of individuals to the I compartment. In this study, the transition of individuals from S to I is conditioned on exposure to disease at a given rate denoted by $S\pi$. Conditions, such as immunity, vaccination, hospitalization, treatment, and self-recovery, have often reassigned individuals from the I compartment to the R compartment. The movement of individuals to R follows the order of Ig, which signifies recovery from the disease. At some point, individuals in R transition to S, thereby making them susceptible individuals that might be recruited into I unless there is a high influence of immunity that mitigates contagion.

The total duration T for a disease outbreak can be denoted into some $t_1, t_2, \ldots t_n$, which is represented in the figure as $t = 0$, $t = 1$, $t = 2 \ldots t = n$ where t_0 or $t = 0$ is a notation for showing when the outbreak of the disease is reported in a given population P. For the SIR model in this study, it is assumed that the population size of P is kept at

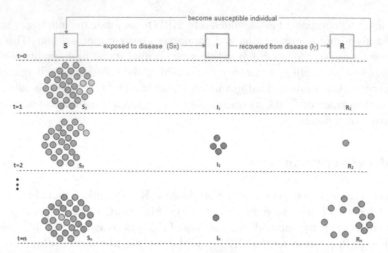

Fig. 1. Illustration of the SIR model in the process of solving optimization problems.

constant c with the implication that birth or death are variables or events that are frozen until duration T is completed. This keeps the size of the population constant, where the sum of individuals includes those in S, I, and R. The optimization process of the SIR models therefore follows that at t_0 or $t = 0$, all individuals in P are represented in S ($size(P) = size(S)$). At time $t = 1$, some individuals in S have been exposed to the disease and now can be infected by the virus, causing the disease after some duration, which is often peculiar to the nature of the disease. At $t = 2$, it can be observed that the exposed individuals in S have now been infected, leading to their reassignment to the I compartment. Meanwhile, at the same time, it is reported that one of the infected cases has recovered and is now assigned to the R compartment. As T increases, the reassignment of individuals across the compartment generates a dynamic process with each instance of the processes having a formation of allocations to S, I, and R with different patterns. The optimization process is resumed with the continuous anatomical changes of each individual whose mutation and displacement within the search space present a useful search pattern.

Table 1 presents an outline of all the parameters and their corresponding definition used to describe the SIR model. We take note of the infection rate π, recovery rate g, contact rate β with infected individuals, and natural death rate G of a population.

Considering the generic nature of the classical SIR model proposed in this study, setting specific values for each of the parameters is not applicable. Moreover, using stochastic models to assign values to the parameters assumes the popular approach used in literature. In this study, we proposed a novel machine learning-based model for the selection and combination of parameter values for all parameters used for the proposed SIRO algorithm. Meanwhile, the mathematical model of the algorithm is first presented and then a description of the learning-based method is also discussed in the following paragraphs.

Table 1. Symbols and definitions of the SIR model parameters

Symbols	Descriptions
π	Infection rate
γ	Recovery rate
Γ	Disease-induced death rate
τ	Natural death rate
β_1	Contact rate of infected
β_2	Contact rate of recovered

2.2 SIRO Model

The entire population of the SIR model is represented in Eq. (1), where at any time t_i, the summation of all individuals in the S, I, and R yields the total population:

$$P = S + I + R \tag{1}$$

We note that the generation of individuals into the S, I, and R compartments is dependent on a system differential equation consisting of three sub-models, as captured in Eqs. (2), (3), and (4) respectively:

$$S_t = \frac{\partial S(t)}{\partial t} = S\pi - I\beta \tag{2}$$

$$I_t = \frac{\partial I(t)}{\partial t} = I + S\beta - Ig \tag{3}$$

$$R_t = \frac{\partial R(t)}{\partial t} = R + gI \tag{4}$$

where S_t, I_t, and R_t are the computed number of individuals allocated to S, I, and R at an arbitrary time t_i, respectively. Updates to these compartments are achieved during every iteration phase of the algorithm during the training or optimization process. Each time there is an update, it implies that individuals have been reassigned to different compartments, and therefore, there is a mutation of individuals according to the operation associated with a compartment. In the population initialization phase, all individuals are appropriated a certain composition as discussed in the following paragraph.

2.3 Basic and Neural Network-Based Initialization Methods

The population initialization and representation for population-based optimization algorithms often assume a stochastic approach or use some recent methods reported in the literature. In this study, a stochastic population initialization method is evolved to incorporate a deep learning approach. First, the population is initialized and represented by the matrix representation in Eq. (5) using a two-dimensional approach, where d is equivalent to the dimension of the optimization problem:

$$X = \begin{bmatrix} x_{1,1} & x_{1,2} & \cdots & x_{1,d-1} & x_{1,d} \\ x_{2,1} & x_{2,2} & \cdots & x_{2,d-1} & x_{2,d} \\ & & \vdots \;\; \vdots \; x_{i,j} \;\; \vdots \;\; \vdots \\ x_{n,1} & x_{n,2} & \cdots & x_{n,d-1} & x_{n,d} \end{bmatrix} \tag{5}$$

The computation for each x_i in X is achieved using the stochastic method represented in Eq. (6), where U and L respectively denote the upper and lower bounds typical in optimization problems with the optimization problem:

$$x_i = L + rand(0,1) * (U - L) \tag{6}$$

The resulting initial solution in X was applied to train a related biology-based and disease propagation optimization algorithm to obtain some final solution. The trajectory of solutions obtained was then curated and applied to train a deep learning model so that the model can learn a suitable solution space that can yield a potential initial solution for training the proposed SIRO algorithm. A long short-term memory (LSTM), which is a kind of recurrent neural network (RNN) and generally classified as deep learning (DL), was considered for this task of learning to generate a suitable initial search space. The architectural representation of LSTM is shown in Fig. 2.

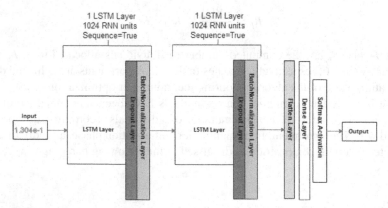

Fig. 2. The design of the proposed LSTM architecture used for generating suitable and optimal initial solutions for the proposed SIRO.

The deep learning architecture consists of two LSTM layers, each of which is followed by a dropout layer and batch-normalization layer, with each layer equivalent to

1024 RNN units. Each layer is followed by flattened and dense layers to produce outputs typical of a generator G whose sequence represents the initial solution space of SIRO. The output from the G is denoted by R, which is a sample raw solution that requires some processing or parsing to allow for it to be used as the search space. This generator and parser are represented in Eq. (7), where $\dim(P)$ denotes the size and dimension of the population as required by the SIRO algorithm:

$$R = parse(G(X, \dim(P))) \tag{7}$$

The parsed initial solution is then used to generate the individuals, which are populated into S in Eq. (8):

$$S = \left\{ s_i \in R \not\! \phi \, d \mid 0 \leq i \leq \dim(P) \right\} \tag{8}$$

where ϕ is an N-ary operator with a clockwise integral operation on R; and d represents the delimiter, which indicates the end and beginning of the previous s_{i-1} from next s_i individual in the population. To ensure that each s_i is kept within the bounds of the problem space, Eq. (9) is applied to amend all s_i in S, where ub and lb are upper and lower bounds, respectively:

$$S = \left\{ s_i \in [amend(S, i)]_{lb}^{ub} \mid 0 \leq i \leq \dim(P) \right\} \tag{9}$$

During iteration for the optimization process, the global best s_{best} individual is obtained by using Eq. (10), which compares s_{cbest} with the previously obtained s_{cbest}:

$$S_{best} = \begin{cases} s_{best}, fits(s_{cbest}) < fits(s_{best}) \\ s_{cbest}, fits(s_{cbest}) \geq fits(s_{best}) \end{cases} \tag{10}$$

The position of every s_i in the search space is used to discover when the displacement of the individual has reached a disease-super-spreading point when such an individual is infected. As a result, at the beginning of every iteration, the current positions of the individuals in I are computed and used to determine when intensification and explorative mechanism of the algorithm occur. In Eq. (11), the computation of the positions for each individual in I is represented:

$$lpos_i^{t+1} = Ipos_i^t + rand(0, 1) * \rho \tag{11}$$

$$M(s) = lrate * rand(0, 1) + M(Ind_{best}) \tag{12}$$

where ρ represents the scale factor of displacement of an individual; $Ipos_i^{t+1}$ and $Ipos_i^t$ are the updated and original position at time t and t + 1, respectively; and $rand(0, 1)$ randomly yields a value in the range 0 and 1.

The value obtained for $Ipos_i^{t+1}$ is assigned to $srate$ when $Ipos_i^{t+1} < 0.5$; otherwise, it is assigned to $lrate$.

2.4 SIRO Neural Network-Based Parameter Selection

Compared with other related methods, SIRO uses a few parameters for the control and optimization process. To take this advantage further, this study proposes a novel deep learning approach that intelligently combines and selects the best parameter-value configuration required to yield a state-of-the-art performance. To achieve this, a pool of parameter-value configurations is generated and applied to optimize some benchmark functions so that the performance of the image of SIRO for each configuration is collected as a dataset for the deep learning model. Using a convolutional neural network (CNN), the datasets are supplied as input for training the model so that the trained model can be used to predict the best configuration appropriate for yielding the best performance of the best SIRO algorithm. In Eq. (13), the modeling of the mapping function, which generates the pool of configurations used in the experimentation, is described:

$$pv(\pi, g, \beta, G) = pv : \{p_1, p_2, p_3, p_4\} \rightarrow \{v_1, v_2, v, v_4\} \tag{13}$$

where pv represents a mapping function whose output is a set C of a unique combination of the SIRO parameters. The elements of C are described as $\{c \in pv(\pi, g, \beta, G)_i | 0 \le i \le N\}$. Each c is equivalent to a combination of parameters and their corresponding generated values so that such mapping of parameter-value is applied for partial training SIRO, and the convergence of the graph is generated and collected as a dataset for training the CNN. The architecture of the CNN used for this training is illustrated in Fig. 3, whereby the completely trained network is then used to predict which best c_i is suitable for fully training SIRO.

The CNN architecture consists of six blocks of convolution-pooling units, where each block contains two convolution layers, one zero-padding layer, and one max pooling layer. A 3 × 3 filter size is used for all layers of convolution in the architecture, though the filter count for each block of convolution-pooling follows 2^k where k ranges between [5, 10]. The architecture is completed using a flattened layer, Dense layer, and Dropout layer, and the SoftMax activation function is used for classification purposes.

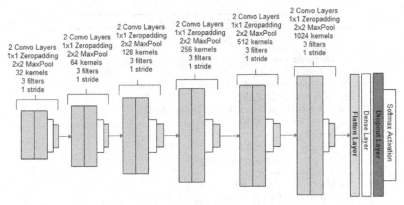

Fig. 3. The design of the convolutional neural network (CNN) architecture applied for learning solution pattern of SIRO to obtain the best configuration and combination of parameter values

2.5 SIRO Algorithms and Computational Complexity

In Algorithm 1, the pseudo-code of the proposed SIRO metaheuristic algorithm is presented. The algorithm lists all the procedures described in previous paragraphs and outlines how the optimization process is achieved during the iterative process. Input to the algorithm includes the number of iterations required for training the model, population size N, and dataset required to train the CNN model for obtaining the best c_i to be used for training SIRO. The expected output is the combination of all the best solutions obtained for each iteration and also the final global best. Line 1 of the algorithm shows the initialization of the S, I, and R compartments and also the container for storing the best solutions. In order for the LSTM model to generate the best initial solution for training SIRO, a stochastically generated initial solution X is supplied as input for training the LSTM model. Then, the trained LSTM model generates the learned-solution space for SIRO for full training. Meanwhile, Lines 5–7 display the application of the CNN model to generate the best configuration of the parameter value set suitable for obtaining the values to use on all parameters of SIRO. The fitness values of all individuals in S are obtained and the global best solution is assigned in Lines 8–9. The iterative training of the SIRO algorithm is listed in Lines 10–31, and the return values are in Line 32. The generation and mutation of individuals into the I compartment are seen in Lines 12–21, and updates of individuals in all three compartments are exhibited in Lines 22–25. Lastly, the update of the current and global best solution for each iteration is reported in Lines 26–30.

Algorithm 1: SIRO metaheuristic algorithm

Result: global Best, solutions
Input: iter, N, dataset
Output: sols, gbest

1 $S, I, R, sols \leftarrow \emptyset$;
2 $X \leftarrow createPopulation(N, S)$ using Eq.6;
3 $lstm = buildLSTM()$;
4 $S = lstm.trainGenS(X, N)$ using Eq.9;
5 $cnn = buildCNN()$;
6 $model = cnn$;
7 $params = \Delta \circ \{\pi, \Gamma, \gamma, \beta\}$;
8 $\pi, \Gamma, \gamma, \beta \leftarrow apply(cnn, params)$;
9 $S = fits(S, asc)$;
10 $gbest, cbest, I[0] \leftarrow S[0]$;
11 **while** $e \leq iter$ **do**
12    $lpos \leftarrow position(I)$ using Eq.11;
13   **for** $i \leftarrow 1$ to $len(I)$ **do**
14      $newI \leftarrow \emptyset$;
15   **if** $lpos_i < 0.5$ **then**
16       $tmp \leftarrow rand(0, Eq.3 \times I \times srate, \gamma, \beta)$;
17   **end**
18   **else**
19       $tmp \leftarrow rand(0, Eq.3 \times I \times lrate, \gamma, \beta)$;
20   **end**
21      $newI + \leftarrow tmp$;
22   **end**
23    $I + \leftarrow newI$;
24    $r \leftarrow rand(0, Eq.4 \times I, \gamma), R + \leftarrow r$;
25    $I + \leftarrow I - r$;
26    $S + \leftarrow r$;
27    $cbest = fits(S)$;
28   **if** $cbest > gbest$ **then**
29      $gbest = cbest$;
30      $sols \leftarrow gbest$;
31   **end**
32 **end**
33 **return** gbest, sols;

The complexity of the proposed SIRO algorithm can be computed by partitioning the algorithm into three segments, namely the initialization, optimization, and result return stages. The first stage, i.e. the initialization stage, has Lines 5, 7, and 8, which will yield a computational analysis of $O(n)$ and all other lines showing evidence of basic operations evaluated to $O(1)$. Computationally speaking, it is clear that $O(n) + O(n) + O(n) = O(n)$ by the rule of summation algorithm analysis, where n is the population size of the search space. The nested loop seen in Lines 11–31 has a computational analysis equivalent to $O(m * n)$ for the optimization stage, where m in this case is the number of iterations during the optimization process. The third stage which returns the result of the optimization process can be equated to $O(1)$. Therefore, the complexity of the entire algorithm may be evaluated by summing $O(n) + O(m * n) + O(1)$, which in turn yields $O(m * n)$. Note that for simplification, may be defined as $O(n^2)$.

In the next section, detailed experimentation of the proposed SIRO algorithm is presented along with a thorough discussion of the results. For the implementation of SIRO, we combined the solution-space initialization, parameter fine-tuning, and optimization process of the SIR model, as described in Algorithm 1.

3 System and Parameter Configuration

The experimentation was conducted using the Google Colab platform, which has 12 GB memory, 100 GB disk size, Python 3 backend, and a GPU. Additional experiments were performed on the Google Cloud compute engine with specific configurations. The parameter values for the SIRO algorithm were derived from ranges and step variables, with certain parameters computed from randomly generated models. Each parameter had a grid of ten candidate values for the deep learning model to select from. The details of the parameter configurations can be found in Table 2. The parameters and candidate values were used to train the deep learning model for SIRO. In the next subsection, we discuss the application of IEEE CEC 2017 functions to rigorously test the resulting SIRO configuration. Moreover, various metaheuristic algorithms were considered, spanning human-based, evolutionary-based, swarm-based, biology-based, physics-based, and math-based approaches. This includes the Arithmetic Optimization Algorithm (AOA), Artificial Bee Colony (ABC), Aquila Optimizer (AO), Archimedes Optimization Algorithm (ArchOA), Ebola Optimization Search Algorithm (EOSA), Firefly Algorithm (FA), Invasive Weed Optimization (IWO), and Memetic Algorithm (MA).

Table 2. List of SIRO parameters with corresponding notations and range of values

Variable	recruitment_rate	disease_induced_death_rate	contact_rate_infectious	contact_rate_recovered	recovery_rate	natural_death_rate
Notation	π	G	β_1	β_2	g	t
Parameter range	[0.1, 0.9]	$randf(0, 1)$	[0.1, 0.9]	[0.1, 0.9]	$randf(0, 1)$	$randf(0, 1)$
Step	0.1	N/A	0.1	0.1	N/A	N/A
Size	10	10	10	10	10	10

Training SIRO for each parameter-value configuration was achieved using 50 iterations since the aim was to perform partial training for plotting the convergence curve. Therefore, the scatter plots representing the convergence of solutions in the solution space were obtained as an image input for training the CNN model. Figure 4 displays sample scatter plots for the parameter-value configurations listed in Table 3. It can be seen that each plot has a unique representation of the distribution of points in the scatter plot to show how the solutions converge after some training time.

Scatter plot-generated image samples were employed to train the CNN to learn input convergence patterns. The goal was to predict optimal image representations for fully training the SIRO method. After the completion of CNN training, a correctly predicted image with superior convergence was selected, and its corresponding parameter values were extracted. The optimal parameter-value combination for training SIRO was identified as π: 0.3, $\beta 1$: 0.0783, $\beta 2$: 0.3, Γ: 0.3, γ: 0.1914, and τ: 0.1272.

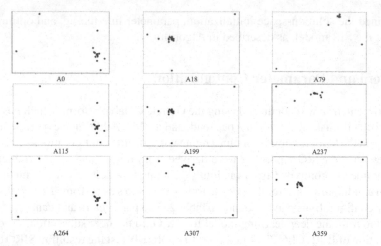

Fig. 4. Scatter plot illustrating parameter-value configurations used as input for training a CNN in an image-label solution.

3.1 Results and Discussion

Benchmark functions are traditionally used for evaluating metaheuristic algorithms to compare their performances. The standard benchmark functions and IEEE CEC functions have been widely accepted and applied for testing the performance of optimization algorithms. Since the IEEE CEC functions demonstrate some uniqueness and complexity, which are comparable with real-life optimization problems, we focused on the evaluation of the SIRO algorithm using these functions. Ideally, the CEC functions consist of fourteen (14) functions, offering a comprehensive range of computational capabilities. We experimented with 28 hybrid combinations derived from these 14 fundamental CEC functions. These hybrids include variants such as C1–C8 and C10, which undergo shifts (S) relative to their original CEC counterparts. The C9 and C11–C16 represent the shift-rotate (SR) adaptation of the corresponding CEC functions. The functions C17–C22 represent shift variants of a combination of some CEC functions, whereas C23–C28 are hybrids of predefined CEC hybrids. Each of the 28 derived hybrids has 30 dimensions.

Results presented in this section are based on the pre-training of SIRO using deep learning methods and the training and testing of SIRO using the hybrid CEC functions. The section is divided into three subsections so that three major issues are addressed: 1) evaluation of the SIRO parameter selection capability as obtained using the CNN; 2) comparative analysis of SIRO with other similar optimization methods to show if the deep learning aided method is preferable; and 3) a collection of observations into a discussion of findings as revealed from the study. The selected parameter-value configuration was used to fully train SIRO under 500 iterations. In addition to SIRO, other metaheuristic algorithms such as ABC, AO, AOA, ArchOA, EOSA, FFA, IWO, and MA were trained for the purpose of comparative analysis.

Based on the experimental results presented in Table 3, the SIRO algorithm demonstrates exceptional performance across the various hybrid benchmark functions, particularly surpassing in handling complex real-life problems involving shift and rotate operations. Its competitiveness is evident in outperforming other compared methods across different hybrid function types, indicating its suitability for addressing a wide range of optimization challenges. Moreover, SIRO maintains its effectiveness even when confronted with higher levels of complexity, as observed in functions involving combinations of CEC benchmarks. The method's consistent performance across different function categories underlines its utility and confirms its status among top optimization algorithms for tackling complex real-world problems. In summary, it was found that SIRO returned the best performance among all methods for the C1–C8 hybrid categories, while the performance is competitively tied among AO, AOA, EOSA, and ArchOA. A further evaluation of SIRO on C9 and C11–C16 functions also revealed good performance. Recall that these functions present a higher level of complexity than the C1–C8, notwithstanding, the performance of SIRO with this increasing level of complexity is highly encouraging.

Figure 5 depicts the analysis of SIRO's exploration and exploitation patterns. The ideal curves exhibit a smoothly drawn slanted U-shape, with the exploration curve descending from the top-left to bottom-right, and the exploitation curve ascending from bottom-left to top-right. SIRO displayed distinctive and nearly perfect curve patterns, showcasing its exceptional performance in navigating the solution space. In comparison to other algorithms, SIRO demonstrated superior abilities in efficiently finding solutions within both exploration and exploitation phases, affirming its capacity to intelligently avoid local optima. An effective optimization algorithm is characterized by convergence curves aligning into a straight line over iteration or training time. In contrast, if these curves scatter without evident convergence as iterations increase, the optimization method is deemed underperforming. SIRO's categorization among top-performing metaheuristic algorithms, demonstrated by its convergence graph, underscores its suitability for challenges demanding smooth and desirable convergence in the solution space.

Figure 6 illustrates the computational time needed by the optimization methods, including the SIRO algorithm, to execute the distinctive C10 benchmark function. Notably, SIRO outperformed other high-performing algorithms like EOSA and AOA, completing the task in just 0.05 secs, while EOSA and AOA required an average of 0.30 secs, respectively. This highlights SIRO's exceptional efficiency compared to similar methods. The exhaustive experimentation with the SIRO method indicates its competitiveness with state-of-the-art methods in addressing real-life optimization problems. Notably, the use of deep learning for parameter-value configuration improved SIRO's performance, ensuring consistent configuration across instances. This is a significant contribution to the field of optimization domain. SIRO's performance on benchmark CEC functions highlights its utility for real-life optimization problems, especially those akin to C1–C28 functions. In summary, SIRO competes favorably, demonstrating outstanding performance and effective balance between exploration and exploitation processes. Its convergence graphs affirm its relevance for problems demanding smooth convergence of the solution space.

3.2 Analysis of Statistical Results

The Anderson-darling test states that if the p-value > 0.05, then the dataset is normally distributed and the parametric test should be considered, otherwise a non-parametric

Table 3. A comparative analysis of SIRO against other optimization techniques using the hybrid functions of CEC 2017, each characterized by 30 dimensions.

Func.	Metric	ABC	AO	AOA	ArchOA	EOSA	FFA	IWO	MA	SIRO
C1	best	3.6E+03	**1.0E+02**	6.7E+04	**1.1E+02**	1.0E+03	7.9E+03	3.8E+05	2.2E+06	**1.1E+02**
	mean	2.9E+04	5.3E+04	6.7E+04	1.8E+04	4.2E+03	1.4E+04	3.9E+05	2.2E+06	2.1E+03
	std	1.4E+05	1.8E+05	1.5E−11	6.4E+04	2.7E+04	2.6E+04	4.5E+04	6.1E+03	1.6E+04
	worst	1.8E+06	2.1E+06	6.7E+04	6.4E+05	5.5E+05	5.7E+05	6.2E+05	2.3E+06	2.6E+05
C2	best	2.9E+05	2.7E+02	4.0E+06	**2.0E+02**	7.5E+05	5.1E+05	3.6E+08	1.3E+08	**2.0E+02**
	mean	1.8E+07	4.5E+07	4.0E+06	1.1E+07	1.7E+06	1.9E+07	3.9E+08	1.3E+08	1.0E+06
	std	1.1E+08	1.6E+08	0.0E+00	1.1E+08	5.5E+06	1.3E+08	1.1E+08	4.5E+06	5.3E+06
	worst	1.2E+09	1.6E+09	4.0E+06	1.6E+09	7.2E+07	1.4E+09	1.1E+09	1.6E+08	1.0E+08
C3	best	3.0E+02	3.0E+02	1.0E+06	3.0E+02	3.1E+02	1.9E+03	1.7E+04	1.9E+04	4.6E+02
	mean	9.6E+02	6.9E+03	1.0E+06	7.0E+02	2.0E+03	2.7E+03	1.7E+04	1.9E+04	8.0E+03
	std	4.5E+03	6.4E+03	0.0E+00	2.8E+03	6.9E+03	5.1E+02	1.5E+03	0.0E+00	1.4E+05
	worst	5.1E+04	2.6E+04	1.0E+06	2.4E+04	5.9E+04	5.2E+03	2.4E+04	1.9E+04	3.1E+06
C4	best	4.0E+02	4.0E+02	4.0E+02	4.0E+02	4.0E+02	4.0E+02	4.8E+02	4.1E+02	4.0E+02
	mean	**4.0E+02**	**4.0E+02**	**4.0E+02**	**4.0E+02**	**4.0E+02**	**4.0E+02**	4.8E+02	4.1E+02	**4.0E+02**
	std	5.4E+00	1.5E+01	5.7E−14	6.2E+00	5.6E+00	5.8E+00	5.7E−14	2.0E+00	7.5E+00
	worst	4.6E+02	5.5E+02	**4.0E+02**	4.9E+02	5.2E+02	4.7E+02	4.8E+02	4.4E+02	5.6E+02
C5	best	5.0E+02	**5.0E+02**	**5.0E+02**	**5.0E+02**	**5.0E+02**	5.2E+02	5.2E+02	5.2E+02	**5.0E+02**
	mean	5.0E+02	5.1E+02	5.0E+02	5.1E+02	5.0E+02	5.2E+02	5.2E+02	5.2E+02	5.0E+02
	std	2.9E+00	9.4E+00	1.1E−13	8.8E+00	5.2E−01	3.8E−02	5.9E−04	6.8E−01	2.5E+00
	worst	5.2E+02	5.2E+02	5.0E+02	5.2E+02	5.1E+02	5.2E+02	5.2E+02	5.2E+02	5.2E+02
C6	best	**6.0E+02**	**6.0E+02**	**6.0E+02**	**6.0E+02**	**6.0E+02**	**6.0E+02**	**6.0E+02**	**6.0E+02**	**6.0E+02**
	mean	6.0E+02	6.0E+02	6.0E+02	6.0E+02	6.0E+02	6.0E+02	6.0E+02	6.0E+02	6.0E+02
	std	2.8E−01	5.1E−01	1.1E−13	2.8E−01	7.8E−02	1.8E−01	1.1E−13	3.7E−02	1.1E−01
	worst	6.0E+02	6.0E+02	6.0E+02	6.0E+02	6.0E+02	6.0E+02	6.0E+02	6.0E+02	6.0E+02
C7	best	**7.0E+02**	**7.0E+02**	**7.0E+02**	**7.0E+02**	**7.0E+02**	**7.0E+02**	7.1E+02	**7.0E+02**	**7.0E+02**
	mean	7.0E+02	7.0E+02	7.0E+02	7.0E+02	7.0E+02	7.0E+02	7.1E+02	7.0E+02	7.0E+02
	std	2.7E+00	3.1E+00	1.1E−13	1.2E+00	1.6E+00	1.5E+00	5.8E−01	7.8E−02	1.4E+00
	worst	7.3E+02	7.3E+02	**7.0E+02**	7.1E+02	7.2E+02	7.2E+02	7.2E+02	**7.0E+02**	7.3E+02
C8	best	**8.0E+02**	**8.0E+02**	**8.0E+02**	**8.0E+02**	**8.0E+02**	8.1E+02	8.3E+02	8.1E+02	**8.0E+02**
	mean	8.0E+02	8.1E+02	8.0E+02	8.1E+02	8.0E+02	8.2E+02	8.3E+02	8.1E+02	8.0E+02
	std	3.2E+00	3.7E+00	0.0E+00	7.8E+00	2.7E+00	2.8E+00	8.3E−02	6.4E−01	2.1E+00
	worst	8.3E+02	8.2E+02	**8.0E+02**	8.4E+02	8.2E+02	8.4E+02	8.3E+02	8.1E+02	8.4E+02
C9	best	**9.1E+02**	**9.0E+02**	**9.0E+02**	**9.0E+02**	**9.0E+02**	9.1E+02	9.4E+02	9.1E+02	**9.0E+02**
	mean	9.1E+02	9.0E+02	9.0E+02	9.0E+02	9.0E+02	9.1E+02	9.4E+02	9.1E+02	9.0E+02
	std	1.8E+00	3.9E+00	1.1E−13	6.4E+00	2.2E+00	1.8E+00	1.3E−01	9.4E−01	2.5E+00
	worst	9.3E+02	9.2E+02	**9.0E+02**	9.3E+02	9.5E+02	9.3E+02	9.4E+02	9.2E+02	9.4E+02
C10	best	1.3E+03	**1.0E+03**	1.1E+03	**1.0E+03**	1.1E+03	1.3E+03	2.0E+03	1.6E+03	**1.0E+03**
	mean	1.3E+03	1.2E+03	1.1E+03	1.1E+03	1.1E+03	1.3E+03	2.0E+03	1.6E+03	1.0E+03

(*continued*)

Table 3. (*continued*)

Func.	Metric	ABC	AO	AOA	ArchOA	EOSA	FFA	IWO	MA	SIRO
	std	6.8E+01	1.6E+02	4.5E−13	1.4E+02	4.2E+01	7.6E+01	4.5E−13	3.5E+01	9.6E+01
	worst	2.1E+03	1.7E+03	1.1E+03	1.7E+03	2.1E+03	1.7E+03	2.0E+03	1.8E+03	1.9E+03
C11	best	1.2E+03	**1.1E+03**	1.2E+03	**1.1E+03**	**1.1E+03**	1.2E+03	2.0E+03	1.8E+03	**1.1E+03**
	mean	1.3E+03	1.3E+03	1.2E+03	1.3E+03	1.1E+03	1.3E+03	2.0E+03	1.8E+03	1.1E+03
	std	4.4E+01	1.7E+02	2.3E−13	2.5E+02	3.8E+01	7.2E+01	4.6E+00	1.7E+01	6.8E+01
	worst	1.8E+03	2.0E+03	1.2E+03	2.1E+03	1.9E+03	1.6E+03	2.1E+03	1.8E+03	1.9E+03
C12	best	**1.2E+03**	**1.2E+03**	**1.2E+03**	**1.2E+03**	**1.2E+03**	**1.2E+03**	**1.2E+03**	**1.2E+03**	**1.2E+03**
	mean	1.2E+03	1.2E+03	1.2E+03	1.2E+03	1.2E+03	1.2E+03	1.2E+03	1.2E+03	1.2E+03
	std	3.2E−01	8.1E−01	2.3E−13	4.6E−01	6.8E−01	1.4E−01	5.2E−01	0.0E+00	7.2E−01
	worst	1.2E+03	1.2E+03	1.2E+03	1.2E+03	1.2E+03	1.2E+03	1.2E+03	1.2E+03	1.2E+03
C13	best	**1.3E+03**	**1.3E+03**	**1.3E+03**	**1.3E+03**	**1.3E+03**	**1.3E+03**	**1.3E+03**	**1.3E+03**	**1.3E+03**
	mean	1.3E+03	1.3E+03	1.3E+03	1.3E+03	1.3E+03	1.3E+03	1.3E+03	1.3E+03	1.3E+03
	std	1.5E−01	2.6E−01	4.5E−13	1.5E−01	3.6E−01	1.9E−01	2.3E−13	5.2E−02	3.8E−01
	worst	1.3E+03	1.3E+03	1.3E+03	1.3E+03	1.3E+03	1.3E+03	1.3E+03	1.3E+03	1.3E+03
C14	best	**1.4E+03**	**1.4E+03**	**1.4E+03**	**1.4E+03**	**1.4E+03**	**1.4E+03**	**1.4E+03**	**1.4E+03**	**1.4E+03**
	mean	1.4E+03	1.4E+03	1.4E+03	1.4E+03	1.4E+03	1.4E+03	1.4E+03	1.4E+03	1.4E+03
	std	6.8E−02	6.3E−01	2.3E−13	3.9E−01	6.3E−01	6.7E−01	2.3E−13	1.8E−02	1.3E−01
	worst	1.4E+03	1.4E+03	1.4E+03	1.4E+03	1.4E+03	1.4E+03	1.4E+03	1.4E+03	1.4E+03
C15	best	**1.5E+03**	**1.5E+03**	**1.5E+03**	**1.5E+03**	**1.5E+03**	**1.5E+03**	**1.5E+03**	**1.5E+03**	**1.5E+03**
	mean	1.5E+03	1.5E+03	1.5E+03	1.5E+03	1.5E+03	1.5E+03	1.5E+03	1.5E+03	1.5E+03
	std	2.9E−01	6.1E+00	0.0E+00	6.4E−01	1.7E+00	3.4E−01	4.2E−02	2.6E−01	5.3E−01
	worst	1.5E+03	1.6E+03	1.5E+03	1.5E+03	1.5E+03	1.5E+03	1.5E+03	1.5E+03	1.5E+03
C16	best	**1.6E+03**	**1.6E+03**	**1.6E+03**	**1.6E+03**	**1.6E+03**	**1.6E+03**	**1.6E+03**	**1.6E+03**	**1.6E+03**
	mean	1.6E+03	1.6E+03	1.6E+03	1.6E+03	1.6E+03	1.6E+03	1.6E+03	1.6E+03	1.6E+03
	std	7.4E−02	9.9E−02	0.0E+00	1.5E−01	6.3E−02	4.1E−02	2.3E−13	9.3E−02	1.6E−02
	worst	1.6E+03	1.6E+03	1.6E+03	1.6E+03	1.6E+03	1.6E+03	1.6E+03	1.6E+03	1.6E+03

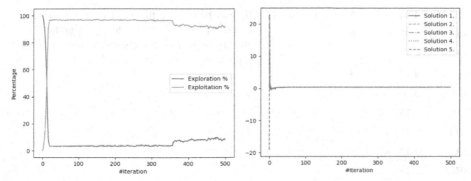

Fig. 5. The left plot illustrates the exploration and exploitation of SIRO, while the right plot shows the convergence trajectory of the first five (5) solutions in the solution space of SIRO.

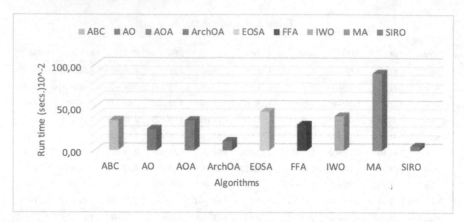

Fig. 6. Computational time comparison of SIRO and other optimization methods on benchmark function C10.

test should be employed. However, the initial test result obtained, with a p-value of 3.53e−25 shows that the data do not follow normal distribution. Therefore, Friedman's test which is a non-parametric statistical test was selected for the comparative study of SIRO with other algorithms. The non-parametric Friedman's test was conducted to obtain the rank SIRO in comparison with the other algorithms. According to Friedman's test, the algorithms are not equally effective. The proposed algorithm SIRO ranked 1 as the best algorithm, while other well-performing algorithms such as EOSA, ArchOA, and AOA are ranked, 2, 3, and 4 respectively. In Table 4, the ranks of the compared algorithms based on Friedman's rank test with a 95% level of confidence are illustrated.

Table 4. Friedman's rank test of proposed SIRO with considered algorithms Null hypothesis (H_0): All the algorithms are equally effective.

Method	Mean rank	Rank	p-value	Conclusion
ABC	4.84	5	1.318e−17	H_0 may be rejected for $\alpha = 1\%$ since p-value = 1.318e−17 < 0.01. At 1% level of significance, effectiveness of several algorithms are not equal considering type − I error of 1%
AO	5.02	6		
AOA	4.48	4		
ArchOA	4.23	3		
EOSA	3.50	2		
FFA	5.13	7		
IWO	7.45	9		
MA	7.13	8		
SIRO	3.23	1		

4   Conclusion and Future Work

This paper introduces a machine-learning approach to optimize parameter configurations for the SIRO algorithm. It utilizes an LSTM model to broaden the search space and explores different SIRO-parameter schemes through a parameter grid. SIRO is trained using permutations of these values, generating image representations for a CNN to identify optimal parameter values. Fully trained SIRO excels in IEEE CEC 2017 benchmark test functions, outperforming similar methods. Analysis of exploration-exploitation and convergence demonstrates SIRO's competitiveness. The approach enables consistent parameter values, challenging stochastic methods' dependency. Future research could extend benchmarks and assess SIRO's robustness against challenging functions and real-world problems. Moreover, SIRO demonstrates versatility in addressing challenging computer vision problems, including medical image classification, along with complex machine learning tasks like deep-leaning parameter optimization and feature selection.

References

1. Seyyedabbasi, A.: A reinforcement learning-based metaheuristic algorithm for solving global optimization problems. Adv. Eng. Softw. **178**, 103411 (2023)
2. Talbi, E.G.: Machine learning into metaheuristics: a survey and taxonomy. ACM Comput. Surv. (CSUR) **54**(6), 1–32 (2021)
3. Dokeroglu, T., Sevinc, E., Kucukyilmaz, T., Cosar, A.: A survey on new generation metaheuristic algorithms. Comput. Ind. Eng. **137**, 106040 (2019)
4. Hussain, K., Mohd Salleh, M.N., Cheng, S., Shi, Y.: Metaheuristic research: a comprehensive survey. Artif. Intell. Rev. **52**(4), 2191–2233 (2019)
5. Wang, L., Cao, Q., Zhang, Z., Mirjalili, S., Zhao, W.: Artificial rabbits optimization: a new bio-inspired meta-heuristic algorithm for solving engineering optimization problems. Eng. Appl. Artif. Intell. **114**, 105082 (2022)
6. Zhao, S., Zhang, T., Ma, S., Chen, M.: Dandelion Optimizer: a nature-inspired metaheuristic algorithm for engineering applications. Eng. Appl. Artif. Intell. **114**, 105075 (2022)
7. Wolpert, D.H., Macready, W.G.: No free lunch theorems for optimization. IEEE Trans. Evol. Comput. **1**(1), 67–82 (1997)

Investigation of the Benefit of Extracting Patterns from Local Optima to Solve a Bi-objective VRPTW

Clément Legrand[1]([✉]) [iD], Diego Cattaruzza[2] [iD], Laetitia Jourdan[1] [iD], and Marie-Eléonore Kessaci[1] [iD]

[1] Univ. Lille, CNRS, Centrale Lille, UMR 9189 CRIStAL, 59000 Lille, France
{clement.legrand4.etu,laetitia.jourdan,marie-eleonore.kessaci}@univ-lille.fr
[2] Univ. Lille, CNRS, Inria, Centrale Lille, UMR 9189 CRIStAL, 59000 Lille, France
diego.cattaruzza@centralelille.fr

Abstract. Hybridizing learning and optimization often improves existing algorithms in single-objective optimization. Indeed, high-quality solutions often contain relevant knowledge that can be used to guide the heuristic towards promising areas. Learning from the structure of solutions is challenging in combinatorial problems. Most of the time, local optima are considered for this task since they tend to contain more relevant structural information. If local optima generally contain more interesting information than other solutions, producing them requires a time-consuming process. In this paper, we study the benefits of learning from local optima during the execution of a multi-objective algorithm. To this end, we consider a hybridized MOEA/D (a multi-objective genetic algorithm) with a knowledge discovery mechanism adapted to the problem solved and we conduct experiments on a bi-objective vehicle routing problem with time windows. The knowledge discovery mechanism extracts sequences of customers from solutions. The results show the benefit of using different strategies for the components of the knowledge discovery mechanism and the efficacy of extracting patterns from local optima for larger instances. An analysis of speed-up performance gives deeper conclusions about the use of local optima.

Keywords: Combinatorial Optimization · Multi-Objective Optimization · Online Learning · Genetic Algorithm · Local Search

1 Introduction

It is known that optimization and learning have a good synergy [8,21], especially in combinatorial optimization. Knowledge discovery (KD) mechanisms consist of an extraction step, where knowledge is discovered from solutions, and an injection step, exploiting the knowledge to guide the algorithm towards new regions of the exploration space.

Due to the success of the use of KD to solve single-objective problems [1], it seems natural to consider the integration of KD into multi-objective combinatorial optimization problems (MoCOPs) [7]. Such problems are frequent in the industry where decision-makers are interested in optimizing several conflicting objectives at the same time. Each objective reflects a different point of view on the problem. More precisely, in a MoCOP, the quality of a solution is evaluated according to multiple criteria that are generally conflicting. In that context, a solution may not be better than another one.

Moreover, the structure of the solutions themselves is important. In single-objective problems, the learning tasks are often performed on local optima [1, 2,10], which tend to contain more relevant structural information. Although obtaining local optima is a time-consuming task, it is known that using a local search as a mutation operator in evolutionary algorithms improves their performance [11,13]. The local search is commonly applied following a probability to reduce the computational overhead induced, bringing more diversity to the solutions found.

In this paper, we investigate the benefit of extracting knowledge from local optima in a multi-objective context. More precisely, we solve a bi-objective vehicle routing problem with time windows (bVRPTW), where total cost and waiting time are simultaneously minimized. To that aim, we consider a MOEA/D [28] hybridized with a KD mechanism [14], which already proved its efficiency, and we propose four related variants, each one with a different strategy for extraction and injection.

The remainder of the paper is organized as follows. In Sect. 2 MoCOPs are formalized. Section 3 presents a brief review of existing works on learning from solutions to optimization problems. After an overview of MOEA/D, Sect. 4 presents how the learning is integrated into MOEA/D and describes four related variants. Section 5 presents the problem studied, the benchmark, and the tuning of the variants. Section 7 gives the experimental protocols followed to answer the following question: Are local optima the only kind of solution from which we should learn? In Sect. 8 our results are presented and discussed. Finally, Sect. 9 concludes and presents perspectives for this work.

2   Multi-objective Optimization

Many logistic problems can easily be considered as multi-objective problems since many challenges have to be tackled simultaneously. These challenges may be linked to economic or environmental issues. Formally, *Multi-objective Combinatorial Optimization Problems* (MoCOPs) are defined as follows [7].

$$(MoCOP) = \begin{cases} Optimize \ F(x) = (f_1(x), f_2(x), \ldots, f_n(x)) \\ s.t. \ x \in \mathcal{D}, \end{cases} \tag{1}$$

where n is the number of objectives ($n \geq 2$), x is the vector of decision variables, $\mathcal{D}$ is the (discrete) set of feasible solutions and each objective function $f_i(x)$ has to be optimized (i.e. minimized or maximized). In multi-objective optimization, the objective function F defines a so-called objective space denoted by $\mathcal{Z}$. Each solution $x \in \mathcal{D}$ is associated with a unique point $F(x)$ in $\mathcal{Z}$.

The notion of *dominance* is defined as follows: a solution x dominates a solution y, in a minimization context, if and only if for all $i \in [1 \ldots n]$, $f_i(x) \leq f_i(y)$ and there exists $j \in [1 \ldots n]$ such that $f_j(x) < f_j(y)$. This criterion induces a partial order on the set of feasible solutions since two solutions can be incomparable (i.e. no one dominates the other one).

Then a set of non-dominated solutions is called a *Pareto front*. A feasible solution $x^* \in \mathcal{D}$ is called *Pareto optimal* if and only if there is no solution $x \in \mathcal{D}$ such that x dominates x^*. Resolving a MoCOP involves finding all the Pareto optimal solutions that form the *Pareto optimal set*. The *true Pareto front* of the problem is obtained by plotting the objective function values corresponding to the solutions in the Pareto optimal set.

Over the years, many metaheuristics based on local search techniques or using evolutionary algorithms [5,9,28] have been designed to solve multi-objective problems. Moreover, many tools [17] have been developed to assess and compare the performance of multi-objective algorithms. In this paper, we consider the unary hypervolume (HV) [29]. It is a metric defined relatively to a reference point Z_{ref}, generally $(1.001, \ldots, 1.001)$, and requires that the objectives of the solutions are normalized between 0 and 1. A value slightly higher than 1 (like 1.001) is often preferred for computational issues. This indicator evaluates the accuracy, diversity, and cardinality of the front. Moreover, the HV can be used without knowing the true Pareto front of the problem. Geometrically, it reflects the volume covered by the members of a non-dominated set of solutions. The larger the hypervolume, the better the set of solutions.

3 Learning and Multi-objective Optimization

Hybridizing machine learning methods and metaheuristics is recently quite common to solve combinatorial problems. Indeed the survey [21] reviews different kinds of hybridizations and proposes to classify the different methods according to where the hybridization is performed: at a problem level, a low level, or a high level. When learning is integrated at a low level, the knowledge is extracted from solutions to the problem. This is the integration we are interested in.

Moreover, the hybridization can be realized either *online* or *offline* [8]. The learning is said online when it uses resources generated during the execution. Otherwise, the learning is said offline.

Most of the KD mechanisms are composed of an *extraction* step, where something is learned, and an *injection* step, which uses the extracted knowledge to find new promising solutions. A study of existing works in machine learning and its hybridization with metaheuristics [21] leads to four questions: *What/Where/When/How* is the knowledge extracted/injected?

Question *What* is problem-dependent, since each problem may have a specific relevant knowledge. In our case, this question is related to the structure of the solutions obtained, since we learn patterns present inside them. Question

Where is algorithm-dependent, since the extraction and injection steps have to be integrated into the process of the algorithm. More precisely, the position of the extraction step in the algorithm highly influences what is learned. Indeed, suppose the extraction is performed after a local search. In that case, solutions will have a more interesting structure relevant to the learning, which will lead to an overall improvement of the following iterations. This strategy has been successfully applied to solve single-objective routing problems [1,3]. However, it could also be interesting to learn about solutions that are not local optima to bring more diversity to the search. Question *When* is algorithm-dependent as well and deals with the frequency of applying the extraction and injection steps. Question *How* corresponds to the design of the extraction and injection steps considering the nature of the problem (multi-objective in our case). While question *What* is considered in Sect. 5.2, once the problem has been presented, Sect. 4.2 addresses the three other questions.

4 Hybridization Between Learning and MOEA/D

4.1 MOEA/D

MOEA/D [28], is a genetic algorithm that approximates the Pareto front by decomposing the multi-objective problem into several scalar objective subproblems, as illustrated in Fig. 1. Each iteration of the algorithm optimizes one of the subproblems, by applying a genetic step composed of crossover and mutation operators. The mutation is commonly replaced by a local search [11,13], to intensify the search in a region of the space. Consequently, it generates better solutions for the next crossover step. However, the local search is time-consuming and increases the time allocated to each generation. Thus, given the same time of execution, fewer generations are performed, and consequently, less crossover, leading to potentially less diversity. MOEA/D is a simple algorithm that has already been widely studied in the literature [27]. Moreover, it has already been successfully hybridized with an opposition-based learning mechanism [16], making it a good candidate for our study.

Here, we consider scalar problems defined with a weighted sum of the objectives. More precisely a convex combination of the n objectives is defined by attributing a weight $w_i \in [0,1]$ to the objective f_i such that $\sum_{i=1}^{n} w_i = 1$. Then the fitness of a solution is the following quantity: $g(x|w) = \sum_{i=1}^{n} w_i \cdot f_i(x)$. Thus to generate different Pareto optimal solutions one can use M different weight vectors $w^1, \dots, w^M$. However, in practice, not all the Pareto optimal solutions can be obtained with such aggregations.

MOEA/D solves the i-th subproblem, by using the solutions of its closest neighbors. Indeed, the *neighborhood*, of size m, of a weight vector w^i is defined as the set of its m closest (for the euclidean distance) weight vectors in $\{w^1, \dots, w^M\}$. Then the neighborhood $\mathcal{N}_m(i)$ of the i-th subproblem simply consists of the m subproblems defined with a weight vector belonging to the neighborhood of w^i. In the following, we consider a uniform distribution on the weight vectors, and we assume that is enough to obtain diverse subproblems.

During the execution of MOEA/D, only the best solution found is kept for each subproblem. When a subproblem i is optimized, the genetic step generates a new solution. The crossover occurs with probability p_{cro} and the mutation with probability p_{mut}. Note that, the crossover is realized between two randomly chosen solutions from subproblems of $\mathcal{N}_m(i)$. Moreover, an external archive stores non-dominated solutions found during the search. These solutions are returned once the termination criterion is reached. In MOEA/D, the crossover applied is the partially mapped crossover (PMX) [12]. Only one solution is randomly kept after the crossover, to reduce the computation time. The mutation is a local search, briefly described in Sect. 5.1. In the remainder of the paper, this version of MOEA/D is called $\mathcal{A}$.

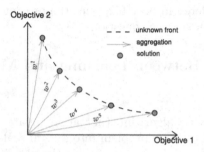

Fig. 1. Decomposition of a bi-objective problem into five scalar problems with weight vectors $w^1, \ldots, w^5$ in MOEA/D. Each subproblem is associated with its current best solution.

4.2   Learning Within $\mathcal{A}$ and Variants

Firstly, we present *How* the KD mechanism works. The KD mechanism integrated within $\mathcal{A}$ is based on the following assumption: close solutions in the objective space share a similar structure. Note that, the *closeness* between two solutions is evaluated according to their objective vector with the Euclidean distance. We present now the notion of *knowledge groups*, as introduced by Legrand et al. [14].

Each knowledge group $\mathcal{G}$ is associated with one of the subproblems defined in $\mathcal{A}$. More precisely $\mathcal{G}_i$ is associated with the subproblem of weight vector w^i. Thus, there are as many knowledge groups as subproblems. Moreover, since the neighborhood of each subproblem is already defined in $\mathcal{A}$, we keep the same neighborhood for the knowledge groups. In other words, if each subproblem has m neighbors, then each solution will belong to m knowledge groups. With this construction, each knowledge group focuses on a part of the objective space.

Once the knowledge groups are defined, they can be used for the extraction and the injection steps. In the following, we answer questions *where* and *when* presented in Sect. 3. The knowledge discovery MOEA/D is presented in Algorithm 1. Briefly, the algorithm follows the MOEA/D framework as described in

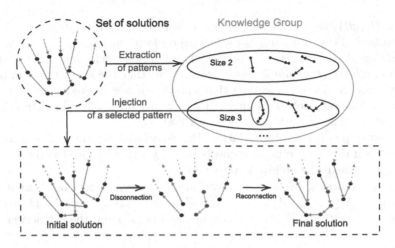

Fig. 2. For each solution belonging to a set of solutions, its patterns are extracted and added to the corresponding knowledge group. A frequent pattern (here of size 3) is selected and injected into a VRPTW solution.

Sect. 4.1. During the initialization phase, an archive is created to keep track of non-dominated solutions found during the execution. An initial random solution is associated with each subproblem, providing an initial population P. For each subproblem i, its $\mathcal{N}(i)$ neighboring subproblems are computed, its initial solution x^i undergoes a LS step with probability p_{mut}, and is then evaluated. The associated knowledge group $\mathcal{G}_i$ is initially empty. Then, while the termination criterion is not achieved, the subproblems are iteratively solved. Considering subproblem i, two neighboring subproblems are selected and their associated solutions undergo a partially mapped crossover to generate a new solution for the subproblem. If the crossover is not applied, the current associated solution is kept. The injection step occurs, with a probability p_{inj}, and the LS is applied to the resulting solution, with probability p_{mut}.

More precisely, to perform the injection step we distinguish between two different strategies $\mathcal{S} = \{int, div\}$. The first one is an intensification strategy ($s = int$), where the search is focused on a specific part of the exploration space. In this case, only one knowledge group is chosen for the injection. More precisely, if the solution x_c is obtained after the crossover for the subproblem i (line 11 of Algorithm 1), then it receives knowledge contained in the knowledge group $\mathcal{G}_i$ that is to say the knowledge group associated with its own subproblem. The other strategy concerns diversification ($s = div$), where the knowledge of all the groups can be used, to favor a larger exploration of the space. More precisely, the solution x_c receives knowledge contained in a randomly chosen knowledge group among all the existing groups.

The extraction step is performed after the genetic step (after line 13 of Algorithm 1), but at some conditions and following a probability p_{ext}. Since we investigate the importance of local optima during learning, we consider two strategies

$q \in Q = \{lo, all\}$, that is the extraction is realized on local optima only or on every solution. If the extraction is performed on local optima only ($q = lo$), then the condition *allowExtraction* is verified if and only if the local search occurred. Otherwise, *allowExtraction* is always verified, so that the extraction can occur on every solution ($q = all$) meaning that knowledge is extracted from all potential solutions found. Note that in both cases, the probability of extraction is considered once the condition has been verified.

The last step concerns the update of the associated solutions of neighboring subproblems if the final solution found is better for them. Once the termination criterion is reached, the archive is returned.

Finally, our design leads to four variants of $\mathcal{A}$, which will be compared in the following: $\mathcal{A}_{int}^{lo}$, $\mathcal{A}_{div}^{lo}$, $\mathcal{A}_{int}^{all}$ and $\mathcal{A}_{div}^{all}$. These variants are called KD variants. Each variant has its specificity concerning the injection and the extraction step.

5   Problem and Related Knowledge

5.1   Vehicle Routing Problems with Time Windows (VRPTW)

The VRPTW [23] is defined on a graph $G = (V, E)$, with the set of vertices $V = \{0, 1, \ldots, N\}$ and $E = \{(i, j) | i, j \in V\}$ the set of arcs. The travel from i to j incurs a travel cost c_{ij} and a travel time t_{ij}. Usually, the cost is computed as the Euclidean distance between the customers, and $t_{ij} = c_{ij}$. A fleet of K identical vehicles with limited capacity Q is based at the depot, represented by vertex 0. The other vertices represent the customers to be served. Each customer has a demand q_i, a time window (TW) $[a_i, b_i]$ during which service must occur, and a service time s_i estimating the required time to perform the delivery. Arriving later than b_i is not allowed here. We recall that a *route* r is an elementary cycle on G that contains the depot and is expressed as a sequence of vertices $r = (v_0, v_1, \ldots, v_{|r|}, v_{|r|+1})$ where $v_0 = v_{|r|+1} = 0$ and vertices $v_1, \ldots, v_{|r|}$ are distinct. The cost c_r of a route r is then given as the sum of traveling costs on arcs used to visit subsequent vertices, that is $\sum_{i=0}^{|r|} c_{v_i, v_{i+1}}$. A solution x is represented as a set of K (possibly empty) routes, that is $x = \{r_1, \ldots, r_K\}$, and its cost is expressed as:

$$f_1(x) = \sum_{k=1}^{K} c_{r_k} \tag{2}$$

When vehicles arrive before a_i, the driver has to wait until a_i to accomplish service incurring a waiting time. We formalize the notion of waiting time as follows. The waiting time W_i at a customer i is given as the maximum between 0 and the difference between the opening of the TW a_i and the arrival time T_i at location i, that is $W_i = \max\{0, a_i - T_i\}$. Note that each route $r = (v_0, v_1, \ldots, v_{|r|}, v_{|r|+1})$ is associated with a feasible (i.e., consistent with traveling times and TWs) arrival time vector $T_r = (T_{v_0}, T_{v_1}, \ldots, T_{v_{|r|}}, T_{v_{|r|+1}})$ and the total waiting time $W_r(T_r)$ on route r, with respect to T_r is given by $W_r(T_r) = \sum_{i=1}^{|r|} W_{v_i}$. Thus the total waiting time of a solution $x = \{r_1, \ldots, r_K\}$

Algorithm 1: The generic $\mathcal{A}_s^q$ framework.

Input: M weight vectors $w^1, \ldots, w^M$ and the size m of each neighborhood. s denotes the strategy followed for the injection. q refers to the nature of the solutions that undergo the extraction.

Output: The external archive A

/* Initialization   */

1 $A \leftarrow \emptyset$

2 $P \leftarrow$ random initial population (x^i for the i-th subproblem)

3 **for** $i \in \{1, \ldots, M\}$ **do**

4      $\mathcal{N}(i) \leftarrow$ indexes of the m closest weight vectors to w^i

5      $x^i \leftarrow LS(x^i)$

6      $Obj^i \leftarrow F(x^i)$

7      $\mathcal{G}_i \leftarrow \emptyset$

/* Core of the algorithm   */

8 **while** *not stopping criterion satisfied* **do**

9   **for** $i \in \{1, \ldots, M\}$ **do**

10          $(k, l) \leftarrow$ select randomly two indexes from $\mathcal{N}(i)$

11          $x_c \leftarrow$ Crossover(x^k, x^l)

12          $x_{inj} \leftarrow$ Injection(x_c, s)

13          $x' \leftarrow LS(x_{inj})$

14   **if** $allowExtraction(x', q)$ **then**

15              $\mathcal{K} \leftarrow$ Extraction(x')

16              $\mathcal{G}_1, \ldots, \mathcal{G}_M \leftarrow$ update with $\mathcal{K}$

/* Updating the neighbors   */

17   **for** $j \in \mathcal{N}(i)$ **do**

18   **if** $g(x'|w^j) \leq g(x^j|w^j)$ **then**

19                  $x^j \leftarrow x'$

20                  $Obj^j \leftarrow F(x')$

21          $A \leftarrow$ Update(A, x')

22 **return** A

---

on a graph G, given a time arrival vector for each route in the solution, i.e., $T_x = (T_{r_1}, \ldots, T_{r_K})$, is given by the following formula:

$$f_2(x) = \sum_{k=1}^{K} W_{r_k}(T_{r_k}) \qquad (3)$$

The VRPTW calls for the determination of at most K routes such that the traveling cost is minimized and the following conditions are satisfied: (a) each route starts and ends at the depot, (b) each customer is visited by exactly one route, (c) the sum of the demands of the customers in any route does not exceed Q, (d) time windows are respected. In the remainder of the paper, the problem considered is a Bi-Objective VRPTW (bVRPTW), where functions f_1 (Eq. 2) and f_2 (Eq. 3) are simultaneously minimized.

Usually, when dealing with the VRPTW, the number of vehicles is minimized first and then the total transportation cost is minimized. However, considering these objectives leads to the creation of a Pareto front with few solutions, which is not an interesting situation for the KD mechanism described. Furthermore, the number of vehicles and the total transportation cost tend to be positively correlated, while it is the opposite for the total waiting time and the total transportation cost [6]. Hence, we decided to minimize the total waiting time instead of the number of vehicles.

A solution to the problem is represented as a permutation of the customers. To evaluate such a solution, we use the algorithm `split` developed by [26]. The evaluation always provides a feasible solution (i.e. respecting capacities and time windows).

Concerning the LS performed in $\mathcal{A}$ (cf. Sect. 4.1), we implemented the same operators as described in [18]: `Relocate`, `Swap`, and `2-opt*`. These simple operators are largely used in local search algorithms for routing problems since they can produce a large neighborhood. The `Relocate` operator moves customer i from its current position to another location (possibly on the same route). The `Swap` operator exchanges in the solution the position of two customers. The `2-opt*` operator generalizes the 2-opt (that is an exchange of two arcs in the same route), by involving different routes. Note that, a move is accepted only if the solution remains feasible. A Randomized Variable Neighborhood Descent is applied for exploration [20], where the order of the neighborhood operators is kept during descent but shuffled each time the LS is applied. For each operator, customers are shuffled, and iteratively, all the moves involving the current customer are tried until the best one is found. The moves considered for one customer are reduced using a granularity parameter δ [22]. The granularity defines the furthest (for the Euclidean distance) possible neighbor to consider in a move. Finally, the operator is iterated until no customers can be moved, and the next operator is applied similarly. Once the three operators have been applied, the solution is returned. To perform an efficient exploration of the neighbors, we use incremental evaluation as defined in [25].

5.2 Pattern Injection Local Search

This section is dedicated to answering question *What* from Sect. 3. In the field of routing problems, a learning mechanism called Pattern Injection Local Search (PILS) has recently been introduced by Arnold et al. [1]. This mechanism is an optimization strategy that uses frequent patterns from high-quality solutions, to explore high-order local-search neighborhoods. PILS has been hybridized with the Hybrid Genetic Search (HGS) of Vidal et al. [26] and the Guided Local Search (GLS) of Arnold and Sörensen [2] to solve the Capacitated Vehicle Routing Problem (CVRP) with good results.

PILS is based on an extraction step, focused on the patterns of the current solution, and an injection step, which brings diversity to the solution by adding some patterns learned.

Given a solution x of the problem, patterns are sequences of consecutive customers in a route and PILS extracts all the patterns of x with a size between 2 and $MaxSize$, a user-defined parameter. For instance, a route $r = (0, v_1, \ldots, v_{|r|}, 0)$, contains $max(|r| - k + 1, 0)$ patterns of size k (e.g. (v_1, v_2) is a pattern of size two). The depot is not considered inside patterns. The extracted patterns are added to the corresponding knowledge group by incrementing their frequency of appearance.

At some point, PILS tentatively injects $N_{Injected}$ patterns in the current solution x. Only improving patterns (for the current subproblem) are kept in the solution, assuming that the solution remains feasible. The selection of a pattern is performed as follows: first, the knowledge group is selected (at random if several possibilities are available), then the size of the pattern is randomly chosen among the possible sizes of patterns, and in the corresponding subset one pattern is randomly chosen among the $N_{Frequent}$ most frequent patterns of the same size. When all patterns have been selected, they are injected one by one according to the steps presented in Fig. 2. Firstly arcs incident to a node of the pattern are removed to form the pattern. This step creates several pieces of routes, that are finally reconnected to form a feasible solution. Note that, because of time windows, we do not consider reversed patterns in our mechanism.

6 Experimental Setup

6.1 The Solomon's Benchmark

We use Solomon's instances [19] to evaluate the performances of the five variants presented in Sect. 4.2. The set contains 56 instances divided into three categories according to the type of generation used, either R (random), C (clustered), or RC (random-clustered). The generation R (23 instances) randomly places customers in the grid, while the generation C (17 instances) tends to create clusters of customers. The generation RC (16 instances) mixes both generations. Each category is itself divided into two classes 1 or 2 according to the width of time windows. Instances of class 2 have wider time windows than instances of class 1, meaning that instances 1 are more constrained. All 56 instances exist in three sizes: 25, 50, and 100. However, instances of size 25 and 50 are restrictions of instances of size 100. For our experiments, we do not consider instances of size 25 since they are too small. Although this set was created to evaluate single-objective algorithms, it is also used in the literature to evaluate the performances of multi-objective algorithms.

6.2 Setup and Tuning

As mentioned above, each algorithm is tuned to find the best setting of the parameters. However, the instances for this tuning phase have to be different from the ones used to evaluate the final performance of the tuned algorithms. Thus, we generated 96 new instances of sizes 50 and 100, by using the method described by Uchoa et al. [24], to mimic Solomon's instances.

Up to 10 parameters are used in the variants considered: 5 parameters are related to $\mathcal{A}$. Those are the number of subproblems to solve (M), the size of the neighborhood for each subproblem (m), the granularity during the local search (δ), the probability of local search (p_{mut}), and the probability of crossover (p_{cro}). The other 5 parameters are specific to the learning. Those are the probability of extraction (p_{ext}), the maximum size of patterns extracted ($MaxSize$), the probability of injection (p_{inj}), and the number of patterns injected ($N_{Injected}$) among the maximum number of most frequent patterns ($N_{Frequent}$). The configurations obtained with irace [15] are available as supplementary materials.

The following experiments are performed on two computers "Intel(R) Xeon(R) CPU E5-2687W v4 @ 3.00 GHz", with 24 cores each, in parallel (with slurm). The variants have been implemented using the jMetalPy framework [4].

7   Experimental Design

In the following, three batches of experiments are considered. The first batch evaluates the impact of the integration of the KD into $\mathcal{A}$, by comparing the variants with the same parameters. The second batch evaluates the performance of the KD variants with their best parameters. The last batch analyses the speed-up of the KD variants compared with $\mathcal{A}$ to reach a target hypervolume.

For the first batch of experiments, we define a new configuration considering the configurations obtained for all variants with irace. We take the 5 parameters of $\mathcal{A}$ and for the KD parameters, we compute the mean of the corresponding parameters of the KD variants, and we round the result, to conserve the same precision. Each variant is executed 30 times on each instance of Solomon's benchmark (56 instances of size 50, and 56 instances of size 100). For each algorithm, the k-th run of an instance is executed with the seed $10 \times (k - 1)$, so that, all algorithms are compared with the same seeds. The termination criterion is fixed to $N \times 6$ s, where N is the size of the instance. The results are compared using the hypervolume, since the true Pareto fronts of the instances are not known.

For the second batch of experiments, we proceed similarly to the first batch, except that we use the configurations returned by irace for each variant.

Finally, for the third batch of experiments, we evaluate the speed of the learning variants to reach 95% of the mean hypervolume returned by $\mathcal{A}$. The termination criterion is slightly different from the other batches since we consider the value of the hypervolume to be reached. Except for that change, the remaining of the experiment is similar to the previous one. Note that, for all the experiments, we use the same values to normalize the objectives of all variants. These values are obtained with the first experiment and are simply the best and worst values obtained among all the executions. It allows an easy computation of the hypervolume during the execution of the algorithm.

8   Experimental Results

This section presents a synthesis of the results obtained. The tables report the average values calculated over all instances for each class and each size of the problems. The results for all instances are given as supplementary materials.

Table 1 shows the mean gain, in percentage, obtained with all KD variants when A is the reference algorithm. First one can see that all gains are strictly positive, meaning that the KD variants return better results, in general, than A. Bold results represent the maximum gain. However, when using Friedman and pairwise Wilcoxon tests, the KD variants are statistically equivalent most of the times. The KD variants are much more efficient on instances of class 1 since the gain is always higher than the gain for instances of class 2. We recall that instances of class 2 contain wider TW and thus are less constrained.

Table 1. Mean gain (%) obtained with the KD variants with respect to A, when they all have the same parameters.

Class	Size	A_{int}^{all}	A_{div}^{all}	A_{int}^{lo}	A_{div}^{lo}
C1	50	**52.6**	52.3	52.2	52.2
R1	50	24.4	25.5	25.1	**26.3**
RC1	50	43.1	43.1	**43.9**	43.7
C2	50	12.2	11.8	**12.8**	12.7
R2	50	5.7	6.3	**6.5**	5.7
RC2	50	20.5	20.7	21.2	**21.3**
C1	100	232.2	**234.9**	234.2	234.2
R1	100	83.9	85.1	83.2	**85.3**
RC1	100	145.0	**148.0**	143.3	145.9
C2	100	**70.9**	70.5	69.7	70.3
R2	100	25.4	25.1	**26.7**	25.0
RC2	100	34.2	**38.0**	36.1	35.7

Table 2. Mean gain (%) obtained with the KD variants with respect to A, when the parameters are given by irace.

Class	Size	A_{int}^{all}	A_{div}^{all}	A_{int}^{lo}	A_{div}^{lo}
C1	50	**55.1**	55.1	53.1	53.3
R1	50	35.9	**36.5**	30.5	26.2
RC1	50	58.1	**59.6**	53.5	50.8
C2	50	16.0	**16.5**	10.4	7.8
R2	50	**14.0**	13.9	9.7	9.2
RC2	50	28.6	**29.0**	26.1	24.4
C1	100	227.2	239.1	235.0	**249.3**
R1	100	126.1	143.0	140.9	**144.1**
RC1	100	164.1	203.2	209.6	**222.6**
C2	100	41.6	61.7	69.9	**72.4**
R2	100	43.0	56.0	58.4	**60.7**
RC2	100	35.2	57.2	65.9	**72.7**

Table 2 shows the mean gain obtained when the parameters are chosen by irace for each variant. It leads to an overall improvement of the results of Table 1 (except for the algorithms learning on local optima on some C2 and R1 instances). More importantly, we see that learning on local optima is mainly beneficial for instances of size 100 (on average the variant A_{div}^{lo} returns the best results on these instances). However, it is outperformed by the other KD variants on instances of size 50. On instances of size 50, it seems more interesting to learn from any solutions found. Indeed the algorithm may be easily get stuck on the same local optima, since the instance is easier to solve, which does not add useful information to the mechanism. Table 3 gives the gaps with the best-known solutions for the total cost objective. The results highlight that our KD variants return much better results than $\mathcal{A}$. However, in some instances (e.g., RC1 of size 100), the gaps obtained are still high ($> 2\%$), meaning that in a multi-objective context, it is harder to find the optimal value of each objective. Moreover, larger gaps are obtained on instances of size 100, showing scalability issues.

Table 4 shows the speed-up of the KD variants to reach 95% of the mean hypervolume returned by $\mathcal{A}$. The table shows that we reach an average speed-up of 73.5% (resp. 64.1%) on instances of size 50 (resp. 100) compared to $\mathcal{A}$, leading to a significant improvement. By considering the results in Table 2, we observe that $\mathcal{A}_{div}^{all}$ reaches good results faster than $\mathcal{A}_{div}^{lo}$ on instances of size 100, but then slows down and is outperformed by $\mathcal{A}_{div}^{lo}$. On instances of size 50, it is the intensification variant $\mathcal{A}_{int}^{all}$, which is the fastest variant on average.

Table 3. Mean gap (%) obtained regarding the total cost objective. The gaps are computed with the optimal value known for each instance.

Class	Size	$\mathcal{A}$	$\mathcal{A}_{int}^{all}$	$\mathcal{A}_{div}^{all}$	$\mathcal{A}_{int}^{lo}$	$\mathcal{A}_{div}^{lo}$
C1	50	10.86	0.07	0.08	0.40	0.40
R1	50	3.09	0.90	0.92	1.23	1.43
RC1	50	7.41	1.72	1.56	2.17	2.54
C2	50	2.15	0.35	0.35	0.58	0.55
R2	50	4.56	2.37	2.40	2.66	2.65
RC2	50	7.31	1.52	1.22	1.97	1.95
C1	100	35.03	5.13	3.79	4.02	2.22
R1	100	10.44	5.15	4.43	4.51	4.37
RC1	100	18.84	10.31	8.11	7.67	7.03
C2	100	5.67	2.64	1.24	0.54	0.38
R2	100	9.58	7.81	6.58	6.15	5.71
RC2	100	12.02	11.00	7.61	6.24	5.04

Table 4. Mean gain (%) obtained, in terms of speed-up, with the learning-variants with respect to $\mathcal{A}$, to reach 95% of the mean hypervolume returned by $\mathcal{A}$.

Class	Size	$\mathcal{A}_{int}^{all}$	$\mathcal{A}_{div}^{all}$	$\mathcal{A}_{int}^{lo}$	$\mathcal{A}_{div}^{lo}$
C1	50	**85.0**	79.9	84.4	80.8
R1	50	73.7	70.4	**76.7**	74.3
RC1	50	**84.7**	81.7	83.5	83.9
C2	50	64.5	**64.7**	59.2	55.5
R2	50	**65.8**	58.9	65.6	63.6
RC2	50	78.6	72.6	**78.9**	77.9
C1	100	73.4	**81.2**	76.2	69.9
R1	100	55.7	**65.9**	62.6	54.4
RC1	100	68.1	**75.6**	73.5	68.6
C2	100	59.9	**70.0**	68.0	62.5
R2	100	53.4	**58.2**	56.4	46.6
RC2	100	57.0	61.9	**62.7**	57.0

9 Conclusion

Exploiting knowledge in multi-objective optimization requires taking into account the absence of order between incomparable solutions. In this paper, we presented a MOEA/D algorithm enhanced with a knowledge discovery step that extracts knowledge from the representation of solutions and integrates this knowledge within other solutions. The question raised and discussed in this paper concerns the impact of extracting knowledge from local optimum solutions only. It led to the design of four variants of MOEA/D, depending on the strategy followed during the injection step and depending on the solutions used during extraction. In particular, the variant $\mathcal{A}_{div}^{all}$ (resp. $\mathcal{A}_{div}^{lo}$) extracts the knowledge from all potential (resp. local optima) solutions found, and follows a diversification strategy during the injection, meaning that we use knowledge from all groups to favor larger exploration of space. We conducted experiments on Solomon's instances of the bi-objective VRPTW and we showed the benefit of exploiting knowledge to better optimize solutions. Additionally, extracting patterns from local optima ($\mathcal{A}_{div}^{lo}$), especially for larger instances, is preferable to obtain better solutions. The investigation of the speed-up reveals that $\mathcal{A}_{div}^{all}$ converges faster towards good solutions (i.e., when the learning is not focused on local optima only). In practice, it means that focusing on local optima solutions only is not a necessity to quickly achieve good performances. Future works should consider an adaptive mechanism to control the learning on local optima only or not. Finally, other combinatorial problems should be considered like the bi-objective flow-shop, to know if similar conclusions are reached.

Supplementary Materials

The detailed results for each instance and the source code, can be found here: https://gitlab.univ-lille.fr/clement.legrand4.ctu/knowledge-discovery.

References

1. Arnold, F., Santana, Í., Sörensen, K., Vidal, T.: PILS: exploring high-order neighborhoods by pattern mining and injection. Pattern Recogn. **116**, 107957 (2021)
2. Arnold, F., Sörensen, K.: Knowledge-guided local search for the vehicle routing problem. Comput. Oper. Res. **105**, 32–46 (2019)
3. Barbalho, H., Rosseti, I., Martins, S.L., Plastino, A.: A hybrid data mining grasp with path-relinking. Comput. Oper. Res. 40(12), 3159–3173 (2013)
4. Benitez-Hidalgo, A., Nebro, A.J., Garcia-Nieto, J., Oregi, I., Del Ser, J.: jMetalPy: a python framework for multi-objective optimization with metaheuristics. Swarm Evol. Comput. **51**, 100598 (2019)
5. Blot, A., Marmion, M., Jourdan, L.: Survey and unification of local search techniques in metaheuristics for multi-objective combinatorial optimisation. J. Heuristics **24**(6), 853–877 (2018)
6. Castro-Gutierrez, J., Landa-Silva, D., Pérez, J.M.: Nature of real-world multi-objective vehicle routing with evolutionary algorithms. In: 2011 IEEE International Conference on Systems, Man, and Cybernetics, pp. 257–264. IEEE (2011)

7. Coello Coello, C.A., Dhaenens, C., Jourdan, L.: Multi-objective combinatorial optimization: problematic and context. In: Coello Coello, C.A., Dhaenens, C., Jourdan, L. (eds.) Advances in Multi-Objective Nature Inspired Computing. SCI, vol. 272, pp. 1–21. Springer, Heidelberg (2010). https://doi.org/10.1007/978-3-642-11218-8_1
8. Corne, D., Dhaenens, C., Jourdan, L.: Synergies between operations research and data mining: the emerging use of multi-objective approaches. Eur. J. Oper. Res. 221(3), 469–479 (2012)
9. Deb, K., Pratap, A., Agarwal, S., Meyarivan, T.: A fast and elitist multiobjective genetic algorithm: NSGA-II. IEEE Trans. Evol. Comput. 6(2), 182–197 (2002)
10. Khalil, E., Dai, H., Zhang, Y., Dilkina, B., Song, L.: Learning combinatorial optimization algorithms over graphs. In: Advances in Neural Information Processing Systems, vol. 30 (2017)
11. Knowles, J.D.: Local-search and hybrid evolutionary algorithms for Pareto optimization. Ph.D. thesis, University of Reading Reading (2002)
12. Kora, P., Yadlapalli, P.: Crossover operators in genetic algorithms: a review. Int. J. Comput. Appl. 162, 10 (2017)
13. Land, M.W.S.: Evolutionary algorithms with local search for combinatorial optimization. University of California, San Diego (1998)
14. Legrand, C., Cattaruzza, D., Jourdan, L., Kessaci, M.-E.: Improving neighborhood exploration into MOEA/D framework to solve a bi-objective routing problem. Int. Trans. Oper. Res. (2023)
15. López-Ibáñez, M., Dubois-Lacoste, J., Cáceres, L.P., Birattari, M., Stützle, T.: The irace package: iterated racing for automatic algorithm configuration. Oper. Res. Perspect. 3, 43–58 (2016)
16. Ma, X., et al.: MOEA/D with opposition-based learning for multiobjective optimization problem. Neurocomputing 146, 48–64 (2014)
17. Riquelme, N., Von Lücken, C., Baran, B.: Performance metrics in multi-objective optimization. In: 2015 Latin American computing conference (CLEI), pp. 1–11. IEEE (2015)
18. Schneider, M., Schwahn, F., Vigo, D.: Designing granular solution methods for routing problems with time windows. Eur. J. Oper. Res. 263(2), 493–509 (2017)
19. Solomon, M.M.: Algorithms for the vehicle routing and scheduling problems with time window constraints. Oper. Res. 35(2), 254–265 (1987)
20. Subramanian, A., Uchoa, E., Ochi, L.S.: A hybrid algorithm for a class of vehicle routing problems. Comput. Oper. Res. 40(10), 2519–2531 (2013)
21. Talbi, E.-G.: Machine learning into metaheuristics: A survey and taxonomy. ACM Comput. Surv. (CSUR) 54(6), 1–32 (2021)
22. Toth, P., Vigo, D.: The granular tabu search and its application to the vehicle-routing problem. INFORMS J. Comput. 15(4), 333–346 (2003)
23. Toth, P., Vigo, D.: Vehicle Routing: Problems, Methods, and Applications. SIAM (2014)
24. Uchoa, E., Pecin, D., Pessoa, A., Poggi, M., Vidal, T., Subramanian, A.: New benchmark instances for the capacitated vehicle routing problem. Eur. J. Oper. Res. 257(3), 845–858 (2017)

25. Vidal, T., Crainic, T.G., Gendreau, M., Prins, C.: A hybrid genetic algorithm with adaptive diversity management for a large class of vehicle routing problems with time-windows. Comput. Oper. Res. **40**(1), 475–489 (2013)
26. Vidal, T., Crainic, T.G., Gendreau, M., Prins, C.: A unified solution framework for multi-attribute vehicle routing problems. Eur. J. Oper. Res. (2014)
27. Xu, Q., Xu, Z., Ma, T.: A survey of multiobjective evolutionary algorithms based on decomposition: variants, challenges and future directions. IEEE Access **8**, 41588–41614 (2020)
28. Zhang, Q., Li, H.: MOEA/D: a multiobjective evolutionary algorithm based on decomposition. IEEE Trans. Evol. Comput. **11**(6), 712–731 (2007)
29. Zitzler, E., Thiele, L., Laumanns, M., Fonseca, C.M., Da Fonseca, V.G.: Performance assessment of multiobjective optimizers: an analysis and review. IEEE Trans. Evol. Comput. **7**(2), 117–132 (2003)

A Memetic Algorithm for Large-Scale Real-World Vehicle Routing Problems with Simultaneous Pickup and Delivery with Time Windows

Ethan Gibbons and Beatrice Ombuki-Berman[✉]

Department of Computer Science, Brock University, St. Catharines, Canada
bombuki@brocku.ca

Abstract. The vehicle routing problem with simultaneous pickup and delivery with time windows (VRPSPDTW) is an important variant of the vehicle routing problem which has received considerable attention among researchers in the last decade. The vast majority of solution methodologies for the VRPSPDTW have been applied to synthetic problem instances that bear little resemblance to routing problems found in the real world. Recently, 20 large-scale VRPSPDTW instances based on real customer data from the transportation company known as JD Logistics became publicly available as a new benchmark VRPSPDTW problem set.

In this paper, a memetic algorithm (MA), referred to as MA-BCRCD, is proposed for use on these real-world instances. The MA prioritizes efficient search and utilizes a crossover method which is shown to be more effective than that of the previous MA approach (known as MATE) applied to this set. MA-BCRCD finds new best known solutions for all 20 instances. It also performs better on average for all instances in comparison to the performance of MATE. The results and analysis provided in this study suggest that further improvements on this problem set are possible both in terms of solution quality and search efficiency.

Keywords: Vehicle routing problem · Memetic algorithm ·
Combinatorial optimization · Industrial application

1 Introduction

The vehicle routing problem with simultaneous pickup and delivery with time windows (VRPSPDTW) is a variant of the well-established vehicle routing problem which features two common constraints faced by various industries involved with the transportation of goods. In many instances it is required as part of a delivery service that drivers not only deliver required goods to customers, but also pick up goods held by customers which may be wasted or unusable [12]. A common scenario is the picking up of empty, reusable containers which companies can collect, process, and reuse for later deliveries [4]. For this reason, as

M. Sevaux et al. (Eds.): MIC 2024, LNCS 14753, pp. 78–92, 2024.
https://doi.org/10.1007/978-3-031-62912-9_8

companies seek to operate with environmentally-friendly goals in mind, research into practical solution methods for the vehicle routing problem with simultaneous pickup and delivery (VRPSPD) becomes more relevant [10].

Another common constraint in transportation logistics is the requirement that goods be delivered to customers within a given time window [6]. This constraint arises in many situations where customers cannot feasibly receive their deliveries outside their given time window or where customer satisfaction may be impacted by early and late deliveries. As a result, the vehicle routing problem with time windows (VRPTW) is one of the most popularly researched variants of the VRP [5]. The VRPSPDTW combines the problem features of the VRPSPD and the VRPTW. Over the last decade, the VRPSPDTW has received a fair amount of attention in the literature. Though the problem was first introduced by Angelelli and Mansini [2] in 2002, most of the attention to this problem came after Wang and Chen [22] released publicly available VRPSPDTW instances which were derived from the popular Solomon instances [20] for the VRPTW. Following Wang and Chen's work, many follow-up studies proposed new metaheuristics which were able to find improved solutions for their instances (for some examples, see [8,11,19,21]).

One such work by Liu et al. [12] proposed a memetic algorithm approach for the VRPSPDTW in 2021. This MA, referred to as MATE, currently outperforms all existing approaches both in terms of number of current best-known-solutions found as well as when compared to each approach one-on-one. As an additional contribution, Liu et al. made the source code of MATE publicly available. More importantly, they released 20 new large-scale VRPSPDTW instances which were derived from real customer data from the company JD Logistics. They applied MATE to these instances to provide initial benchmark solutions and noted that these instances should be more difficult to solve and relevant for researchers, since realistic problem instances are not commonly found in the literature.

More recent approaches to the VRPSPDTW have been published [11,13,23]. Each of these studies attempt to optimize for the Wang and Chen instances and compare their results with previous works. However, the instances from JD Logistics (referred to as the *jd* instances) have not received any attention since their release despite their practical utility. Thus, to close this gap, a primary aim of this paper is to follow up on the approach by Liu et al. by providing a memetic algorithm (referred to as MA-BCRCD) for the real-world *jd* instances. The remainder of this paper is organized as follows: a literature review summarizing previous research on the VRPSPDTW is given in Sect. 2. Section 3 provides required notation and a formulation of the problem at hand. Section 4 describes the components of MA-BCRCD, including the crossover which is a new alteration to the popular Best-Cost-Route-Crossover (BCRC) [16] referred to in this paper as BCRC with route destruction (BCRCD). Simple but effective components and heuristics are also introduced which reduce the number of local searches performed during the search. In Sect. 5, a computational study using both MA-BCRCD as well as the open-source code of MATE on the *jd* instances is reported. As a result of this study, BCRCD is shown to be an improvement over the unaltered BCRC, and MA-BCRCD is shown to outperform MATE on all instances. Section 6 concludes the paper by summarizing the significance of the computational study performed as well as providing possible future directions for research into this problem.

2 Related Works

As mentioned above, most of the studies dealing with the VRPSPDTW have employed their metaheuristics on the instances released by Wang and Chen [22]. For these instances, a multi-objective problem is assumed, where priority is always given to solutions requiring fewer vehicles. Solutions with the same number of vehicles are then judged by the combined total distance of all the routes, and the solution with less total distance is considered better. The few approaches which were not based on these instances only attempt to minimize the total distance. They include the work by Angelelli and Mansini [2] who first introduced the problem and developed a branch-and-cut-and-price algorithm which was able to solve the VRPSPDTW for problem sizes of up to 20 optimally. Mingyong and Erbao [15] approached the problem with a differential evolution method. They successfully used a decimal coding method to adapt DE to this combinatorial problem. Kassem and Chen [9] used a sequential route construction heuristic to build a solution and then applied a simulated annealing approach to improve on the initial solution.

Wang and Chen [22] introduced their co-evolutionary algorithm to tackle the VRPSPDTW using their benchmark problem instances. The GA maintained two populations, one which promoted diversity, while the other housed stronger solutions. They compared their co-evolutionary GA against a typical GA to validate their approach. Wang et al. [21] used the Residual Capacity and Radial Surcharge (RCRS) [7] heuristic to construct an initial solution, followed by a simulated annealing method which is parallelized to allow for a wider search in less time. Hof and Schneider [8] used an adaptive large neighborhood search approach with path-relinking. Their approach is quite intricate, involving several large neighborhood search methods, a number of customer removal and reinsertion methods, learned values which measure the past effectiveness of these operators, and a memory of elite solutions that are used to improve the current solution through path re-linking. The authors applied their algorithm to several variants of the VRPSPD, including that with time windows. Shi et al. [19] used a two-stage hybrid approach with variable neighborhood search and tabu search. Due to the multi-objective nature of the Wang-Chen instances, they proposed a unique evaluation function for the purpose of reducing the number of vehicles first before attempting to reduce travel distance.

As already mentioned, Liu et al. [12] developed MATE for the VRPSPDTW. They applied the RCRS heuristic for constructing initial solutions using evenly distributed parameter configurations in order to increase initial population diversity. For crossover, they applied the method used by Wang and Chen and originally developed in [1] while altering the crossover's reinsertion phase by implementing regret insertion [18]. They dubbed their crossover the route-assembly with regret-insertion crossover, or RARI. For a local search procedure, they applied typical local search operators such as *or-opt*, *swap*, *two-opt*, and *two-opt**. In addition, they applied the relatedness-removal large neighborhood operator [8] to perturb solutions once the local search had reached a local minimum.

More recent approaches include Wu and Gao [23], who developed the first ant colony algorithms for the VRPSPDTW where one ant solution is constructed using typical solution construction methods, and the remaining solutions are built using various neighborhood generating operators. Liu, F. et al. [11] employed a late acceptance hill climbing strategy with a multi-armed bandit algorithm for intelligent neighborhood selection. Liu, Z. et al. [13] applied a sparrow search algorithm which is a swarm intelligence paradigm wherein different population members serve different roles to guide the search. The approaches in [11] and [13] contribute several new best known solutions for the Wang and Chen instances. However, none of these studies tested their approach on the realistic instances introduced in [12].

3 VRPSPDTW Problem Formulation

The following formulation is equivalent to the formulation given by Liu et al. in their paper [12], though the notation is somewhat different.

Given a set of customers $N = \{1, 2, ..., n\}$, a depot 0, a set of J identical vehicles with a cargo capacity of Q. The objective of the VRPSPDTW is to assign routes to each vehicle such that each customer in N is visited and serviced by exactly one vehicle in an efficient manner. The problem can be represented using a complete graph G with vertex set $V = \{0\} \bigcup N$. Each customer $i \in V$ has the following attributes: p_i denotes the amount of goods to be picked up at customer i, d_i denotes the amount of goods to be delivered from the depot to customer i, a_i is the earliest time that customer i can begin being serviced by a vehicle, b_i is the latest time i can begin being serviced, and s_i is the length of time it takes for a vehicle to service i. The pair $[a_i, b_i]$ denotes the time window of each customer. E denotes the arc set of G, where each arc has two values $c_{i,j}$ and $t_{i,j}$. The former represents the distance a vehicle must travel to get from node i to j in V while the latter represents the time it takes to travel from node i to j.

A solution to the VRPSPDTW, denoted with $S = \{R_1, R_2, ..., R_K\}$, consists of K routes, where $K \leq J$. A route $R_k = (i_0, i_1, i_2, ..., i_{H_k}, i_{H_k+1})$ is an ordered sequence of customers to be visited by vehicle k, where H_k is the number of customers assigned in route k and where $i_0 = i_{H_k+1} = 0$. For convenience when talking about a single route, the k subscript is omitted from the notation. Each route has a cost associated with it, which is the sum of the edges in the route,

$$C_R = \sum_{h=0}^{H} c_{i_h, i_{h+1}}. \tag{1}$$

A feasible route requires that the sum of the delivery request quantities (denoted as D_R) is less than the capacity of the vehicle, i.e. $D_R = \sum_{h=1}^{H} d_{i_h} \leq Q$. The load l_{R_k,i_h} (l_{R,i_h}) of a vehicle just after servicing customer i_h consists of the remaining deliverable goods to be delivered after customer i_h, as well as the goods which have already been picked up by the vehicle. Thus we have

$$l_{R,i_h} = \sum_{x=1}^{h} p_{i_x} + \sum_{x=h+1}^{H} d_{i_x}. \tag{2}$$

For a route to be feasible, the load of a vehicle at any point in the route must never be higher than the vehicle capacity, i.e., $l_{R,i_h} \leq Q$ for $1 \leq h \leq H$.

In addition to the capacity constraints, routes also must be feasible with regards to time constraints. Let $B_{R_k,i}$ $(B_{R,i})$ be defined as the time that a vehicle k would start servicing customer i given a complete solution. A feasible route requires that all customers be visited before their closing time window, i.e. $B_{R_k,i} \leq b_i$ for all customers in route k. The opening time window has slightly different properties. A vehicle may arrive at a customer i before its opening time window a_i, but if it does, it must wait until the time a_i before it can start servicing the customer. The sequential values of $B_{R_k,i}$ must be calculated using these potentially required waiting times and the time it takes to service each customer. For a route R_k, the values of $B_{R,i}$ can be calculated as follows:

$$
\begin{aligned}
B_{R,i_0} &= 0, \\
B_{R,i_{H+1}} &= B_{R,i_H} + s_{i_H} + t_{i_H,i_{H+1}}, \\
B_{R,i_h} &= \max \left(B_{R,i_{h-1}} + s_{i_{h-1}} + t_{i_{h-1},i_h}, a_{i_h} \right) \quad \text{for } h = 1, \ldots, H
\end{aligned}
\tag{3}
$$

The time window for the depot is a special case where $a_0 = 0$, $b_0 = T$, and where T is the length of the working day. All vehicles must return to the depot before the depot's closing time window T.

The instances from [12] are weighted multi-objective problem instances. The two objectives are to reduce the number of vehicles used in a solution and to reduce the total distance travelled by all the vehicles. The weights for these objectives are denoted as u_1 and u_2, respectively. For the sake of the problem formulation, let $x_{i,j}^k$ be a binary variable which equals 1 if customer j is visited by vehicle k immediately after customer i in route R_k, and 0 otherwise.

For all possible solutions S and their corresponding K routes, the objective is the following:

$$\min_{S} \left(u_1 \cdot K + u_2 \cdot \sum_{k=1}^{K} C_{R_k} \right) \tag{4}$$

such that:

$$K \leq J \tag{5}$$

$$i_0 = i_{H_R+1} = 0 \qquad \forall R \in S \tag{6}$$

$$\sum_{i \in V} \sum_{k=1}^{K} x_{i,j}^k = 1 \qquad \forall j \in N \tag{7}$$

$$D_{R_k} \leq Q \qquad \forall k \in \{1, 2, \dots, K\} \tag{8}$$

$$l_{R_k, i_h} \leq Q \qquad \begin{array}{l} \forall k \in \{1, 2, \dots, K\}, \\ \forall h \in \{1, 2, \dots, II_{R_k}\} \end{array} \tag{9}$$

$$a_{i_h} \leq B_{R_k, i_h} \leq b_{i_h} \qquad \begin{array}{l} \forall k \in \{1, 2, \dots, K\}, \\ \forall h \in \{1, 2, \dots, H_{R_k}\} \end{array} \tag{10}$$

$$B_{R_k, i_{H_k}+1} \leq T \qquad \forall k \in \{1, 2, \dots, K\}. \tag{11}$$

The objective function 4 seeks to minimize the weighted sum of the NV and TD, with the weights to be determined for the instance at hand. Constraint 5 implies that there is a maximum number of routes that a solution can have. Constraint 6 ensures that each route begins and ends at the depot. Constraint 7 ensures that each customer is visited by exactly one vehicle exactly one time. Constraint 8 requires that the required amount to be delivered to a route does not exceed a vehicle's capacity. Constraint 9 ensures that the load of a vehicle never exceeds its capacity throughout its route, with Eq. 2 being used to determine the current load of each vehicle. Constraint 10 ensures that the servicing of every customer begins within that customer's time window, with Eq. 3 used to determine the start of the service time for each customer in a route. Finally, constraint 11 requires that all vehicles return to the depot before the end of the day.

4   Memetic Algorithm for the VRPSPDTW

Memetic algorithms are a class of population-based metaheuristics which imitate the evolutionary concept of natural selection, where good solutions are more likely to survive and influence or participate in the production of new solutions. A population of solutions are initialized with some heuristic, and the solutions improve via local search or are replaced by new solutions constructed through a crossover method. This occurs over several generations until further improvements cannot be found. A memetic algorithm may or may not have a mutation operator, which perturbs a solution in an unpredictable fashion in order to improve population diversity.

Our proposed MA was developed with the intention of reducing the use of computationally expensive construction heuristics and local search operators while still performing an effective search. Algorithm 1 provides the overall framework of the MA. There are three parameters for this MA: n_{pop} which determines the population size, f_{search} which determines how many generations pass before a local search step is applied, and p_{search} which determines the proportion of solutions which undergo the local search step. The changing of these parameters has a large affect on the run time of the MA.

Algorithm 1: Memetic Algorithm for the VRPSPDTW

Input: VRPSPDTW Instance, n_{pop}, f_{search}, p_{search}.

1   $U \leftarrow$ set of n_{pop} empty solutions.
2   Initialize solutions in U.
3   Perform local search on $n_{pop} * p_{search}$ solutions in U.
4   Maintain elitism.
5   $g \leftarrow 1$.
6   **while** *termination condition is not met* **do**
7      $U_{new} \leftarrow$ perform k-tournament selection on U ($k = 3$).
8   **for** *each pair of solutions* (P_1, P_2) *in* U_{new} **do**
9          $O_1, O_2 \leftarrow$ perform crossover using solutions (P_1, P_2).
10   O_1 and O_2 replace P_1 and P_2 in U_{new}.
11   **end**
12      $U \leftarrow U_{new}$.
13   **if** g **mod** $f_{search} = 0$ **then**
14   Perform local search on $n_{pop} * p_{search}$ solutions in U.
15   **end**
16   Maintain elitism.
17      $g \leftarrow g + 1$.
18 **end**
19 **return** *Best solution found.*

4.1   Solution (Chromosome) Representation and Initialization

For GAs, some crossover and mutation methods which are well-suited for a single-route VRP (TSP) cannot sensibly work on a VRP solution unless the solution is represented as a long tour with [3] or without [17] route delimiters. If route delimiters are not used, then some decoding method must be applied to partition the tour into separate routes [16].

The crossover used in this MA works intuitively with a direct solution representation as described in the problem formulation. Thus, each solution in the population is a collection of discrete routes where the order of routes does not matter. After the initialization process, each route maintains feasibility throughout the search.

Since some of the instances for the problem at hand are large, a simple initialization process is used to save time. For a population U of n_{pop} empty solutions, a single solution is built using a Nearest Neighbor (NN) method which works as follows: as long as there are customers needing to be assigned to a route, create an empty route and append the depot to it. Append the unrouted customer which is nearest to the previous customer (or depot) to the end of the current route and repeat until the nearest customer cannot be appended feasibly. Once this happens, repeat the process with a new route. The remaining $(n_{pop} - 1)$

solutions get initialized using a Random Neighbor (RN) method, which works similarly to NN except that random neighbors are appended to routes instead of nearest neighbors. For RN, a new route is started each time the randomly chosen customer cannot be appended to the end of the current route.

4.2 Crossover

The crossover used in this MA is based on the Best-Cost-Route-Crossover (BCRC) originally introduced by Ombuki et al. in [16]. Given two parent solutions P_1 and P_2, BCRC works by choosing a random route R^1 from P_1 and a random route R^2 from parent P_2. The customers which are in R^1 are removed from P_2 and vice versa. Then, the incomplete solutions are completed by inserting the missing customers in some order back into the solutions using cheapest insertion. That is, for the next customer c to be reinserted, every possible position in every route (including the route and position from which c was removed) is checked to see if c can be feasibly inserted, and the route and position which increases the cost of the solution the least is chosen as the new position for c. If c cannot be placed anywhere, a new route is created containing only c.

We propose a new alteration of the BCRC for use in this MA. In addition to removing the customers in R^1 from P_2 and vice versa, also remove the entirety of R^1 from P_1 (and R^2 from P_2). We refer to this altered crossover as BCRC with route destruction (BCRCD). Since part of the weighted objective function of the jd instances is to have fewer vehicles, this route destruction step was added with the hope that completed offspring would have fewer routes than their parents.

In addition to the route destruction step, another alteration is made to ensure solutions remain feasible throughout the search. Since the jd instances do not always satisfy the triangle inequality, removing customers from routes can render them infeasible with regards to time windows. Thus, if the removal of a customer from a route would render that route infeasible, that customer is not removed. Figure 1 shows an example of these alterations to the BCRC.

It should be noted that the order of which customers should be reinserted first via cheapest insertion is not specified in [16]. There are a number of heuristics which could be used to decide the order of customer insertions [8], but for this implementation of BCRCD, the order of insertion is randomized for simplicity.

4.3 Local Search

Because exhaustive local search descents are very computationally expensive, two parameters are used to limit how often the local search step is performed. First, since BCRCD performs a neighborhood search and is less expensive than a full local search descent, it may be beneficial to have a higher frequency of crossover steps in comparison to local search steps. Thus f_{search} is proposed as a user-defined parameter to find a better balance between these two components. That is, after every f_{search} generations of crossover occur, a greedy local search descent is applied to a proportion of the population. The second user-defined parameter p_{search} determines the proportion of solutions which undergo local search descent at every local search step. Three common local neighborhood generators are used in the local search step: *swap*, *relocate*, and *two-opt** [8,12,19,21,23]. The *relocate* operator (sometimes referred to as *or-opt* [12]) involves removing a customer i and reinserting that customer somewhere else in the solution. For this implementation, *relocate* is also used for a subsequence of two consecutive customers, where the subsequence is removed and reinserted elsewhere in the solution. The *swap* operator involves two subsequences of customers. Similarly to *relocate*, either subsequence can be of length 1 or 2. The two subsequences of customers are swapped in the solution. Either of these two operators can be applied to alter either one or two routes. Finally, *two-opt** is employed which always involves two different routes R^1 and R^2. In *two-opt**, the two routes are split into two, and the earlier part of R^1 is attached to the later part of R^2, and vice versa.

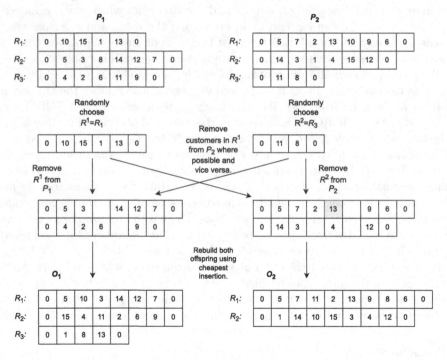

Fig. 1. An example of Best-Cost-Route-Crossover with Route Destruction (BCRCD) with guaranteed feasibility for the VRPSPDTW. In this example, customer 13 in P_2 is not removed, since its removal would cause the remaining partial route to be infeasible.

The greedy local search descent works as follows: for a solution undergoing local search, generate all neighboring solutions possible from applying any of the above operators once. Whichever neighboring solution has the best improvement, keep the corresponding changes to the solution. Repeat this process until no further improvements can be made. The feasibility and cost-savings of each generated neighborhood are evaluated in constant time using the method proposed by Liu et al. [12] for the VRPSPDTW.

To determine which members of the population get to undergo local search descent, three simple heuristics are used. With $n_{search} = n_{pop} * p_{search}$, one of the following three heuristics is chosen with equal probability to choose n_{search} solutions for local search every f_{search} generations:

1. Pick the best n_{search} solutions in terms of objective function,
2. Pick n_{search} solutions randomly,
3. Sort solutions by fitness score, then pick every (n_{pop}/n_{search})-th solution in the sorted list.

5 Computational Study and Experimental Analysis

5.1 Problem Instances from JD Logistics

Liu et al. describe the derivation of the jd instances introduced in [12]. The distribution operations of JD Logistics involves both the delivery of purchased goods to customers, but also the collection of defective goods or goods in need of maintenance within a predefined time window per customer. To derive these instances, 3000 real customer requests were collected and then sampled to generate instances with customer sizes of either 200, 400, 600, 800, or 1000 customers. 4 instances of each size were generated for a total of 20 instances. The instances all have tight capacity and time window constraints, so solutions will consist of many small routes. For example, in our experimentation, final solutions usually consisted of around 40 routes for the 200 customer instances and around 200 routes for the 1000 customer instances.

The objective function for these instances is a weighted-sum objective, where both reducing the total number of vehicles used and reducing the distance travelled are objectives. The specific weights u_1 and u_2 were chosen to accurately reflect operational costs faced by JD Logistics.

A major complicating factor of these instances is that the distances and times between customers are given explicitly. This is in contrast to many classic VRP benchmark instances which map each customer on a 2D euclidean plane, where for any two customers i and j, the distance between them is equal to the time it takes to travel from one to the other, i.e. $c_{i,j} = t_{i,j}$. In addition, these values are calculated using euclidean distance, so the triangle inequality always holds. However, these assumptions are obviously not realistic, and solution methods which rely on these assumptions may not do so well in real-world applications.

5.2 Experimental Setup

The computational experiments reported in this section involve both the MA introduced in this paper (referred to as MA-BCRCD) as well as the open source code for MATE released by Liu et al. The source code for MA-BCRCD was written in Java, while the code for MATE was written in C++.

All reported runs were run using on Linux machines using an Intel i7-9700 CPU at 3.00 GHz. Since the jd instances are large, algorithm can run a long time before converging. Liu et al. set a time limit of 2 h for each run in their experiments with MATE. We adopt this time limit as the only stopping criteria for all experiments shown. Each combination of algorithm configuration and problem instance is run 30 times. The average fitness scores shown are the average of the final solutions found in the 30 runs, while the best scores shown are the best found solutions out of all the runs.

5.3   Comparing BCRCD with Other Crossovers

In order to introduce BCRCD, we compare its performance to BCRC without the route destruction step as well as two other state-of-the-art crossovers. One is RARI, which was used in MATE. The other is referred to as RCX, which was used recently in a many-to-many bike-sharing re-balancing problem [14]. These two crossovers were implemented in our MA as described in [12] and [14], respectively.

During initial experimentation, it was found that changes the population n_{pop} and proportion of population searched p_{search} caused similar changes to performance regardless of the crossover used, and a small value for p_{search} was crucial to the success of the MA, especially for the larger instances. However, the frequency of search f_{search} was very sensitive both to the crossover used as well as the size of the instance. Only BCRCD was found to benefit from this feature regardless of instance size, while BCRC and RCX only benefited from an increase in f_{search} for larger instances. Thus the parameter values used in the crossover comparison are as follows: $n_{pop} = 50$ for all crossovers and all instances, $p_{search} = 0.1$ for instances of size 600 or less, and $p_{search} = 0.06$ for larger instances. For BCRCD, $f_{search} = 20$ for all instances. All other crossovers have $f_{search} = 1$ with the following exceptions: BCRC has $f_{search} = 20$ for instances with 1000 customers, and RCX has $f_{search} = 20$ for instances with 800 and 1000 customers.

The results in Table 1 show the excellent performance of BCRCD compared to the other crossovers. The MA using BCRCD had the best average score and the best found solution for every instance. Importantly, these results show that BCRCD reliably outperforms BCRC without the route destruction step. By examining the percent difference in average score, it can be seen that BCRC starts off weak for the smaller instances, but the gap between its performance and BCRCD narrows for the larger instances. However, both RCX and RARI become less competitive as instance sizes grow. These crossovers tend to require a large number of customer re-insertions to complete offspring solutions, while BCRCD only removes at most two routes from each parent for reinsertion. For the jd instances, two routes is a very small proportion of each solution. Thus, it is likely the case that RARI and RCX change too much of each parent solution and run too slowly to be competitive for the larger jd instances.

Table 1. The average and best final scores for the *jd* instances using the proposed MA using different crossovers. For BCRC, RARI and RCX, the difference of the average run compared to that of BCRCD is given as a percentage. The best average solution is highlighted in grey, while the best solution found is bolded.

Instance	BCRCD Avg	Std	Best	BCRC Avg	% Diff.	Std	Best	RCX Avg	% Diff.	Std	Best	RARI Avg	% Diff.	Std	Best
200_1	65363	368	**64413**	66933	2.4	520	65950	66206	1.29	640	65051	66769	2.15	561	65756
200_2	65322	330	**64753**	67085	2.7	727	65685	65811	0.75	395	64935	67213	2.89	606	66132
200_3	66442	329	**65679**	68530	3.14	651	67371	66916	0.71	523	65791	67944	2.26	642	66917
200_4	65188	296	**64620**	67032	2.83	590	65641	65223	0.05	233	64776	66611	2.18	733	65365
400_1	118972	501	**117812**	121790	2.37	1097	119310	122318	2.81	784	120398	122153	2.67	1024	120155
400_2	124312	685	**123110**	126853	2.04	759	125116	126850	2.04	827	125135	127391	2.48	1452	125002
400_3	118207	500	**117043**	120982	2.35	923	119164	120880	2.26	1153	119112	121195	2.53	1133	118996
400_4	121321	647	**120255**	123739	1.99	925	122275	124222	2.39	922	122621	124227	2.4	1306	122540
600_1	179504	967	**178113**	182571	1.71	1218	180001	187127	4.25	1171	184902	184680	2.88	1784	180463
600_2	184160	1227	**182052**	187904	2.03	1575	185292	191778	4.14	1359	188772	190155	3.26	1510	186025
600_3	182760	926	**180797**	187303	2.49	1268	185058	191053	4.54	1276	188814	187697	2.7	1690	184935
600_4	183431	857	**182006**	186347	1.59	1177	183850	191121	4.19	1510	187411	187866	2.42	2057	184204
800_1	213206	954	**211125**	215719	1.18	1592	212677	218199	2.34	1170	216020	220226	3.29	1568	217185
800_2	210938	711	**209559**	214664	1.77	980	212602	217643	3.18	1289	215467	217964	3.33	1179	214567
800_3	213562	1193	**211602**	216623	1.43	1304	214025	217788	1.98	1317	215256	220115	3.07	1517	215023
800_4	208114	1064	**205902**	211335	1.55	1503	208539	214449	3.04	1260	211921	215940	3.76	1426	212844
1000_1	312110	1385	**309891**	316273	1.33	1787	311856	321706	3.07	1593	318993	322824	3.43	2241	319716
1000_2	309818	1910	**306018**	313684	1.25	1551	311364	319431	3.1	1499	315752	320988	3.61	2321	316747
1000_3	310085	1612	**307275**	316150	1.96	1659	313016	320529	3.37	1684	316459	322499	4	2569	317852
1000_4	307405	1412	**304769**	312512	1.66	2192	308259	318281	3.54	1172	316323	318925	3.75	2417	316166
Median:					1.98%				2.93%				2.89%		

Table 2. The average and best final scores using MA-BCRCD compared against our own re-runs of MATE with and without the use of crossover. Asterisks indicate which MATE re-run performed better on average and in terms of the better solution found. The published results of MATE in [12] are included for reference.

Instance	MA-BCRCD Avg	Std	Best	MATE Re-run Avg	% Diff	Std	Best	MATE Re-run without Crossover Avg	% Diff	Std	Best	MATE Avg	% Diff	Std	Best
200_1	65363	368	**64413**	66196	1.27	288	65479	65991*	0.96	241	65411*	66097	1.12	292	65106
200_2	65322	330	**64753**	65956*	0.97	392	65002*	66171	1.13	505	65002*	66038	1.1	422	65012
200_3	66442	329	**65679**	67141*	1.05	176	66625*	67190	1.13	254	66625*	67090	0.98	332	65980
200_4	65188	296	**64620**	65921	1.12	365	65113*	65898*	1.09	198	65390	65851	1.02	326	64747
400_1	118972	501	**117812**	123085	3.46	335	122413	122892*	3.29	423	121901*	123261	3.61	446	122319
400_2	124312	685	**123110**	127832*	2.83	392	127169*	127962	2.94	366	127169*	128091	3.04	410	126887
400_3	118207	500	**117043**	122163	3.35	459	121260	121905*	3.13	429	121023*	122306	3.47	682	120130
400_4	121321	647	**120255**	124928	2.97	360	124098*	124910*	2.96	310	124265	125242	3.23	359	124517
600_1	179504	967	**178113**	183648*	2.31	525	182442*	183862	2.43	606	182837	184119	2.57	608	182504
600_2	184160	1227	**182052**	188546*	2.38	684	187221*	188681	2.45	672	187221*	188950	2.57	645	187236
600_3	182760	926	**180797**	188083	2.91	664	186546*	187847*	2.78	519	186546*	188050	2.89	621	186644
600_4	183431	857	**182006**	187832*	2.4	603	186805*	187881	2.43	549	186958	188110	2.55	790	186289
800_1	213206	954	**211125**	215260	0.96	695	213410	214985*	0.83	630	213181*	214634	0.67	561	213661
800_2	210938	711	**209559**	213916	1.41	493	213148	213646*	1.28	717	211664*	213276	1.11	292	212752
800_3	213562	1193	**211602**	215315*	0.82	685	213358*	215548	0.93	572	214467	214870	0.61	318	214126
800_4	208114	1064	**205902**	211164*	1.47	564	209702*	211302	1.53	496	210026	210845	1.31	429	209431
1000_1	312110	1385	**309891**	314526	0.77	791	312171	314444*	0.75	1062	311087*	314914	0.9	1096	312606
1000_2	309818	1910	**306018**	311419	0.52	845	309249	311081*	0.41	899	308205*	311718	0.61	1153	309158
1000_3	310085	1612	**307275**	313201*	1	867	311449*	313424	1.08	669	311640	313989	1.26	981	311377
1000_4	307405	1412	**304769**	310574*	1.03	971	307707	310610	1.04	1131	306263*	311415	1.3	943	308816
Median:					1.34%				1.29%				1.28%		
Total:				10*			12*	10*			13*				

5.4   Comparing MA-BCRCD with MATE with and Without Crossover

Since MATE is open source, it is easy to perform follow-up experiments on the results from Liu et al. [12]. We took advantage of this code in two ways. First, since the only stopping criteria for both algorithms is the 2 h time limit, the environment in which the algorithm runs might significantly affect the quality of the final solutions found. Running on a slower machine may result in worse solutions. Because of this, we ran MATE in the same environment as we did MA-BCRCD to provide as fair a comparison as possible.

Second, it is possible that the poor results of using RARI in our MA is due to other algorithmic components in the MA such as the simple initialization procedure, and that RARI might work well in a different memetic algorithm. Luckily, the source code of MATE includes a configurable setting where crossover is not applied at all during the search. To further test the effectiveness of RARI, we also ran MATE without the use of crossover to see if there would be any significant difference in solution quality.

For MA-BCRCD, the parameters used in the comparison against MATE are the same values as what were used for the crossover comparison. MATE has three user-defined parameters: The parameters ω_1 and ω_2 determine the lower and upper bounds of the number of customers which are removed during MATE's large search operator, and N in this case is the population size. We ran MATE using the same parameter choices as was used for the jd instances in [12], which are: $N = 36$, $\omega_1 = 0.2$, and $\omega_2 = 0.4$.

Table 2 shows all the experiments run with the source code of MATE, and the results are compared to MA-BCRCD. For reference, the published results from [12] are included as well. Both the comparison between MA-BCRCD and the MATE re-run as well as the comparison between MATE with and without crossover need to be examined.

From these results, we see that MA-BCRCD outperformed the MATE re-run on all instances, with some improvements reaching up to more than 3% on average. For more than half of the instances, the average score of MA-BCRCD was better than the best score found by MATE.

When comparing MATE with and without crossover, no clear advantage can be detected when using RARI. In terms of number of better average solutions, the runs with crossover has 10 better averages, while the runs without crossover also has 10 better averages. Comparing the best solutions found, the results show that using RARI yielded 12 better solutions while using no crossover gave 13 better. In addition, the median percent average difference from MA-BCRCD was slightly worse for MATE with crossover at 1.34% against the 1.29% median difference from the runs without crossover.

It is important to note that MATE currently outperforms all existing approaches which have been applied to the popular Wang and Chen instances both when compared one-on-one and in terms of number of best-known solutions found. This fact in conjunction with the results discussed above suggest that the ways in which these instances differ from the Wang and Chen instances (size, objective function, road network complexity, etc.) warrant special consideration from researchers if they are to be solved effectively.

6   Conclusion

In this study, the vehicle routing problem with simultaneous pickup and delivery with time windows was discussed, and a memetic algorithm was proposed for tackling the problem. The proposed MA was designed to address properties specific to the large real-world jd instances from Liu et al. [12]. In particular, the MA was designed with search efficiency in mind by employing a simple initialization procedure, a fast yet effective crossover based on the popular BCRC, and judicious use of a local search procedure. The altered BCRC was shown to be well suited for these instances, and the MA was able to strongly outperform the previous solution method applied to them.

Further work should be done to be able to solve the jd instances more efficiently. Because distances and times between customers are given explicitly, more intelligent clustering techniques should be applied to allow for efficient divide-and-conquer approaches which could greatly reduce algorithm run times. Some characteristics of the jd instances were not exploited in this study, but for example, taking advantage of tight time window constraints could simplify the search further.

Acknowledgement. This research was supported financially by the Natural Sciences and Engineering Research Council of Canada (NSERC).

References

1. Alvarenga, G.B., Mateus, G.R., De Tomi, G.: A genetic and set partitioning two-phase approach for the vehicle routing problem with time windows. Comput. Oper. Res. **34**(6), 1561–1584 (2007)
2. Angelelli, E., Mansini, R.: The vehicle routing problem with time windows and simultaneous pick-up and delivery. In: Klose, A., Speranza, M.G., Van Wassenhove, L.N. (eds.) Quantitative Approaches to Distribution Logistics and Supply Chain Management. LNE, vol. 519, pp. 249–267. Springer, Heidelberg (2002). https://doi.org/10.1007/978-3-642-56183-2_15
3. Baker, B.M., Ayechew, M.: A genetic algorithm for the vehicle routing problem. Comput. Oper. Res. **30**(5), 787–800 (2003)
4. Battarra, M., Cordeau, J.F., Iori, M.: Pickup-and-delivery problems for goods transportation, chapter 6. In: Vehicle Routing: Problems, Methods, and Applications, 2nd edn., pp. 161–191. SIAM (2014)
5. Braekers, K., Ramaekers, K., Van Nieuwenhuyse, I.: The vehicle routing problem: state of the art classification and review. Comput. Ind. Eng. **99**, 300–313 (2016)
6. Desaulniers, G., Madsen, O.B., Ropke, S.: The vehicle routing problem with time windows, chapter 5. In: Vehicle Routing: Problems, Methods, and Applications, 2nd edn., pp. 119–159. SIAM (2014)

7. Dethloff, J.: Vehicle routing and reverse logistics: the vehicle routing problem with simultaneous delivery and pick-up: Fahrzeugeinsatzplanung und redistribution: Tourenplanung mit simultaner auslieferung und rückholung. OR-Spektrum **23**, 79–96 (2001)
8. Hof, J., Schneider, M.: An adaptive large neighborhood search with path relinking for a class of vehicle-routing problems with simultaneous pickup and delivery. Networks **74**(3), 207–250 (2019)
9. Kassem, S., Chen, M.: Solving reverse logistics vehicle routing problems with time windows. Int. J. Adv. Manuf. Technol. **68**(1–4), 57–68 (2013)
10. Koç, Ç., Laporte, G., Tükenmez, İ: A review of vehicle routing with simultaneous pickup and delivery. Comput. Oper. Res. **122**, 104987 (2020)
11. Liu, F., et al.: A hybrid heuristic algorithm for urban distribution with simultaneous pickup-delivery and time window. J. Heuristics 1–43 (2023)
12. Liu, S., Tang, K., Yao, X.: Memetic search for vehicle routing with simultaneous pickup-delivery and time windows. Swarm Evol. Comput. **66**, 100927 (2021)
13. Liu, Z., et al.: A new hybrid algorithm for vehicle routing optimization. Sustainability **15**(14), 10982 (2023)
14. Lu, Y., Benlic, U., Wu, Q.: An effective memetic algorithm for the generalized bike-sharing rebalancing problem. Eng. Appl. Artif. Intell. **95**, 103890 (2020)
15. Mingyong, L., Erbao, C.: An improved differential evolution algorithm for vehicle routing problem with simultaneous pickups and deliveries and time windows. Eng. Appl. Artif. Intell. **23**(2), 188–195 (2010)
16. Ombuki, B., Ross, B.J., Hanshar, F.: Multi-objective genetic algorithms for vehicle routing problem with time windows. Appl. Intell. **24**, 17–30 (2006)
17. Prins, C.: A simple and effective evolutionary algorithm for the vehicle routing problem. Comput. Oper. Res. **31**(12), 1985–2002 (2004)
18. Ropke, S., Pisinger, D.: An adaptive large neighborhood search heuristic for the pickup and delivery problem with time windows. Transp. Sci. **40**(4), 455–472 (2006)
19. Shi, Y., Zhou, Y., Boudouh, T., Grunder, O.: A lexicographic-based two-stage algorithm for vehicle routing problem with simultaneous pickup-delivery and time window. Eng. Appl. Artif. Intell. **95**, 103901 (2020)
20. Solomon, M.M.: Algorithms for the vehicle routing and scheduling problems with time window constraints. Oper. Res. **35**(2), 254–265 (1987)
21. Wang, C., Mu, D., Zhao, F., Sutherland, J.W.: A parallel simulated annealing method for the vehicle routing problem with simultaneous pickup-delivery and time windows. Comput. Ind. Eng. **83**, 111–122 (2015)
22. Wang, H.F., Chen, Y.Y.: A genetic algorithm for the simultaneous delivery and pickup problems with time window. Comput. Ind. Eng. **62**(1), 84–95 (2012)
23. Wu, H., Gao, Y.: An ant colony optimization based on local search for the vehicle routing problem with simultaneous pickup-delivery and time window. Appl. Soft Comput. 110203 (2023)

Tabu Search for Solving Covering Salesman Problem with Nodes and Segments

Takafumi Matsuura[✉][iD]

Nippon Institute of Technology, 4-1 Gakuendai, Miyashiro-machi,
Minamisaitama-gun, Saitama 345-8501, Japan
matsuura@nit.ac.jp

Abstract. We have already mathematically formulated a Covering Salesman Problem with Nodes and Segments (CSPNS). In CSPNS, a node distribution is given. In CSPNS, some nodes are selected from the given set, and a tour is constructed that visits these nodes. An objective of CSPNS is to identify the shortest tour that covers all given nodes by segments and nodes in the tour. In this study, to find good near-optimal solutions within a reasonable time frame, we propose a heuristic method using tabu search.

Keywords: Covering Problem · Tabu Search · Heuristic Method

1 Introduction

Large-scale natural disasters have occurred almost every year in Japan. Efforts are being made to protect lives and minimize damage by strengthening river embankments, improving sewage systems, and seismic retrofitting. However, preventing all damage is impossible. Therefore, to rescue as many people as possible in a disaster, it is important to quickly assess the damage and identify those in need of rescue as soon as possible.

In recent years, the performance of drones has significantly improved. Compared to traditional aircraft and helicopters that were used for information gathering, drones require less preparation time. Therefore, drones are widely used for gathering information on damage situations. However, drones have a shorter continuous flight time than airplanes and helicopters, thus, it is necessary to conduct efficient search activities within a limited flight time.

To decide efficient flight route of a drone, we have already proposed a combinatorial optimization problem called "Covering Salesman Problem with Nodes and Segments (CSPNS)." In the CSPNS, a node distribution is provided. The objective of the CSPNS is to identify the shortest tour constructed by a subset of the given nodes set such that a node not on the tour is within a radius r of any node or segment on the tour. When the number of nodes is less than 50, an optimal solution of CSPNS is found by general purpose mixed integer solver such as Gurobi Optimization [1, 3] However, when the number of nodes increases, the

M. Sevaux et al. (Eds.): MIC 2024, LNCS 14753, pp. 93–99, 2024.
https://doi.org/10.1007/978-3-031-62912-9_9

optimal solution of CSPNS cannot be found in a reasonable time frame. Thus, we have already propose local search methods [1]. Although the method quickly constructs the solutions, the obtained solutions are local optimal solutions.

To find good near-optimal solutions for combinatorial optimization problems, various meta-strategies have been proposed such as the simulated annealing [4], the genetic algorithm [5], and the tabu search [6–8]. In this study, we propose a heuristic method by using the tabu search. In the method, the tabu search controls an execution of three local search methods: removing method, adding method, and two types of swapping methods. From results of numerical simulation, the proposed method shows good performances for CSPNS.

2   Covering Salesman Problem with Nodes and Segments

A set of a depot and nodes $N = \{0, 1, ..., n\}$, euclidean distance d_{ij} between node i and node j, perpendicular distance c_{ijk} between node i and segment jk, and a covering distance $r > 0$ are given. In the set N, node 0 is the depot where the drone takes off and lands. If the distance d_{ij} is less than or equal to r, the node i can cover the node j. If the perpendicular distance c_{ijk} is less than or equal to r, the edge j-k can cover the node i. The constraint conditions of CSPNS are as follows:

1. The drone starts from the depot (node 0) and goes back to the depot.
2. The drone can visit each node at most once.
3. All nodes must be covered by either nodes or edges on the tour.

An objective of CSPNS is to identify the shortest tour time which satisfies the constant conditions. Figure 1 shows a graphical example of CSPNS. In CSPNS, nodes included on the tour are referred to as visited nodes, while nodes not included in the tour are called unvisited nodes.

One of the problems that is most similar to CSPNS is covering salesman problem (CSP) [2]. The CSP is also given a set of nodes, and the goal is to find the shortest tour that covers all the given nodes. In CSPNS, the given nodes can be covered by both visited nodes and segments within the tour. In CSP, however, the given nodes can only be covered by visited nodes.

3   Proposed Method

To quickly find good near-optimal solutions, we propose a heuristic method by using tabu search. The procedure of the proposed method is as follows:

1. CSPNS is solved as a TSP. That is, a tour that visits all given nodes is constructed. The construction method constructs an initial tour by randomly visiting nodes.
2. The initial tour is improved by Lin-Kernighan algorithm.
3. If a node on the tour can be removed and still cover all nodes, it is removed from the tour.

4. Until a local optimal solution is obtained, the travel time of tour is improved by five local search methods.
5. To find good near-optimal solutions, tabu search explores the solution space of CSPNS.

3.1 Local Search Method

To find local optimal solutions of CSPNS, the proposed method uses five local search methods.

The removing method removes a visited node from a tour if all nodes are covered by a new tour after removing the visited node. The inserting method inserts a visited node between an edge in the tour. The swapping method swaps visited nodes in the tour. The exchanging nodes method exchanges a visited node with an unvisited node. The exchanging edges method exchanges an edge in the tour with an edge constructed by unvisited nodes. The adding method inserts an unvisited node between an edge in the tour. When an unvisited node is inserted between edges in the tour, the travel time increases. However, when the number of nodes in the tour is small, it is difficult to generate feasible neighboring solutions, or the number of such solutions decreases. To address this issue, the adding method is used only when executing the tabu search.

3.2 Tabu Search

Tabu search is one of the powerful meta-strategies for solving combinatorial optimization problems. In the proposed method by using tabu search, the neighborhood solutions of the current solution are generated by six local search methods described in Sect. 3.1. Then, to prevent periodic solution search, five tabu lists are used (Table 1).

The procedure of the proposed method by using tabu search is described as follows:

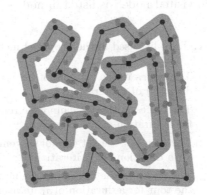

■: depot ●: visited nodes ●: unvisited nodes

Fig. 1. Covering Salesman Problem with Nodes and Segments ($n = 150$ and $r = 50$)

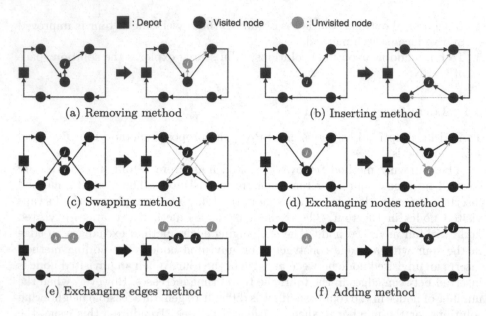

Fig. 2. Local search methods used in the proposed method

Step 1: Exchanging nodes method by tabu search The neighborhood solutions of the current tour is generated by the exchanges nodes methods. If visited node i and unvisited node j is exchanged, visited node i is listed in additional tabu list and unvisited node j is listed in removal tabu list (Fig. 2).

Step 2: Inserting method by tabu search The neighborhood solutions of the current tour is generated by the inserting method. If visited node i is inserted into an edge, visited node i is listed in insertion tabu list.

Step 3: Swapping method by tabu search The neighborhood solutions of the current tour is generated by the inserting method. If visited node i is inserted into an edge, visited node i is listed in node exchanged tabu list.

Table 1. Tabu list used in the proposed method

Additional tabu list	The nodes in the list cannot be add in the tour for a duration of τ_a iteration
Removal node tabu list	The nodes in the list cannot be removed from the tour for a duration of τ_{rn} iteration
Removal edge tabu list	The edges in the list cannot be removed from the tour for a duration of τ_{re} iteration
Insertion tabu list	The nodes in the list cannot be inserted into the edge in the tour for a duration of τ_i iteration
Node exchange tabu list	The pairs of nodes in the list cannot be exchanged for a duration of τ_n iteration

Step 4: Exchanging edges method by tabu search The neighborhood solutions of the current tour is generated by the exchange edges method. If edge i-j in the tour and edge k-l not in the tour are exchanged, edge k-l is listed in removal edge tabu list.

Step 5: Adding method by tabu search The neighborhood solutions of the current tour is generated by the adding method. If node k is added into edge i-j, node k is listed in additional tabu list.

Step 6: Removing method by tabu search The neighborhood solutions of the current tour is generated by the removing method. If node i is removed from the tour, node i is listed in removal tabu list.

The proposed method repeats Steps 1 to 6 until the termination condition is satisfied. If the new best solution is found in each step, the local search methods described in Sect. 3.1 are applied until a local optimal solution is obtained.

4 Simulations and Results

To investigate performances of the proposed method, we generate benchmark instances by using DIMACS [9] which is one of the benchmark problems for the traveling salesman problem. The number of nodes including the depot is 101 and 151. The nodes are uniformly distributed in the 10^4[m] $\times$ 10^4[m] square and seed for making the instances is set to 1. The covering range r is set to 25 and 50 m. The speed of the drone is $s = 30$[km/h]. The travel time t_{ij}[s] between nodes i and j is calculated by the following equation:

$$t_{ij} = \left\lfloor d_{ij} \times \frac{3,600[\text{s}]}{s \times 1,000[\text{m}]} \right\rfloor.$$

The tabu tenure is set to several values. In the simulation, we used a Mac Stdio (Apple M1 Ultra) with 64 GB of memory running Mac OS X 12.6. The termination condition is after $n/10$[s] from the start of the propose method.

Table 2. Results of the proposed method

n	r	OPT	Local Search		Tabu Search			
			Ave.	Best	Ave.	Best	IR	tabu tenure
100	25	957	990.25	963	977.35	961	1.32%	$\tau_{rn} = 14, \tau_a, \tau_{re}, \tau_i, \tau_n = 4$
	50	820	885.40	860	848.38	828	4.36%	$\tau_{rn} = 14, \tau_a, \tau_{re}, \tau_i, \tau_n = 7$
150	25	1089	1135.72	1115	1117.30	1100	1.65%	$\tau_{rn} = 15, \tau_a, \tau_{re}, \tau_i, \tau_n = 2$
	50	888	972.48	937	923.17	904	5.34%	$\tau_{rn} = 13, \tau_a, \tau_{re}, \tau_i, \tau_n = 9$

Table 2 shows the results of the proposed method. The column of OPT shows the optimal travel time. By using the formulation of the CSPNS [1] and a mixed-integer programming solver, such as Gurobi Optimization [3], we can obtain an optimal solution. By using 16 threads, the CPU time required to obtain the optimal solution was 2330.43 s for 100 nodes and 27864.00 s for 150 nodes when the covering range is 50 m. "Local Search" shows the results obtained by up to step 4 of the proposed method described in Sect. 3. "Tabu Search" represents the results of the proposed method after executing tabu search. The columns of Ave. and Best represent the average and best results obtained in 30 trials, respectively. The column of IR represents an improvement rate of the average travel time by the proposed method compared to the average travel time of the local search methods. The column of tabu tenure indicates the tenures at which the average and best travel times were obtained.

From Table 2, although the proposed method cannot find the optimal solution, the good near-optimal solutions can be found. The gaps between the optimal solution and the best solution obtained by the proposed method are approximately 1%.

From the results of IR, it is observed that the improvement rate is higher when the covering distance r is larger. This results suggests that the proposed method is effective for CSPNS, where the coverage factor is important.

In this simulation, $\tau_a, \tau_{re}, \tau_i$, and τ_n are set to the same value and only τ_{rn} was set to a different value because the proposed method has many tabu lists. From the results of tabu tenure, it can be observed that better solutions are obtained when the tabu tenure is longer for τ_a, τ_{re}, and τ_i compared to τ_n.

5   Conclusion

This study proposed a heuristic method by using tabu search for find good near-optimal solutions for covering salesman problem with nodes and segments. From the computational results, although the proposed method uses simple local search methods and the tabu search, it obtains good solutions. In the future work, it is important to develop an effective adjustment method of tabu tenure. It is also important to compare the performance of other meta-strategies, such as genetic algorithms, ant colony algorithms, simulated annealing methods, and others.

Acknowledgments. This study was funded by JSPS KAKENHI (grant number JP22K04602).

References

1. Takafumi, M., Takayuki, K.: Covering salesman problem with nodes and segments. Am. J. Oper. Res. **7**(4), 249–262 (2017)
2. Current, J.R., Schilling, D.A.: The covering salesman problem. Transp. Sci. **23**(3), 208–213 (1989)
3. Gurobi Optimization. https://www.gurobi.com/
4. Kirkpatrick, S., Gelatt, C.D., Vecchi, M.P.: Optimization by simulated annealing. Science **220**, 671–680 (1983)

5. Holland, J.H.: Adaptation in Natural and Artificial Systems. University of Michigan Press, Ann Arbor (1975)
6. Glover, F.: Tabu search I. ORSA J. Comput. **1**(3), 190–206 (1989)
7. Glover, F.: Tabu search II. ORSA J. Comput. **2**, 4–32 (1990)
8. Glover, F., Taillard, E.: A user's guide to tabu search. Ann. Oper. Res. **41**(1), 1–28 (1993)
9. 8th DIMACS Implementation Challenge: The Traveling Salesman Problem. http://dimacs.rutgers.edu/archive/Challenges/TSP/

GRASP with Path Relinking

VNS with Path Relinking
for the Profitable Close-Enough Arc
Routing Problem

Miguel Reula[✉][iD], Consuelo Parreño-Torres[iD], Anna Martínez-Gavara[iD],
and Rafael Martí[iD]

University of Valencia, Valencia, Spain
miguel.reula@uv.es

Abstract. Arc Routing Problems typically deal with traversing a set
of connecting edges or arcs in a network at the minimum possible cost.
In this paper, we target the close enough model in which clients can be
served from relatively close arcs, addressing some practical situations,
such as inventory management or automated meter reading. We propose a heuristic to maximize the sum of profits of the clients served
(penalized with the distance traveled). Our solving procedure, based on
the VNS methodology, incorporates efficient search strategies to obtain
high-quality solutions in short computational times, as required in practical applications. We study its improvement by coupling the method
with Path Relinking as a post-processing. Our experimentation over a
benchmark of previously reported instances shows the good performance
of the heuristics as compared with a previous GRASP.

Keywords: Metaheuristics · Logistics · Inventory Management ·
Close-Enough · Arc Routing

1 Introduction

In the dynamic field of logistics and operations research, Arc Routing Problem
(ARP) continuously evolve, mainly driven by new methods and technologies.
This paper deals with a specific variant of the classical ARP, based on the inter-
play between profit maximization and service delivery, that addresses the modern
needs where service provision does not require on-site presence but is achieved
by being sufficiently close to the customer.

Recent advancements in radio frequency identification (RFID) technology
have significantly transformed the traditional ARP framework. These advancements enable remote execution of tasks that previously required direct interaction, thus redefining the concept of "service" in ARP models. In this context,
servicing clients is possible by traversing specific arc families assigned to them,
a concept known as the Close-Enough Arc Routing Problem (CEARP). Furthermore, these new models typically optimize a balance between incurred costs
(related to distance or time) and profits (associated to the provided service). An

M. Sevaux et al. (Eds.): MIC 2024, LNCS 14753, pp. 103–109, 2024.
https://doi.org/10.1007/978-3-031-62912-9_10

ARP is deemed "profitable" when it strategically maximizes the net difference between the profits from service provision and the associated travel costs.

Our study focuses on the Profitable Close-Enough Arc Routing Problem (PCEARP), where the required arcs for traversal are not predetermined. In this model, servicing a client involves traversing any one of a set of predefined arcs that are sufficiently close to the client's location. This approach provides a flexible framework to meet the evolving needs of remote services. The PCEARP selectively targets clients offering higher profits and does not necessitate servicing all clients. This flexibility introduces additional complexity to the problem's resolution.

Given a strongly connected digraph $G = (V, A)$, where V is the set of vertices, and A the set of arcs, we consider $d_{ij} \geq 0$ as the distance (or length) of arc $(i, j) \in A$, and vertex 1 denotes the depot. The set of clients $\mathbb{H}$, is not necessarily located in the vertices or arcs of the graph. Each client $c \in \mathbb{H}$ receives service if any of its associated arcs $(H_c \subseteq A)$ is traversed. We also define the profit $p_c \geq 0$ of client c. This profit is collected (only once) if the client is serviced. The Profitable Close-Enough Arc Routing Problem (PCEARP) consists in finding a route on G that starts and ends at the depot. Its objective is to maximize the difference between the sum of the profits collected in the route, and its total length.

The PCEARP encompasses elements of both the CEARP—applicable when all client-associated profits are substantial—and the Profitable Routing Problem with Profits (PRPP)—relevant when each required arc corresponds to an individual client. The PCEARP is an NP-hard problem, suitable for a variety of practical applications. One of the most important applications is found in the supply chain management; in particular in inventory management in warehouses. The application of PCEARP models in this setting optimizes drone routes for inventory checks, enhancing efficiency, accuracy, and safety. Another practical application is in automatic meter reading, where RFID technology enables remote data collection for utility meters [7,9].

2   Previous GRASP Approaches

The PCEARP was initially approached using a branch-and-price algorithm for a related problem, with the variant focusing on profit generation identified by [1] in their pricing problem. This led to the development of a GRASP heuristic [6], named GRASP_BP, tailored for solving the minimization component of the pricing challenge. This method is noteworthy for its rapid computational times, a critical feature for algorithms repeatedly applied in branch-and-price frameworks. In GRASP_BP, a constructive algorithm is utilized in each iteration to build a solution, which is then improved upon using a local search algorithm, with the best-profit route being the final output. Building upon the foundation laid by GRASP_BP, in [2] the authors developed and implemented the GRASP_IT, a modified version of the original heuristic. This GRASP_IT, while slightly slower, significantly outperforms GRASP_BP in terms of solution quality. A short description of the general method follows.

Construction Phase: At each step the method computes, for each required arc a not in the route, its value ψ_a, as the change in the objective function if a is inserted in route r (in the best possible position). It is expected that new clients are now served and therefore the sum of profits will probably increase, but at a price that comes from the increment in the total distance to visit them (to traverse the arcs serving them). In mathematical terms, $f(r') = f(r) + \psi_a$.

The Restricted Candidate List (RCL) is built with arcs with good evaluation, specifically those with the highest ψ_a values. The method then randomly selects one of these arcs. Let ψ_{max} and ψ_{min} be the maximum and minimum respectively of the ψ_a values for all the required arcs not in the current route. Then,

$$RCL = \{a \in A_R \setminus A_R(r) : \psi_a \geq \alpha(\psi_{max} - \psi_{min}) + \psi_{min}\},$$

where the parameter $\alpha = 0.9$ balances the greediness and the randomization in the selection from RCL. The selected arc is added to the route, and its value updated.

Improvement Phase: The method applies a post-processing procedure to improve each constructed solution by exploring its neighborhood. In particular, it consists of a destroy-and-repair method [3] that first removes some arcs from the route, and then add new arcs to improve the resulting solution. It is indeed more complex than the standard local search usually applied in GRASP, in an effort to obtain improved outcomes.

3   A New Heuristic Algorithm Based on VNS

We propose a MultiStart Variable Neighborhood Search (MS-VNS) algorithm that extends the Variable Neighborhood Search (VNS) metaheuristic, integrating it into a multi-start framework to enhance its ability to diversify the search process. MS-VNS employs four different neighborhood structures, each based on swap mechanisms where required arcs in a route are exchanged with others not in the solution. The neighborhoods are: N_{1-1}, swapping one required arc with another not in the solution; N_{2-1}, swapping two required arcs with one not in the solution; N_{1-2}, swapping one required arc with two not in the solution; and N_{2-2}, exchanging two arcs with two different arcs. These neighborhoods use a first improvement strategy and incorporate shortest paths to potentially serve non-served clients.

The core of MS-VNS is the Variable Neighborhood Descendent (VND) algorithm, which systematically changes the neighborhood structure. VND begins with a solution and explores different neighborhoods in a predefined order. If an improvement is found in a neighborhood, the search returns to the first neighborhood; otherwise, it proceeds to the next. This process continues until no further improvements can be found in any neighborhood, indicating that the solution is locally optimal across all considered neighborhoods.

The MS-VNS algorithm itself starts with an initial best solution and iterates over a set time limit. Each iteration generates a new solution using a constructive algorithm, which is then locally improved through VND. The Shake method, a key component of MS-VNS, diversifies the search by randomly altering parts of the current solution. This method selects a random arc in the route and removes a percentage of consecutive arcs, then reconstructs the route. The percentage of removed arcs increases iteratively if no improvements are found, up to a maximum limit. The VND algorithm is then applied to improve the shaken solution. We refer the reader to [8] for a detailed description of this component.

As expected, the algorithm's performance is influenced by the maximum time allowed for the search and the largest neighborhood to be explored. The time limit is reactive and is adjusted based on the improvements found during the search. Initially set at 60 s, the time limit increases if a better solution is found, allowing more time for further improvements.

3.1 The Path Relinking Post-processing

Path Relinking (PR) was initially proposed as a mechanism for long-term memory within tabu search, but it can be applied as a post-processing to improve the solutions obtained with any metaheuristic [5]. This strategy intensifies the search by generating new solutions through paths in the neighborhood space, starting with an *initiating solution* and moving towards the *guiding solutions*. PR selects moves to introduce attributes from the guiding solutions. The moves chosen during the relinking process are different from those moves during a "normal" local search because the relinking moves do not use the change of the objective function as the guiding principle.

Path relinking can be thought of as a constrained neighborhood search, where the search is limited to explore the solutions in the neighborhood with characteristics of the guiding solution. The selected neighborhood will determine the set of solutions visited by path relinking. For example, consider a solution represented by a permutation and consider two neighborhoods, swap and insert. In swap $(\pi^1, \ldots, \pi^i, \ldots, \pi^j, \ldots, \pi^n)$ and $(\pi^1, \ldots, \pi^j, \ldots, \pi^i, \ldots, \pi^n)$ are neighbors because π^i and π^j swap their positions whereas in insert $(\pi^1, \ldots, \pi^{i-1}, \pi^i, \ldots, \pi^n)$ and $(\pi^1, \ldots, \pi^{i-1}, \pi^j, \pi^i, \ldots, \pi^n)$ are neighbors since π^j is inserted in position i. In line with the VNS methodology described in the previous section, we consider these two neighbors in our PR for the PCEARP.

4   Computational Experiments and Conclusions

In our computational testing, we consider the 396 PCEARP benchmark instances [2] categorized into *Albaida, Madrigueras,* and *Random* of up to 400 vertices, available at [4]. Three profit scenarios were defined for each instance, based on servicing 60%, 80%, and 90% of clients, respectively. All the algorithms have been implemented in C++ and run on an OpenStack virtualization platform, supported by several blade servers with two 18-core Intel Xeon Gold 5220 processors, running at 2.2 GHz, and has 384 GBytes of RAM.

We compare our proposal, MS-VNS, with two previous GRASP heuristics, namely GRASP_BP [1], designed for the pricing part of the min-max close enough arc routing problem, adapted for PCEARP as a maximization problem, and GRASP_IT [2], is an iterative algorithm that combines constructive and local search heuristics, improving the construction of solutions for better quality bounds in branch-and-cut algorithms, particularly effective for smaller instances. We do not include in the tables the PR results due to the space limitations, and considering that in its current implementation it only provides a marginal improvement in the VNS results.

Table 1. Solutions on 30 runs (with 60 s per instance).

	#Inst.	MS-VNS		GRASP_IT		GRASP_BP	
		of_{avg}	CPU(s)	of_{avg}	CPU(s)	of_{avg}	CPU(s)
Albaida	72	851.3	2.5	848.1	1.7	715.3	0.0
Madrigueras	72	1201.6	13.9	1163.8	4.6	829.8	0.1
Random50	36	1024.4	9.2	1018.9	17.8	901.5	0.1
Random75	36	1470.5	19.7	1463.7	23.8	1296.0	0.5
Random100	36	1679.6	26.5	1629.2	32.4	1286.8	0.9
Random150	36	2245.8	39.1	2141.9	37.7	1653.8	3.2
Random200	36	2451.3	43.7	2262.7	46.4	1894.1	9.3
Random300	36	2554.6	48.8	2231.6	49.2	1746.2	18.8
Random400	36	3611.6	54.0	3304.7	55.7	2887.3	30.1
		1740.3	**24.9**	**1643.3**	**25.1**	**1341.4**	**5.7**

Table 1 shows the average objective function values and the average running time over the 30 replicates conducted with each algorithm on each instance. The MS-VNS heuristic clearly outperforms GRASP_IT and GRASP_BP in solution quality, with an average value of 1740.3, significantly better than the 1643.3 of GRASP_IT and 1341.4 of GRASP_BP.

Although not reported in the tables here, it is worth mentioning that MS-VNS average worst value across replications is 1656.2, improving the best mean scores of both GRASP_IT and GRASP_BP. Its robustness is further indicated by lower standard deviations (40.5) compared to GRASP_IT (50.3) and GRASP_BP (86.9). Despite GRASP_BP being the fastest with an average time of 5.7 s, MS-VNS and GRASP_IT, averaging around 25 s, offer a better balance between solution quality and computational time.

In Table 2, we conduct a comparison of the top performers in terms of the objective function, focusing on the mean time, the GAP(%) relative to the best known solution, and the number of best known solutions obtained. The effectiveness of the MS-VNS is further validated through its comparison with GRASP_IT across 9 benchmark datasets. Indeed, MS-VNS secures the best solutions in 287 out of the 396 instances tested, markedly outdoing GRASP_IT,

Table 2. Comparison of MS-VNS and GRASP_IT w.r.t best known solutions.

	#Inst.	MS-VNS		GRASP_IT	
		GAP(%)	# BestSol	GAP(%)	# BestSol
Albaida	72	0.00	72	0.01	71
Madrigueras	72	0.15	69	1.10	50
Random50	36	0.01	35	0.26	29
Random75	36	0.00	36	1.16	22
Random100	36	0.10	33	1.16	17
Random150	36	0.50	20	3.67	7
Random200	36	1.62	12	7.01	5
Random300	36	5.71	7	13.76	3
Random400	36	10.74	3	15.17	0
		1.72	**287**	**4.04**	**204**

which achieves only 204 best known solutions. Additionally, MS-VNS demonstrates superior performance in terms of the average gap relative to the best known solution, recording a significantly lower gap of 1.72%, compared to GRASP_IT's 4.04%. Additionally, considering the 362 instances with optimum known, our MS-VNS is able to match the optimal value in around 80% of them.

As a final conclusion, we can state that our experimentation with VNS clearly shows its superiority with respect to previous heuristics, and additionally allowed us to identify its limitations in terms of matching the optimal (or best known) value in large instances, which motivated the inclusion of PR as a post-processing to refine the results. We are working on some promising PR variants to further improve our results. This research is partially supported by ERDF -A way of making Europe- PID2021-125709OB-C21 MCIN/AEI.

References

1. Bianchessi, N., Corberán, Á., Plana, I., Reula, M., Sanchis, J.M.: The min-max close-enough arc routing problem. Eur. J. Oper. Res. **300**(3), 837–851 (2022)
2. Bianchessi, N., Corberán, Á., Plana, I., Reula, M., Sanchis, J.M.: The profitable close-enough arc routing problem. Comput. Oper. Res. **140**, 105653 (2022)
3. Corberán, Á., Plana, I., Reula, M., Sanchis, J.M.: A matheuristic for the distance-constrained close-enough arc routing problem. TOP **27**, 312–326 (2019)
4. Corberán, Á., Plana, I., Reula, M., Sanchis, J.M.. PCEARP Instances. https://www.uv.es/plani/instancias.html. Accessed May 2023
5. Laguna, M., Martí, R.: GRASP and path relinking for 2-layer straight line crossing minimization. INFORMS J. Comput. **11**(1), 44–52 (1999)
6. Laguna, M., Martí, R., Martínez-Gavara, A., Pérez-Peló, S., Resende, M.: 20-years of GRASP with path relinking. Eur. J. Oper. Res. (2024, forthcoming)
7. Renaud, A., Absi, N., Feillet, D.: The stochastic close-enough arc routing problem. Networks **69**, 205–221 (2017)

8. Reula, M., Martí, R.: Heuristics for the profitable close-enough arc routing problem. Expert Syst. Appl. **230**, 120513 (2023)
9. Sinha Roy, D., Defryn, C., Golden, B., Wasil, E.: Data-driven optimization and statistical modeling to improve meter reading for utility companies. Comput. Oper. Res. **145**, 105844 (2022)

Meta-Heuristics for Preference Learning

A Simulated Annealing Algorithm to Learn an RMP Preference Model

Yann Jourdin[1(✉)], Arwa Khannoussi[2], Alexandru-Liviu Olteanu[3], and Patrick Meyer[1(✉)]

[1] IMT Atlantique, Lab-STICC, UMR CNRS 6285, 29238 Brest, France
{yann.jourdin,patrick.meyer}@imt-atlantique.fr
[2] IMT Atlantique, LS2N, UMR CNRS 6004, 4430 Nantes, France
arwa.khannoussi@imt-atlantique.fr
[3] Lab-STICC, UMR 6285, CNRS, Université Bretagne Sud, Lorient, France
alexandru.olteanu@univ-ubs.fr

Abstract. Multiple Criteria Decision Aiding (MCDA) provides preference models and algorithms to assist decision-makers (DMs) in their decision-making tasks. The preference models are characterized by preference parameters which can be learned through preference learning algorithms from holistic judgments given by the DM. Here, we use Simulated Annealing (SA) to learn the parameters of the Ranking based on Multiple Reference Profiles (RMP) model and its simpler variant SRMP. Extensive experiments demonstrate that our proposal outperforms existing methods in terms of both calculation time and accuracy.

Keywords: multi-criteria decision aiding · reference profiles · preference learning · simulated annealing

1 Introduction

Multiple Criteria Decision Aiding (MCDA) helps decision-makers (DMs) make more informed decisions among a set of alternatives through the use of preference models. This work focuses on "ranking" problems, whose objective is to order the alternatives from the most preferred to the less preferred. More precisely, we study a specific preference model, namely Ranking based on Multiple Reference Profiles (RMP) [8] and its simpler variant Simple RMP (SRMP). Both use reference profiles to compare alternatives two by two to build a total preorder.

The model parameters must be chosen to best represent the DM's preferences expressed through pairwise comparisons of alternatives, requiring preference elicitation algorithms. For RMP models, a SAT (Boolean satisfiability problem) formulation [1] has been proposed, while a mixed integer linear program [7] and evolutionary metaheuristics [4,6] have been proposed to learn SRMP models. However, computation time for the SAT formulation of [1] increases exponentially with the number of criteria, making it hard to handle large instances. Next to that, to the best of our knowledge, metaheuristics have not been applied to the

M. Sevaux et al. (Eds.): MIC 2024, LNCS 14753, pp. 113–119, 2024.
https://doi.org/10.1007/978-3-031-62912-9_11

elicitation of the more generic RMP model. Finally, the evolutionary algorithms used in [4] and [6] require the calibration of a lot of hyperparameters.

To address these issues, we propose in this article a simulated annealing algorithm [5] to learn both RMP and SRMP models. Results show that the proposed method outperforms the genetic algorithm (GA) of [4] on SRMP models, especially on large instances, while handling larger problems than the SAT formulation on RMP models.

2   Ranking Based on Multiple Reference Profiles (RMP)

As a ranking model, the RMP preference model yields a total preorder $\succsim$ ($\succ$ (resp. $\sim$) represents the asymmetric (resp. symmetric) part of $\succsim$) on the set $\mathcal{A}$ of n alternatives, each described through a set of m criteria $\mathcal{C} = \{1, \ldots, m\}$. Let $a_j \in \mathbb{R}$ be the performance of $a \in \mathcal{A}$ on criterion $j \in \mathcal{C}$, i.e., $a \equiv (a_j)_{j \in \mathcal{C}}$. Without loss of generality, we suppose here that higher performances correspond to preferred evaluations. The preference parameters of an RMP model are:

- $\mathcal{P} = \{p^i, 1 \leq i \leq k\}$, a set of k reference profiles, $p^i \equiv (p_j^i)_{j \in \mathcal{C}}$, where p_j^i is the performance of profile p^i on criterion j. Profiles dominate each other, i.e., $p_j^i \geq p_j^{i+1} \forall j \in \mathcal{C}$.
- $\trianglerighteq$, the importance relation on criteria coalitions, i.e., a total preorder on the powerset $2^\mathcal{C}$ of $\mathcal{C}$ and monotonic with respect to set inclusion. This parameter is equivalent to a capacity μ (a non-negative set function monotonic with respect to set inclusion), with $A \trianglerighteq B \iff \mu(A) \geq \mu(B) \quad \forall A, B \in 2^\mathcal{C}$ [1].
- σ, a permutation of the indexes of the profiles, representing a lexicographic order on the profiles which stands for the importance of the reference points.

In the RMP model, at first, for every alternative $a \in \mathcal{A}$ and every profile $p^i \in \mathcal{P}$, the set of criteria $c(a, p^i) = \{j \in \mathcal{C} | a_j \geq p_j^i\}$ containing all criteria on which a is at least as good as p^i is computed. Then, given these criteria coalitions, an outranking relation $\succsim_i$ for each profile $p^i \in \mathcal{P}$ is created as follows: $a \succsim_i b \iff c(a, p^i) \trianglerighteq c(b, p^i) \forall a, b \in \mathcal{A}$. Finally, the alternatives are ranked through a total pre-order $\succsim$ defined by taking into account the k outranking relations $\succsim_i$ according to the lexicographic order σ, i.e., $a \succ b \iff \exists i', \forall i < i'$, $a \sim_{\sigma(i)} b$ and $a \succ_{\sigma(i')} b$, otherwise $a \sim b$.

A particular case of RMP is SRMP, in which the importance relation $\trianglerighteq$ is defined using additive weights, i.e., for $A, B \in 2^\mathcal{C}$, $A \trianglerighteq B \iff \sum_{j \in A} w_j \geq \sum_{j \in B} w_j$, where w_j is the weight of criterion j, and $\sum_{j \in \mathcal{C}} w_j = 1$.

3   A Simulated Annealing Algorithm to Learn RMP/SRMP Models

The goal of the preference inference algorithm that we propose is to determine the parameters of RMP and SRMP models, given *holistic judgments* expressed by a DM on pairs of alternatives. These take the form of strict preference ($\succ$) and indifference ($\sim$) statements on pairs of alternatives from $\mathcal{A}$. Let $\mathcal{D}$ be the set of such statements. Our proposal, described in 1, follows the classical steps of a simulated annealing algorithm which takes as input $\mathcal{D}$, along with an initial and final temperature T_0 and T_f, a cooling factor α and an initial model M_0 (in the form of its parameters), and returns the best known model M_{best}.

For each type of parameters, we design a specific neighborhood. For the profiles ($\mathcal{P}$), we pick a random profile p and a random criterion j, and replace p_j randomly with one of the precomputed midpoints between two consecutive alternatives' evaluations on that criterion. For the lexicographic order, we randomly switch 2 profiles in σ. For the importance relation $\unrhd$, if the sought model is RMP, we pick randomly a subset of criteria A and replace the value of $\mu(A)$ with a random value between 0 and 1 from a uniform distribution, while respecting the monotonic property of μ with respect to set inclusion. If however the model is SRMP, we pick randomly a criterion j and replace w_j with a value from a uniform distribution on $[w_j - \lambda, w_j + \lambda]$, where λ is a hyperparameter.

The fitness of a model M is the ratio of pairwise comparisons from $\mathcal{D}$ which are correctly rendered by M to the size of $\mathcal{D}$. Note that at each iteration, in line 4 of Algorithm 1 a neighborhood is randomly chosen, with a probability proportional to the size of the considered neighborhood.

Algorithm 1: Simulated annealing for RMP/SRMP learning

Input: $\mathcal{D}$, T_0, T_f, α, M_0
Output: M_{best}

```
1  M ← M₀ ; M_best ← M₀           # Initialize current and best known model
2  T ← T₀                          # Initialize temperature
3  while T > T_f do                # Check stopping condition
4  |   N ← pick_random_neighborhood()          # Pick a neighborhood
5  |   M' ← neighbor(M, N)                      # Create a neighbor model
6  |   if exp((fitness(M', D) − fitness(M, D))/T) > random(0, 1) then
7  |   |   M ← M' ; M_best ← M'     # Update current and best known model
8  |   |   if fitness(M_best, D) = 1 then       # Check 100% fitness
9  |   |   |   return M_best                    # Return best known model
10 |   T ← αT                                   # Update temperature
11 return M_best                                # Return best known model
```

---

4 Numerical Analysis

We choose to study problems with $m \in \{3, 4, 5, 6, 7, 11, 15\}$ criteria. For each of those problem sizes, 50 pairs of train $\mathcal{A}_{i,m}^{train}$ ($|\mathcal{A}_{i,m}^{train}| = 500$) and test $\mathcal{A}_{i,m}^{test}$ ($|\mathcal{A}_{i,m}^{test}| = 500$) datasets are generated ($i \in \{1, \ldots, 50\}$). The performances of the alternatives on the m criteria are drawn randomly from a uniform distribution on $[0,1]$. Next to that, we consider both RMP and SRMP ground truth models with $1 \leq k^* \leq 4$ reference profiles. For each k^* and each problem size m we randomly generate 50 such models $M_{i,m}^*$.

The parameters of a ground truth model $M_{i,m}^*$ are randomly generated as follows: for the profiles, for each criterion, we draw randomly k^* floats from a uniform distribution on $[0,1]$, and sort them to respect the dominance constraint on the profiles; σ is generated by performing a random permutation on $(1,\ldots,k^*)$; for an RMP model, the capacities are generated in a uniform way according to [3], and, in case of an SRMP model, weights are generated according to [2].

For each of those $7 \cdot 50 \cdot 2 \cdot 4 = 2,800$ problems, $M_{i,m}^*$ is used to generate the ground truth pairwise comparisons $\mathcal{D}_{i,m}^{train}$ drawn randomly among the $\frac{500*499}{2} = 124,750$ possible pairwise comparisons. For $|\mathcal{D}_{i,m}^{train}|$ sizes between 100 and 1,000 are tested with an increment of 100, as well as 2,000. $\mathcal{D}_{i,m}^{train}$ is used by the simulated annealing algorithm to train $\widehat{M}_{i,m}$ (with the same number of profiles as $M_{i,m}^*$). $\widehat{M}_{i,m}$ is then applied on $\mathcal{A}_{i,m}^{train}$ to evaluate its fitness. Finally, $M_{i,m}^*$ and $\widehat{M}_{i,m}$ are applied on $\mathcal{A}_{i,m}^{test}$, and pairwise comparisons from both rankings are compared, to evaluate $\widehat{M}_{i,m}$'s generalization ability on unseen data.

To determine optimal hyperparameters of Algorithm 1 we perform a hyperparameter optimization leading to $T_0 = \frac{1}{|\mathcal{D}_{i,m}^{train}|}$, $T_f = \frac{1}{10}T_0$, $\lambda = 0.1$. $\alpha = 0.9999$ is chosen so that the computation time is comparable to state-of-the-art methods.

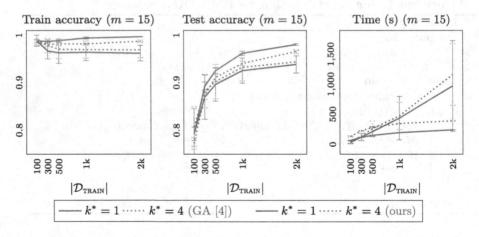

Fig. 1. Average train accuracy, test accuracy and execution time of the GA of [4] (orange) and our proposal (blue) for SRMP models as a function of $|\mathcal{D}_{\mathrm{TRAIN}}|$ for $m = 15$ and $k^* \in \{1,4\}$. Bars represent the limits of the 95% confidence intervals. (Color figure online)

Figure 1 shows, for the learning of SRMP models, the performance of our proposal compared to that of the GA of [4], for a fixed number of criteria ($m = 15$) and two values for k^*. Compared to the GA, we observe that, for our proposal,

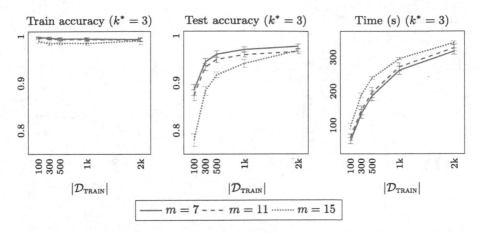

Fig. 2. Performances of our proposal for SRMP models as a function of $|\mathcal{D}_{\text{TRAIN}}|$ for $m \in \{7, 11, 15\}$ and $k^* = 3$. Bars represent the limits of the 95% confidence intervals.

first, average train accuracy is less affected when $|\mathcal{D}_{\text{TRAIN}}|$ increases and second, both average accuracies are higher. Third, for our proposal only, both average accuracies decrease as k^*. Finally, computation times increase with k^*, but depend less on $|\mathcal{D}_{\text{TRAIN}}|$ for our proposal than for the GA. These observations still hold for the other tested values of m. Figure 2 shows the same indicators for our proposal only, while varying m and setting $k^* = 3$. It appears that accuracy decreases with more criteria, while computation time is only slightly affected.

For RMP models, we cannot compare ourselves in every detail to the SAT formulation of [1], as only smaller instances ($m \leq 6$) are used by the authors, and they evaluate their approach using SRMP models. Figure 3 shows the perfor-

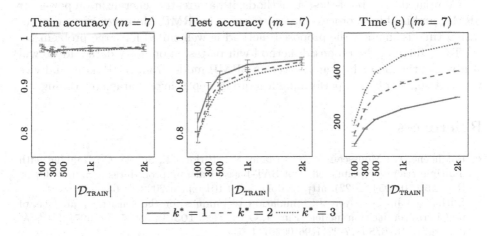

Fig. 3. Performances for an RMP model as a function of $|\mathcal{D}_{\text{TRAIN}}|$ for $m = 7$ criteria and $k^* \in \{1, 4\}$. Bars represent the limits of the 95% confidence intervals.

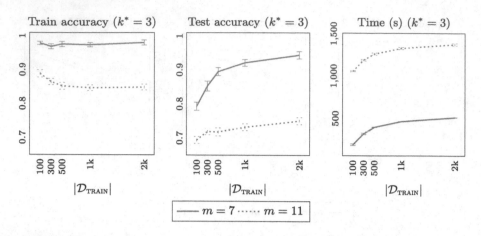

Fig. 4. Performances for RMP models as a function of $|\mathcal{D}_{\text{TRAIN}}|$ for $m \in \{7, 11\}$ criteria and $k^* = 3$. Bars represent the limits of the 95% confidence intervals.

mance evolution for RMP models with $m = 7$ for different k^*. First, both accuracies decrease as k^* increases. Also, as expected, computation time increases with k^*. Figure 4 shows the same indicators, for different m when setting $k^* = 3$. We observe that both train and test accuracies decrease as m increases. Computation time is about three times higher with m going from 7 to 11.

5   Conclusion and Future Work

In this article, we proposed a simulated annealing adapted to the elicitation of parameters of both RMP and SRMP models, and tested it on large instances.

Compared to state-of-the-art methods, it has greater generalization power for SRMP, and is able to process larger instances for RMP. Moreover, since computation time is smaller, the proposed method is well suited for large problems.

To go further, the proposed method will be tested on noisy data and we will study how the algorithm can recover an SRMP model from RMP data, and vice versa. Also, we plan on performing a more in-depth hyperparameter tuning.

References

1. Belahcene, K., Mousseau, V., Ouerdane, W., Pirlot, M., Sobrie, O.: Ranking with multiple reference points: efficient SAT-based learning procedures. Comput. Oper. Res. **150**, 106054 (2022). https://doi.org/10.1016/j.cor.2022.106054
2. Butler, J., Jia, J., Dyer, J.: Simulation techniques for the sensitivity analysis of multi-criteria decision models. Eur. J. Oper. Res. **103**(3), 531–546 (1997). https://doi.org/10.1016/S0377-2217(96)00307-4
3. Grabisch, M., Labreuche, C., Sun, P.: An approximation algorithm for random generation of capacities. Order (2023). https://doi.org/10.1007/s11083-023-09630-0

4. Khannoussi, A., Olteanu, A.L., Meyer, P., Pasdeloup, B.: A metaheuristic for infer-
ring a ranking model based on multiple reference profiles. Ann. Math. Artif. Intell.
(2024). https://doi.org/10.1007/s10472-024-09926-w
5. Kirkpatrick, S., Gelatt, C.D., Jr., Vecchi, M.P.: Optimization by simulated annealing. Science **220**(4598), 671–680 (1983). https://doi.org/10.1126/science.220.4598.
671
6. Liu, J., Ouerdane, W., Mousseau, V.: A metaheuristic approach for preference learning in multicriteria ranking based on reference points. In: Proceeding of the 2nd
DA2PL Workshop, pp. 76–86 (2014)
7. Olteanu, A.L., Belahcene, K., Mousseau, V., Ouerdane, W., Rolland, A., Zheng,
J.: Preference elicitation for a ranking method based on multiple reference profiles.
4OR **20**(1), 63–84 (2022). https://doi.org/10.1007/s10288-020-00468-5
8. Rolland, A.: Reference-based preferences aggregation procedures in multi-criteria
decision making. Eur. J. Oper. Res. **225**(3), 479–486 (2013). https://doi.org/10.
1016/j.ejor.2012.10.013

New VRP and Extensions

Iterative Heuristic over Periods for the Inventory Routing Problem

Katyanne Farias, Philippe Lacomme, and Diego Perdigão Martino[(✉)]

Université Clermont Auvergne, Clermont Auvergne INP, CNRS, Mines Saint-Etienne,
LIMOS, 63000 Clermont-Ferrand, France
katyanne.farias_de_araujo@uca.fr, philippe.lacomme@isima.fr,
diego.perdigao_martino@doctorant.uca.fr

Abstract. Inventory Routing Problems are specially designed to solve transportation problems with inventory management constraints associated. The objective is to serve a set of customers over a finite time horizon, performing product deliveries to meet the demands of customers taking into account constraints regarding inventory levels authorized as well as production capacity. We propose a heuristic method based on an iterative approach that decomposes the original problem into subproblems according to the length of the time horizon. The proposed method is iterative and follows the sequence of periods of the time horizon from the beginning to the end. Therefore, the resolution of a subproblem (with all initial constraints but considering only a part of the time horizon) starts from the resolution of the subproblems for previous periods. The method limits the modification of the solution for periods that has already been considered at an earlier iteration, thus accelerating the resolution. Results shown that our approach is competitive in terms of solution quality and execution time and can provide good solutions for the set of instances considered.

Keywords: Inventory Routing Problem · Iterative algorithm · Heuristic

1 Introduction

The Inventory Routing Problem (IRP) is a multi-period vehicle routing and inventory management problem. The IRP considers a set of customers with deterministic demands per period, a set of homogeneous vehicles with a given capacity and a finite time horizon. The objective is to define the quantities to deliver to the customers, the moment (period) in which the deliveries will take place and the order of visiting customers per period using the available vehicles in such a way that the customer demands are satisfied and that the inventory levels, the capacity of vehicles and supplier production capacity are respected at a minimum-cost total inventory and routing solution cost.

Supervised by authors 1 and 2.

M. Sevaux et al. (Eds.): MIC 2024, LNCS 14753, pp. 123–135, 2024.
https://doi.org/10.1007/978-3-031-62912-9_12

A summary of works dealing with the IRP is presented in [6] and a classical version of the IRP is described by [1]. In our paper, we propose an iterative heuristic over periods based on a Mixed Integer Linear Programming (MILP) model, that is capable of solving the IRP by decomposing its structure in sub-problems according to the time periods in order to facilitate the exploration of the search space and provide high-quality solutions in a reasonable computational time.

An explanation on the Inventory Routing Problem including its mathematical formulation is given by Sect. 2. Then, in Sect. 3, the iterative heuristic approach is introduced and the computational experiments are presented in Sect. 4 as well as the set of instances and parameters used and the results obtained are discussed. Finally, conclusion and perspective work is in Sect. 5.

2   The Inventory Routing Problem

The Inventory Routing Problem is defined on a graph $G = (\mathcal{N}', A)$ in which $\mathcal{N}$ corresponds to the set of n customers $\mathcal{N} = \{1, ..., n\}$ and the node 0 standing for the supplier, with $\mathcal{N}' = \{0\} \cup \mathcal{N}$. Therefore, $A = \{(i,j) : i,j \in \mathcal{N}', i \neq j\}$ is the set of arcs. A time horizon $\mathcal{T} = \{1, ..., H\}$ with H periods is considered and consequently $\mathcal{T}' = \{0\} \cup \mathcal{T}$. A homogeneous fleet of m vehicles is considered, where each presents a capacity B. The cost to travel from i to j is given by $c_{i,j}$ and respect triangular inequalities.

An initial inventory level s_i, $\forall i \in \mathcal{N}'$, is known in advance for the customers and supplier at period 0. The inventory holding costs are given by h_i^t, with $i \in \mathcal{N}'$ and $t \in \mathcal{T}'$. Each customer has a period-independent demand d_i and a maximum and minimum inventory levels allowed $[L_i, U_i]$ per period $t \in \mathcal{T}$.

The sets, variables and data are summarized in Table 1.

The corresponding formulation is presented below and was inspired on the formulation presented by [2].

Objective Function. The objective function (1) aims to minimize the total inventory and transportation cost.

$$\min \sum_{t \in \mathcal{T}'} h_0^t I_0^t + \sum_{i \in \mathcal{N}} \sum_{t \in \mathcal{T}'} h_i^t I_i^t + \sum_{(i,j) \in \mathcal{A}} \sum_{t \in \mathcal{T}} c_{i,j} x_{i,j}^t \tag{1}$$

Inventory Level Constraints. Initially (at period 0), the supplier and customers inventory levels variables I_i^t, $\forall i \in \mathcal{N}'$ and $t \in \mathcal{T}'$, are set to their initial inventory level s_i according to Constraints (2). Then, in Constraints (3), the supplier inventory level is calculated considering the inventory level at the previous period I_0^{t-1}, $\forall t \in \mathcal{T}$, added of its production capacity $r^t, \forall t \in \mathcal{T}$. In Constraints (4), inventory level for each customer $i \in \mathcal{N}$ and period $t \in \mathcal{T}$ considers its precedent inventory level I_i^{t-1} and the amount of products delivered by the supplier

Table 1. Sets, data and variables

Sets/Data	Description
$\mathcal{N}$	Set of customers
$\mathcal{N}'$	Set of customers and the supplier
$\mathcal{A}$	Set of arcs (i,j), where $i, j \in \mathcal{N}'$ with $i \neq j$
$\mathcal{T}$	Set of T discrete time periods from 1 to $|\mathcal{T}|$
$\mathcal{T}'$	Set of T discrete time periods from 0 to $|\mathcal{T}|$
m	Number of vehicles available
B	Capacity of the vehicles
$c_{i,j}$	Distance from $i \in \mathcal{N}'$ to $j \in \mathcal{N}'$, $i \neq j$
r^t	Supplier production at time period $t \in \mathcal{T}$
d_i	Demand of customer $i \in \mathcal{N}$
L_i and U_i	Lower and upper inventory level limits, respectively, for customer $i \in \mathcal{N}$
s_i	Initial inventory level of $i \in \mathcal{N}'$
h_i^t	Inventory holding cost of $i \in \mathcal{N}'$ at time period $t \in \mathcal{T}'$

Variables	Description
$x_{i,j}^t$	Binary variable equal to 1 if arc $(i,j) \in A$ is chosen at $t \in \mathcal{T}$, 0 otherwise
$a_{i,j}^t$	Freight flow passing through arc $(i,j) \in A$ at period $t \in \mathcal{T}$
q_i^t	Quantity delivered to customer $i \in \mathcal{N}$ at period $t \in \mathcal{T}$
I_i^t	Inventory level of $i \in \mathcal{N}'$ at time period $t \in \mathcal{T}'$

and the demands for the current period.

$$I_i^0 = s_i \qquad\qquad \forall i \in \mathcal{N}' \tag{2}$$

$$I_0^t = I_0^{t-1} + r^t - \sum_{i \in \mathcal{N}} q_i^t \qquad\qquad \forall t \in \mathcal{T} \tag{3}$$

$$I_i^t = I_i^{t-1} + q_i^t - d_i \qquad\qquad \forall i \in \mathcal{N} \tag{4}$$

Delivery Quantity Constraints. The quantity q_i^t, $\forall i \in \mathcal{N}, t \in \mathcal{T}$ to be delivered must respect the remaining space available at the customer which is the difference between its previous inventory level and its maximum/minimum storage capacity $(L_i - I_i^{t-1}, U_i - I_i^{t-1} \ \forall i \in \mathcal{N}, t \in \mathcal{T})$ as in Constraints (5) and (6). Constraints (7) requires that the customer receiving products at a time period be visited on one of the routes at the same period.

$$q_i^t \geq L_i - I_i^{t-1} \qquad\qquad \forall i \in \mathcal{N}, t \in \mathcal{T} \tag{5}$$

$$q_i^t \leq U_i - I_i^{t-1} \qquad\qquad \forall i \in \mathcal{N}, t \in \mathcal{T} \tag{6}$$

$$q_i^t \leq \min\{B, U_i\} \sum_{\substack{j \in \mathcal{N}' \\ j \neq i}} x_{i,j}^t \qquad\qquad \forall i \in \mathcal{N}, t \in \mathcal{T} \tag{7}$$

Flow Constraints. Also referred to as degree constraints, these ensure that the entering and leaving flow at each node is equal (Constraints 8). Constraints (9) impose that the number m of available vehicles is respected by regarding the first arc $x_{0,i}^t$ $\forall i \in \mathcal{N}, t \in \mathcal{T}$ in a given route. Constraints (10) ensure that each customer is visited at most once per period since split deliveries are not allowed.

$$\sum_{(i,j)\in\mathcal{A}} x_{i,j}^t = \sum_{(j,i)\in\mathcal{A}} x_{j,i}^t \qquad \forall i \in \mathcal{N}', t \in \mathcal{T} \qquad (8)$$

$$\sum_{i\in\mathcal{N}} x_{0,i}^t \leq m \qquad t \in \mathcal{T} \qquad (9)$$

$$\sum_{\substack{j\in\mathcal{N}'\\ j\neq i}} x_{i,j}^t \leq 1 \qquad \forall i \in \mathcal{N}, t \in \mathcal{T} \qquad (10)$$

Sub-tour Elimination Constraints. In order to avoid sub-tours, variables $a_{i,j}^t$, $\forall i, j \in \mathcal{A}, t \in \mathcal{T}$, are introduced and serve as an increasing counter to ensure that a route starts and ends at node 0, which represents the supplier. All the variables $a_{i,0}^t$, $\forall i \in \mathcal{N}, t \in \mathcal{T}$ returning to the supplier are set to 0 as in constraints (11). Then, the vehicle load along each route is bounded by the vehicle capacity B as expressed in constraints (13).

$$a_{i,0}^t = 0 \qquad \forall(i,j) \in \mathcal{A}, t \in \mathcal{T} \qquad (11)$$

$$\sum_{\substack{j\in\mathcal{N}'\\ j\neq i}} a_{j,i}^t - q_i^t = \sum_{\substack{j\in\mathcal{N}'\\ j\neq i}} a_{i,j}^t \qquad \forall i \in \mathcal{N}, t \in \mathcal{T} \qquad (12)$$

$$a_{i,j}^t \leq Bx_{i,j}^t \qquad \forall(i,j) \in \mathcal{A}, t \in \mathcal{T} \qquad (13)$$

Variables Domain. Lastly, the domain of variables are given by Constraints (14)–(17).

$$x_{i,j}^t \in \{0,1\} \qquad \forall(i,j) \in \mathcal{A}, t \in \mathcal{T} \qquad (14)$$

$$a_{i,j}^t \geq 0 \qquad \forall(i,j) \in \mathcal{A}, t \in \mathcal{T} \qquad (15)$$

$$q_i^t \geq 0 \qquad \forall i \in \mathcal{N}, t \in \mathcal{T} \qquad (16)$$

$$I_i^t \geq 0 \qquad \forall i \in \mathcal{N}', t \in \mathcal{T}' \qquad (17)$$

3   The Iterative Heuristic over Periods

The iterative approach consists of a hybrid heuristic based on the formulation presented in Sect. 2. The method iterates over the time horizon, solving the subproblem composed by all constraints until the current period, and starting the resolution from the partial solution of the previous iteration. Furthermore, the number of modifications to the predefined partial solution is limited by a fixed method parameter.

Let $\mathcal{P}$ be the IRP and $\mathcal{P}^t$, $\forall t \in \mathcal{T}$, the tth subproblem over the $|\mathcal{T}|$ possible subproblems. Solving $\mathcal{P}^t$ means starting the resolution from the previous values found for the routes composition expressed by the variables $x_{i,j}^t$ from periods 1 to $t - 1$ to the current subproblem $\mathcal{P}^t$ and solve it by allowing a degree of freedom for the variables x over the algorithm execution of $\mathcal{P}^t$.

The overall idea is illustrated by Fig. 1.

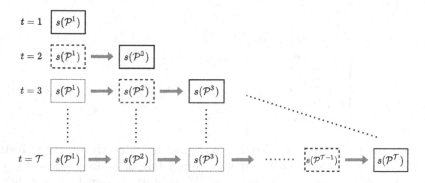

Fig. 1. Iterative heuristic idea

In Fig. 1, we can see the dependance among the subproblems given a period t. For example, when $t - 2$, there is a dependance on subproblem $\mathcal{P}^1$ since it carries a partial solution given as an starting point for $\mathcal{P}^2$. When $t = 3$, subproblem $\mathcal{P}^3$ receives information on $\mathcal{P}^2$ which depends implicitly on $\mathcal{P}^1$. The algorithm iterates over all the time horizon.

It is pertinent to state that subproblems dimension increase according to the iterations over the periods of time, i.e., $|\mathcal{P}^1| < |\mathcal{P}^2| < ... < |\mathcal{P}^{\mathcal{T}-1}| < |\mathcal{P}^{\mathcal{T}}|$ as shown in Fig. 2.

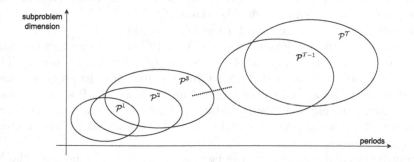

Fig. 2. The iterative heuristic search space evolution

The idea behind that is to explore partially at each iteration the neighbour-hood that is expressed by the variables representing each period of the time horizon. It allows the solver to know implicitly a logical sequence based on the chronological period order. This practice is usually explored by constraint programming algorithms [3,4].

To do so, for a given period $t \in \mathcal{T}$, we consider variables $\bar{x}_{i,j}^t$ as the values obtained when solving the subproblem $\mathcal{P}^{t-1}$. Then, the distance from their corresponding variables $x_{i,j}^t$ in subproblem $\mathcal{P}^t$ is calculated by Eqs. (18). Therefore, Eqs. (18) are linearized by Constraints (19).

$$\delta_{i,j}^{t'} = |\bar{x}_{i,j}^{t'} - x_{i,j}^{t'}| \qquad \forall i,j \in \mathcal{N}', t' \in \{1,...,t-1\} \qquad (18)$$

$$\eta = \begin{cases} \delta_{i,j}^{t'} \geq \bar{x}_{i,j}^{t'} - x_{i,j}^{t'} \\ \delta_{i,j}^{t'} \geq x_{i,j}^{t'} - \bar{x}_{i,j}^{t'} \\ \delta_{i,j}^{t'} \leq \bar{x}_{i,j}^{t'} + x_{i,j}^{t'} \\ \delta_{i,j}^{t'} \leq 2 - \bar{x}_{i,j}^{t'} - x_{ij}^{t'} \end{cases} \qquad \forall i,j \in \mathcal{N}', t' \in \{1,...,t-1\} \qquad (19)$$

Only x variables are considered due to the fact that deactivating or activating an arc in a given period means removing completely or adding a quantity q at a customer. Also, once the resolution of $\mathcal{P}^t$ is started from known values of variables x until the $t-1$, the q values can be more easily defined whether an arc is used or not in a route. Nevertheless, from the subproblem $\mathcal{P}^{t-1}$ to $\mathcal{P}^t$, a starting point on the q variables is added without a degree of freedom.

Then, the degree of freedom is expressed by

$$\sum_{(i,j)\in\mathcal{A}} \sum_{t'\in\{1,...,t-1\}} \delta_{i,j}^{t'} \leq \left\lceil \hat{x} \cdot \frac{\Delta}{100} \right\rceil \qquad (20)$$

in which $\hat{x}$ corresponds to the number of non-zero variables $x_{i,j}^t$ obtained from subproblem $\mathcal{P}^{t-1}$, i.e., all variables for which $x_{i,j}^t = 1$ and Δ to the percentage of changes allowed at each subproblem $\mathcal{P}^t$ based on previous activated arcs. In other words, when $\Delta = 0$, no tolerance is allowed whereas in case when $\Delta = 100$, all the previous activated arcs could be changed.

Algorithm 1 provides the steps for the iterative approach.

Algorithm 1 starts by defining the sets $\bar{X}$ and $\bar{Q}$ and setting them to empty (line 1) as well as initialyzing the elapsed time $elpTime$ to zero (line 2). Then, the first subproblem $\mathcal{P}^1$ is created taking into account only the problem data for the first period (line 3) and by solving it (line 4) by an MILP solver for model (1)–(17). Next, the elapsed time is updated (line 5). If a feasible solution exists (lignes 6 to 24), the associated x and q variables are added to sets $\bar{X}$ and $\bar{Q}$, respectively (lines 8 and 9).

From the second period until the last period of the time horizon, the loop from lines 10 to 24 adds at each iteration the solution obtained previously. The tth subproblem is created (line 11) and since the previous solution found at period $t-1$ is available (note that for the first period, no previous solution can be added to subproblem $\mathcal{P}^1$), Constraints (19) are added taking into account the Δ ratio and the set of linear constraints η given by Inequalities 19 (line 12).

Algorithm 1: Iterative algorithm

Data: problem *data* from Section 2
degree of freedom Δ
solver time limit *timeMax*
Result: a feasible solution $s(\mathcal{P}^{\mathcal{T}})$ for the $\mathcal{P}^t$ subproblem

1 $\bar{X}, \bar{Q} \leftarrow \varnothing$
2 $elpTime \leftarrow 0$
3 Create the first subproblem $\mathcal{P}^1$
4 $s(\mathcal{P}^1) \leftarrow \text{solve}(data, \mathcal{P}^1, timeMax)$
5 Update elapsed time $elpTime$
6 **if** $s(\mathcal{P}^1)$ is feasible **then**
7 $\quad$ $\bar{x}^1_{i,j}, \bar{q}^1_i \leftarrow \text{getValues}(s(\mathcal{P}^1))$
8 $\quad$ $\bar{X} \leftarrow \bar{x}^1_{i,j}$
9 $\quad$ $\bar{Q} \leftarrow \bar{q}^1_i$
10 $\quad$ **for** each period $t \in \mathcal{T} \setminus \{1\}$ **do**
11 $\quad\quad$ Create the tth subproblem $\mathcal{P}^t$
12 $\quad\quad$ $\mathcal{P}^t \leftarrow \text{addHint}(\bar{X}, \bar{Q}, \eta, \Delta)$
13 $\quad\quad$ $timeMax \leftarrow timeMax - elpTime$
14 $\quad\quad$ $s(\mathcal{P}^t) \leftarrow \text{solve}(data, \mathcal{P}^t, timeMax)$
15 $\quad\quad$ Update elapsed time $elpTime$
16 $\quad\quad$ **if** $s(\mathcal{P}^t)$ is feasible **then**
17 $\quad\quad\quad$ $\bar{x}^t_{i,j}, \bar{q}^t_i \leftarrow \text{getValues}(s(\mathcal{P}^t))$
18 $\quad\quad\quad$ $\bar{X} \leftarrow \bar{X} \cup \bar{x}^t_{i,j}$
19 $\quad\quad\quad$ $\bar{Q} \leftarrow \bar{Q} \cup \bar{q}^t_i$
20 $\quad\quad$ **else**
21 $\quad\quad\quad$ stop
22 $\quad\quad$ **end**
23 $\quad\quad$ $\bar{X}, \bar{Q} \leftarrow \varnothing$
24 $\quad$ **end**
25 **else**
26 $\quad$ stop
27 **end**
28 **return** $s(\mathcal{P}^H)$

Here, the hint concerns adding the x variables according to the degree of freedom provided by Δ and a starting point for q values.

The *timeMax* is updated according to the previous iterations elapsed time (line 13). Then, the tth subproblem is solved and the elapsed time $elpTime$ is updated (line 15). If a feasible solution exists (lines 16 to 19), the solution is extracted (line 17) and added to $\bar{X}$ (line 18) and $\bar{Q}$ (line 19).

At last, the solution found for the last subproblem $\mathcal{P}^H$ is returned (line 28).

4   Computational Experiments

The experiments were conducted on a AMD EPYC 7452 32-Core processor computer with 512GB of RAM memory using C++ programming language and the

MILP solver IBM ILOG CPLEX Optimization Studio version 22.1.1.0 was used to solve the model.

The experiments were conducted on a set of classical IRP instances proposed by [1,5] and several experiments varying the maximum authorized distance for the x variables were also explored and are presented below.

For each instance and method, an one-hour time limit was set to the solver. It means that for the iterative method, solving an instance up to a given period t means accumulating the elapsed time from period 1 to t since the subproblem $\mathcal{P}^t$ reuses information from the previous subproblems solved. The same is not valid for the classical approach since there is no period-dependence to solve the subproblems and then the time limit is set to one hour for each instance.

Δ values were set to $\Delta = \{0, 10, 20, ..., 90, 100\}$ in order to analyse the performance to solve the problem. These values do not change over the algorithm execution.

4.1   Instances

The set of instances was initially provided by [1] for the single vehicle IRP and later, [5] adapted these instances to contemplate the multi-vehicle homogeneous vehicle case up to 5 per period.

In total, the set is composed of 160 instances that are characterized by a time horizon with 3 (100 instances) or 6 (60 instances) periods of time and low (50/30 instances) or high (50/30 instances) inventory holding costs. For those with 3 periods, customers are up to 50 and for 6 periods, at most 30 customers are considered. For each dimension and type, 5 different instances are presented. Table 2 gives the instances classification.

Table 2. Classical IRP instances classification

Size		N		T	h
160	100	50	{5, 10, ..., 50}	3	[0.01; 0.05]
		50	{5, 10, ..., 50}	3	[0.1; 0.5]
	60	30	{5, 10, ..., 30}	6	[0.01; 0.05]
		30	{5, 10, ..., 30}	6	[0.1; 0.5]

The instances files are available at https://perso.limos.fr/~diperdigao/MIC2024/instances. File names are identified by *absAnB.dat* in which *A* corresponds to the instance type and *B* to its number of customers.

4.2   Results

Tables 3, 4, 5 and 6 present the results fo a specific set of instances when $\Delta = 50$ and 2 and 3 vehicles. These were selected due to their performance among the

Table 3. Results for $\Delta = 50$ and 3 vehicles

instance				Iterative heuristic		MILP model		gap(%)	ratioTime
type	name	n	H	UB	time(s)	UB	time(s)		
low	abs1n5.dat	5	3	1826.67	0	1430.51	0	22	1
low	abs1n10.dat	10	3	2894.32	1	2732.61	6	6	5
low	abs1n15.dat	15	3	3073.35	10	2783.77	90	9	9
low	abs1n20.dat	20	3	3913.45	146	3605.72	3600	8	25
low	abs1n25.dat	25	3	4312.25	107	3503.3	3600	19	34
low	abs1n30.dat	30	3	4764.89	434	4252.9	3600	11	8
low	abs1n35.dat	35	3	4836.19	3600	4099.58	3600	15	1
low	abs1n40.dat	40	3	5459.64	3600	4624.12	3600	15	1
low	abs1n45.dat	45	3	5309.61	1891	4537.3	3600	15	2
low	abs1n50.dat	50	3	5769	2178	5430	3600	6	2
high	abs1n5.dat	5	3	2695.15	0	2298.73	0	15	3
high	abs1n10.dat	10	3	5666.43	1	5506.09	4	3	4
high	abs1n15.dat	15	3	6480.80	10	6242.9	112	4	11
high	abs1n20.dat	20	3	8506	92	8165.42	3600	4	39
high	abs1n25.dat	25	3	9711.60	71	8893.88	3600	8	51
high	abs1n30.dat	30	3	13464.80	891	12908.9	3600	4	4
high	abs1n35.dat	35	3	13125.20	2604	12445	3600	5	1
high	abs1n40.dat	40	3	15045.40	3600	14229.8	3600	5	1
high	abs1n45.dat	45	3	15443.10	1539	14771	3600	4	2
high	abs1n50.dat	50	3	16292.90	1384	15926	3600	2	3
							average	9	10.28

whole instances set. We compare the values obtained with [7] that, for the best of our knowledge, provide the best bounds for the classical instances.

The results from Table 3 indicate that on average, the iterative method is 10 times faster than direct resolution, with an average gap of 9%. However, a detailed analysis of the results on a per-instance basis reveals significant differences among instances. There are 5 instances (instances low abs1n20, low abs1n25, high abs1n15, high abs1n20, high abs1n25) that exhibit a 10-fold acceleration in computation time, and one of them (instance abs1n25.dat) shows a 19% gap. The most unfavorable case is observed for the instance low abs1n5.dat, for a time horizon with 3 periods, with a 22% cost gap and a ratio of 1 in terms of computation time. Conversely, the most favorable instance is high abs1n20.dat, for a time horizon with 3 periods, with only a 4% cost gap and a computation time ratio of 39. On Tables 4 and 6, when we compare the Iterative approach with the best known literature Upper bounds from [7], we note that the results are competitive in terms of time convergence even if a gap is observed in some cases.

Table 4. Results for $\Delta = 50$ and 3 vehicles

instance				Iterative heuristic		Manousakis et al. (2021)		gap(%)	ratioTime
type	name	n	H	UB	time(s)	UB	time(s)		
low	abs1n5.dat	5	3	1826.67	0	1430.51	1	22	1
low	abs1n10.dat	10	3	2894.32	1	2732.61	21	6	21
low	abs1n15.dat	15	3	3073.35	10	2783.77	23	9	2
low	abs1n20.dat	20	3	3913.45	146	3605.72	196	8	1
low	abs1n25.dat	25	3	4312.25	107	3503.38	83	19	1
low	abs1n30.dat	30	3	4764.89	434	4251.64	1069	11	2
low	abs1n35.dat	35	3	4836.19	3600	4080.60	2463	16	1
low	abs1n40.dat	40	3	5459.64	3600	4532.84	13369	17	4
low	abs1n45.dat	45	3	5309.61	1891	4537.30	29437	15	16
low	abs1n50.dat	50	3	5769	2178	6017.66	42832	−4	20
high	abs1n5.dat	5	3	2695.15	0	2298.73	0	15	1
high	abs1n10.dat	10	3	5666.43	1	5506.09	13	3	13
high	abs1n15.dat	15	3	6480.80	10	6242.90	16	4	2
high	abs1n20.dat	20	3	8506	92	8165.42	229	4	2
high	abs1n25.dat	25	3	9711.60	71	8893.82	53	8	1
high	abs1n30.dat	30	3	13464.80	891	12098.90	2409	10	3
high	abs1n35.dat	35	3	13125.20	2604	12396.00	3380	6	1
high	abs1n40.dat	40	3	15045.40	3600	14224.10	9173	5	3
high	abs1n45.dat	45	3	15443.10	1539	14771	34500	4	22
high	abs1n50.dat	50	3	16292.90	1384	16115.80	43101	1	31
							average	9	7

Table 5. Results for $\Delta = 50$ and 2 vehicles

instance				Iterative heuristic		MILP model		gap(%)	ratioTime
type	name	n	H	cost	time(s)	cost	time(s)		
low	abs1n5.dat	5	6	3776.52	0.61	3775.68	5.56	0	9
low	abs1n20.dat	20	6	7392.19	3503.28	7669.81	3600	−4	1
low	abs1n25.dat	25	6	8233.73	3328.81	7464.18	3600	9	1
high	abs1n5.dat	5	6	6379.56	0.54	6379.56	2.73	0	5
high	abs1n15.dat	15	6	13105.10	137.734	12624.7	3600	4	26
high	abs1n20.dat	20	6	15583.90	3373.58	15551.4	3600	0	1
high	abs1n25.dat	25	6	16222.10	2516.93	16012.1	3600	1	1
							average	2	6.40

Tables 7 and 8 summarize the results obtained for the classical IRP instances with 3 and 6 periods, respectively. We consider here only the first instance type ($abs1nB.dat$). Column *type* indicates the type of inventory holding costs considered, $gapX(\%)$ the average gap among all the instances up to the Xth period,

Table 6. Results for $\Delta = 50$ and 2 vehicles

instance				Iterative heuristic		Manousakis et al. (2021)		gap(%)	ratioTime
type	name	n	H	cost	time(s)	cost	time(s)		
low	abs1n5.dat	5	6	3776.52	0.61	3775.68	8	0	13
low	abs1n20.dat	20	6	7392.19	3503.28	7388.80	2276	0	1
low	abs1n25.dat	25	6	8233.73	3328.81	7461.55	735	9	0
high	abs1n5.dat	5	6	6379.56	0.54	6379.56	8	0	15
high	abs1n15.dat	15	6	13105.10	137.734	12624.70	148	4	1
high	abs1n20.dat	20	6	15583.90	3373.58	15540.40	3232	0	1
high	abs1n25.dat	25	6	16222.10	2516.93	15954.80	197	2	0
						average		2	4

Table 7. Results for $|T| = 3$

type	Δ	gap2(%)	gap3(%)	timeRatio	σ
low	0	0	20	100.69	202.33
	10	0	19	61.06	168.77
	20	0	17	18.45	40.07
	30	0	13	6.64	19.54
	40	0	15	5.77	16.11
	50	0	12	5.22	15.86
	60	0	12	4.81	15.22
	70	0	12	4.56	15.11
	80	0	11	5.00	16.80
	90	0	10	4.89	17.52
	100	0	10	8.08	25.00
high	0	0	8	78.49	204.18
	10	0	8	57.29	239.40
	20	0	7	22.55	74.27
	30	0	5	5.73	14.85
	40	0	5	4.71	11.19
	50	0	5	4.13	9.30
	60	0	5	3.87	9.04
	70	0	4	3.78	9.53
	80	0	4	3.86	10.80
	90	0	4	4.04	11.66
	100	0	4	5.66	9.18
	avg	0	10	19.06	52.53

timeRatio the ratio between the time required to solve the last period by the classical approach and the accumulated time for the iterative heuristic and at last, σ gives the standard deviation of the *timeRatio* values given a specified Δ.

In Tables 7 and 8, note that the *gap* values can be calculated only from the second period since for the first, both iterative and classical approach are

Table 8. Results for $|\mathcal{T}| = 6$

type	Δ	gap2(%)	gap3(%)	gap4(%)	gap5(%)	gap6(%)	timeRatio	σ
low	0	0	18	21	17	18	402.81	491.65
	10	0	19	12	12	10	36.17	32.06
	20	0	17	11	7	8	15.01	11.51
	30	0	16	8	5	5	11.13	10.97
	40	0	15	6	2	6	4.37	3.58
	50	0	15	0	7	−2	3.94	3.64
	60	0	14	−1	6	−4	2.77	2.93
	70	0	12	−1	4	−4	1.72	1.27
	80	0	12	−3	−1	−2	2.03	1.37
	90	0	10	−4	2	−6	1.57	0.67
	100	0	9	−4	−1	−7	1.43	0.84
high	0	0	7	8	11	11	692.68	1264.34
	10	0	8	4	6	5	33.75	43.56
	20	0	4	1	5	15	17.13	28.21
	30	0	7	2	1	0	8.14	11.55
	40	0	6	−1	4	−2	4.83	6.63
	50	0	6	−1	3	−2	5.43	8.06
	60	0	5	−1	2	−5	3.60	6.88
	70	0	4	−2	2	−5	2.66	4.20
	80	0	4	−1	1	−4	2.08	2.41
	90	0	3	−4	0	−5	1.61	2.11
	100	0	3	−5	−2	−5	1.23	1.42
avg		0	10	2	4	1	57.09	88.18

exactly the same since no previous information is available to be added to the first subproblem in the iterative method.

In the first period, for the majority of the instances considered, no routes are scheduled since customers have enough products in their inventory and, consequently, no quantities are delivered. For the second period, *gap* values are all equals to zero since the problem dimension is easy to solve due to the information provided from the first period. From *gap*3 and on, the problem dimension may increase significantly and it is harder to get optimal solutions.

No matter the instance type and dimension, we can clearly state that when increasing the Δ value, a tendance of reducing the gap is observed since we give to the subproblem associated more possibilities on the arc changes. However, the time grows because the search espace get bigger with the Δ value increase. Note that even in the worst case regarding the time, our iterative approach performs better in average when compared to the classical approach.

Detailed results are available at https://perso.limos.fr/~diperdigao/MIC2024/results.

5   Conclusion

An iterative approach to solve the Inventory Routing Problem was presented in this paper. The method combines the exact resolution based on an MILP formulation for the IRP embedded with an iterative exploration of the search space based on the periods available. At each iteration, the subproblem associated with the current time period is defined, starting from the solution obtained for the precedent subproblem. The experiments were conducted in a set of classical instances from the literature and the method has shown a competitive advantage when comparing the execution time and the gap provided within an one-hour time limit of execution. As future works, we intend to consider incorporating other mechanisms that can take advantage of the problem structure and guide the search exploration towards the optimal solutions.

Acknowledgements. This work was sponsored by a public grant overseen by the French National Research Agency as part of the "Investissements d'Avenir" through the IMobS3 Laboratory of Excellence (ANR-10-LABX-0016) and the IDEX-ISITE initiative CAP 20-25 (ANR-16-IDEX-0001).

References

1. Archetti, C., Bertazzi, L., Laporte, G., Speranza, M.G.: A branch-and-cut algorithm for a vendor-managed inventory-routing problem. Transp. Sci. **41**(3), 382–391 (2007)
2. Archetti, C., Bianchessi, N., Irnich, S., Speranza, M.G.: Formulations for an inventory routing problem. Int. Trans. Oper. Res. **21**(3), 353–374 (2014)
3. Bourreau, E., Gondran, M., Lacomme, P., Vinot, M.: De la programmation linéaire à la programmation par contraintes, p. 360. Ellipses (2019)
4. Bourreau, E., Gondran, M., Lacomme, P., Vinot, M.: Programmation Par Contraintes: démarches de modélisation pour des problèmes d'optimisation. Ellipses (2020)
5. Coelho, L.C., Cordeau, J.F., Laporte, G.: Consistency in multi-vehicle inventory-routing. Transp. Res. Part C: Emerg. Technol. **24**, 270–287 (2012)
6. Coelho, L.C., Cordeau, J.F., Laporte, G.: Thirty years of inventory routing. Transp. Sci. **48**(1), 1–19 (2014)
7. Manousakis, E., Repoussis, P., Zachariadis, E., Tarantilis, C.: Improved branch-and-cut for the inventory routing problem based on a two-commodity flow formulation. Eur. J. Oper. Res. **290**(3), 870–885 (2021)

Combining Heuristics and Constraint Programming for the Parallel Drone Scheduling Vehicle Routing Problem with Collective Drones

Roberto Montemanni$^{(\boxtimes)}$ ⓘ, Mauro Dell'Amico ⓘ, and Andrea Corsini ⓘ

University of Modena and Reggio Emilia, 42122 Reggio Emilia, Italy
{roberto.montemanni,mauro.dellamico,andrea.corsini}@unimore.it

Abstract. Last-mile delivery problems where trucks and drones collaborate to deliver goods to final customers are considered. We focus on settings where a fleet with several homogeneous trucks work in parallel to collaborative drones, able to combine with each other to optimize speed and power consumption for deliveries. A heuristic for the min-max vehicle routing problem is coupled with constraint programming models, leading to an effective method able to provide several state-of-the-art solutions for the instances commonly adopted in the literature.

Keywords: Vehicle Routing · Constraint programming · Heuristics

1 Introduction

The first optimization problems involving distribution with trucks and drones were introduced in Murray and Chu [5], where the concept of a new distribution problem where a truck and a drone make deliveries in a collaborative way, was introduced. In the Parallel Drone Scheduling Traveling Salesman Problem (PDSTSP) there is a truck making a tour to service some customers. In parallel, a set of drones is also employed, and each drone can leave the depot, serve a customer, return to the depot, and repeating several times for different customers. The objective of the optimization is to minimize the makespan required to service all the customers and having all the vehicles back to the depot. Models, exact and heuristic algorithms for the problem have been discussed, e.g., in [3] and [7]. Recently, Amazon Technologies Inc. filed a patent [8] where a new distribution paradigm, taking advantage of a so-called "Collective Drone" (c-drone), is introduced. Multiple drones can be connected and fly as a single drone, to serve a customer. The c-drone is able to transport larger and heavier goods with respect to the single drone, and can also modulate its speed more flexibly in order to increase its range [4]. In the work [6], the authors optimized a problem where collective drones and a truck are used to distribute goods, and the resulting problem is named the PDSTSP-c, where c stands for *collective*. An extension of the problem, where multiple trucks are considered, called PDSVRP-c, was

M. Sevaux et al. (Eds.): MIC 2024, LNCS 14753, pp. 136–142, 2024.
https://doi.org/10.1007/978-3-031-62912-9_13

recently introduced in [4]. An example of a PDSVRP-c instance is provided in Fig. 1. Without considering the use of the drones, the problem reduces to a classic Vehicle Routing Problems (VRP) [10] characterized by a *min-max* objective function calculated over the lengths of the different tours, which translates into completing all the deliveries in the shortest possible time. Both exact and heuristic methods have been presented for this problem, that is normally more difficult than a traditional VRP [1].

We investigate how two Constraint Programming models recently proposed for the PDSVRP-c perform once a solution only using trucks (VRP) is fed to the solver as a hint-solution.

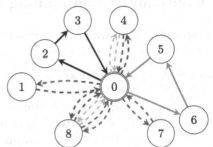

Fig. 1. Example of a PDSVRP-c instance. Node 0 is the depot, while the other nodes are customers. The black and grey continuous arcs represent the tours of the two trucks (0, 2, 3, 0) and (0, 6, 5, 0). The dashed arcs depict instead the missions of the drones, with each colour representing a different one. Notice how for some of the missions multiple drones are collaborating (Color figure online).

2   Problem Description

Given a graph $G(V, E)$ with a set of vertices $V = \{0, 1, \ldots, n\}$, where vertex 0 is the depot and the remaining vertices represent the customers (set $C = V \backslash \{0\}$). A customer i requests a parcel of weight w_i to be delivered to its address from the depot. A set S of s driver-operated delivery trucks, each with unlimited range and capacity, and a set D of m homogeneous drones form the fleet available for deliveries. All the vehicles are based at the depot and the drones have batteries of a given capacity that is installed before each mission. Each truck performs a single delivery tour and no collaboration among trucks is implemented. The deterministic travel time between two vertices $i, j \in V$ is given as t_{ij} for the trucks. The drones instead operate in a back-and-forth fashion from the depot, delivering one parcel at a time. Travel times and maximum ranges of drone missions depend on factors such as the number of drones cooperating and the traveling speed selected. The energy consumption model from [9] is adopted here to calculate battery draining and discharge peaks in order to estimate feasible mission settings. In the configuration considered, characterized by collaborative drones, the weight carried is evenly distributed among the k drones participating in the mission. As described in [6], given a number of drones involved k and a target customer j, the optimal cruise speed that minimizes the time required for the mission itself, while fulfilling the constraints on the batteries (power consumption is used here) can be pre-calculated by inspection. The time τ_j^k required to service customer j with k drones can therefore be pre-calculated as described in [6], with $\tau_i^k = +\infty$ if it is not possible to service customer i with k drones. The set of customers that cannot be serviced by drones is referred to as $C_T \subset C$. Let $C_F = C \backslash C_T$ be the set of customers that can be served by drones, and let q_j and p_j be the minimum and maximum number of drones that can be used to serve $j \in \mathcal{C_F}$. The target of the PDSVRP-c is to find truck tours and drones scheduling that minimize the makespan, i.e. to complete all the deliveries in the shortest possible time, given the resources available.

3 Constraint Programming Models

We present two Constraint Programming models introduced in [4], based on the Google OR-Tools CP-SAT solver [2] and representing the state-of-the-art.

Model M2: The model revolves on the following variables: x_{ij} is 1 if edge (i,j), with $i,j \in V$, is traveled by a truck, 0 otherwise. If $x_{jj} = 1$ means that customer j is served by drones; z_j^k is 1 if k drones serve customer $j \in \mathcal{C_F}$, 0 otherwise; y_{ij} is 1 if vertex a drone serves vertex i right before vertex j by one drone, 0 otherwise; $f_{ij} \in Z^+$ indicates the number of drones serving vertex i right before vertex j; $\overline{T}_j \in R^+$ is the time at which the mission at customer $j \in \mathcal{C_F}$ is completed if the visit is operated by drones; it is the time the truck reaches the customer and the service is started in case the visit is operated by a truck.; α is the completion time of all missions. The model is the following one:

$$(M2) \min \alpha \tag{1}$$
$$s.t. \ \alpha \geq \overline{T}_j + t_{j0}x_{j0}, j \in \mathcal{C} \tag{2}$$
$$x_{jj} = \sum_{q_j \leq k \leq p_j} z_j^k, j \in \mathcal{C_F} \tag{3}$$
$$\text{MultipleCircuit}\begin{pmatrix} i,j \in V, \\ x_{ij} : i \neq 0 \vee j \neq 0, \\ i \in \mathcal{C_T} \Rightarrow j \neq i \end{pmatrix} \tag{4}$$
$$\sum_{j \in \mathcal{C}} x_{0j} \leq s \tag{5}$$
$$x_{ij} \Rightarrow \overline{T}_j \geq T_i + t_{ij}, i \in V, j \in \mathcal{C}, i \neq j \tag{6}$$
$$\sum_{j \in \mathcal{C_F}} f_{0j} \leq m \tag{7}$$
$$\sum_{i \in \mathcal{C_F} \cup \{0\}, i \neq j} f_{ij} = \sum_{q_j \leq k \leq p_j} k z_j^k, j \in \mathcal{C_F} \tag{8}$$

$$\sum_{i \in \mathcal{C_F} \cup \{0\}, i \neq j} f_{ij} = \sum_{l \in \mathcal{C_F} \cup \{0\}, l \neq j} f_{jl}, j \in \mathcal{C_F} \cup \{0\} \tag{9}$$
$$f_{ij} \leq m y_{ij}, i, j \in \mathcal{C_F} \cup \{0\}, i \neq j \tag{10}$$
$$y_{ij} \Rightarrow \overline{T}_j \geq \overline{T}_i + \sum_{q_j \leq k \leq p_j} \tau_j^k z_j^k, \begin{matrix} i \in \mathcal{C_F} \cup \{0\}, \\ j \in \mathcal{C_F}, i \neq j \end{matrix} \tag{11}$$
$$0 \leq f_{ij} \leq m, i, j \in \mathcal{C_F} \cup \{0\}, i \neq j \tag{12}$$
$$x_{ij} \in \{0; 1\}, i, j \in V \tag{13}$$
$$z_j^k \in \{0; 1\}, j \in \mathcal{C_F}, q_j \leq k \leq p_j \tag{14}$$
$$y_{ij} \in \{0; 1\}, i, j \in \mathcal{C_F} \cup \{0\}, i \neq j \tag{15}$$
$$\overline{T}_j \geq 0, j \in V \tag{16}$$
$$m\alpha \geq \sum_{j \in \mathcal{C_F}} \sum_{q_j \leq k \leq p_j} k \tau_j^k z_j^k \tag{17}$$

Constraints (2) state that the total time α to be minimized according to (1), has to be as large as to the time required by the truck and drone missions. Constraints (3) relate x and z variables for each drone-eligible customers; Constraint (4) uses *MultipleCircuit* command of CP-SAT to impose truck tours, while constraint (5) set tos the maximum number of truck tours. Constraints (6) align timing ($\overline{T}$ variables) to tours. Constraints (7)-(9) regulate synchronization among drones (see [4] for details). Constraints (10) and (11) connect y variables with f and z variables, respectively. The remaining constraints (12)-(16) define the domain of the variables. The inequality (17) is not necessary for the validity of the model, but it contributes significantly to tighten the lower bounds, so it is added. The interested reader can refer to [4] for an explanation of the inequality and a formal proof of its validity.

Model M3: The variables of the model are defined starting from those of model M2. Here the x variables are changed to a set of variables w such that $w_{ij}^k = 1$ if edge (i,j) is traveled by truck $k \in S$, 0 otherwise. Moreover, $w_{00}^k = 1$ implies that truck k is not used in the current solution. Variables $\overline{T}$ are substituted by the following variables: T_j represents the completion of the drone-mission to customer $j \in C_F$. The resulting model is as follows:

$$(M3) \quad \min \alpha \qquad (18)$$

$$s.t. \quad \alpha \geq \sum_{i \in V} \sum_{\substack{j \in V, i \neq j}} t_{ij} w_{ij}^k, k \in S \qquad (19)$$

$$\alpha \geq T_j, j \in C_F \qquad (20)$$

$$\sum_{k=1}^{s} (1 - w_{jj}^k) + \sum_{q_j \leq k \leq p_j} z_j^k = 1, j \in C_F \qquad (21)$$

$$\sum_{k=1}^{s} w_{jj}^k = s - 1, j \in C_T \qquad (22)$$

$$Circuit(w_{ij}^k : i, j \in V), k \in S \qquad (23)$$

$$w_{ij}^k \leq 1 - w_{00}^k, k \in S, i, j \in C \qquad (24)$$

$$\sum_{j \in C_F} f_{0j} \leq m \qquad (25)$$

$$\sum_{i \in C_F \cup \{0\}, i \neq j} f_{ij} = \sum_{q_j \leq k \leq p_j} k z_j^k, j \in C_F \qquad (26)$$

$$\sum_{i \in C_F \cup \{0\}, i \neq j} f_{ij} = \sum_{l \in C_F \cup \{0\}, l \neq j} f_{jl}, j \in C_F \cup \{0\} \qquad (27)$$

$$f_{ij} \leq m y_{ij}, i, j \in C_F \cup \{0\}, i \neq j \qquad (28)$$

$$y_{ij} \Rightarrow T_j \geq \overline{T}_i + \sum_{q_j \leq k \leq p_j} \tau_j^k z_j^k, \begin{array}{l} i \in C_F \cup \{0\}, \\ j \in C_F, i \neq j \end{array} \qquad (29)$$

$$0 \leq f_{ij} \leq m, i, j \in C_T \cup \{0\}, i \neq j \qquad (30)$$

$$w_{ij}^k \in \{0; 1\}, k \in S, i, j \in V \qquad (31)$$

$$z_j^k \in \{0; 1\}, j \in C_F, q_j \leq k \leq p_j \qquad (32)$$

$$y_{ij} \in \{0; 1\}, i, j \in C_F \cup \{0\}, i \neq j \qquad (33)$$

$$T_j \geq 0, j \in C_F \cup \{0\} \qquad (34)$$

$$m\alpha \geq \sum_{j \in C_F} \sum_{q_j \leq k \leq p_j} k \tau_j^k z_j^k \qquad (35)$$

The constraints follow the meaning already described for the model $M2$ in Sect. 3. The changes are as follows. Equalities (21) are used to force any each drone-eligible customer has to be services by drones or by a truck. The new constraints (22) is required to force customers in C_T to be service by a truck. Constraints (23), adopting the *Circuit* command from CP-SAT, are defined for each truck k, since the concept of giant-tour seen in the model $CP2$ does not exist here. The constraints (24) state that a truck k can be used only if $w_{00}^k = 1$.

Hint-start: One of the features of CP-SAT is the possibility of passing a (partial) solution to the solver through some values for the variables of the model. The solver takes these settings as suggestions (hints) and potentially improves its

Table 1. Experimental results.

	2 Trucks						3 Trucks					
	VRP	Model M2		Model M3		New bounds	VRP	Model M2		Model M3		New bounds
		[4]	Hint-start	[4]	Hint-start			[4]	Hint-start	[4]	Hint-start	
Instances	UB	[LB, UB]	[LB, UB]	[LB, UB]	[LB, UB]	[LB, UB]	UB	[LB, UB]	[LB, UB]	[LB, UB]	[LB, UB]	[LB, UB]
50-r-e	128	[65, 116]	[61, 116]	[63, 120]	**[69, 112]**	**[69, 112]**	112	[48, 112]	**[62, 104]**	[47, 112]	[52, 108]	**[62, 104]**
53-r-e	128	[77, 112]	[65, 116]	[82, 128]	[78, 112]	[82, 112]	112	[56, 96]	**[64, 96]**	[51, 112]	[56, 104]	**[64, 96]**
66-rc-e	128	[72, 112]	[47, 124]	[73, 136]	[63, 116]	[73, 112]	112	[53, 108]	[47, **100**]	[38, 116]	[**57**, 104]	[**57**, **100**]
67-c-c	56	[38, 52]	[20, **48**]	[31, 52]	[19, 52]	[38, **48**]	56	[27, 52]	[21, 52]	[9, 52]	[12, 52]	[27, 52]
68-rc-c	120	[50, 56]	[36, 104]	[52, 104]	[47, 84]	[52, 56]	84	[39, 56]	[36, 60]	[34, 104]	[**42**, 76]	[**42**, 56]
76-c-c	40	[26, 36]	[16, 36]	[16, 40]	[16, 36]	[26, 36]	28	[18, 24]	[12, 24]	[12, 52]	[16, 24]	[18, 24]
82-c-e	64	[32, 64]	[26, 64]	[17, 64]	[12, 64]	[32, 64]	64	[22, 64]	[**26**, 64]	[8, 64]	[10, 64]	[**26**, 64]
82-rc-c	108	[62, 116]	[32, **100**]	[56, 132]	[54, **100**]	[**62**, **100**]	88	[47, 80]	[34, 84]	[38, 128]	[**48**, 84]	[**48**, 80]
88-c-e	84	[54, 84]	[18, 84]	[58, 112]	[40, 84]	[58, 84]	80	[36, 76]	[18, 76]	[32, 104]	[**39**, 80]	[**39**, 76]
91-r-c	152	[75, 152]	[33, 140]	[75, 160]	[63, **124**]	[75, **124**]	108	[56, 120]	[34, 104]	[42, 148]	[54, **100**]	[56, **100**]
99-rc-c	152	[63, 96]	[26, 136]	[51, 144]	[47, 136]	[63, 96]	96	[47, 64]	[26, 92]	[29, 128]	[41, 88]	[47, 64]
101-r-c	176	[71, 164]	[20, 160]	[53, 152]	[55, 152]	[71, 152]	104	[52, 128]	[20, **96**]	[36, 144]	[44, 100]	[52, **96**]
103-rc-c	128	[69, 124]	[34, 124]	[52, 128]	[49, 128]	[69, 124]	88	[49, 96]	[35, **88**]	[32, 136]	[41, **88**]	[49, **88**]
105-rc-e	140	[65, 136]	[27, **128**]	[57, 148]	[52, **128**]	[65, **128**]	112	[49, 120]	[27, **104**]	[34, 132]	[40, 108]	[49, **104**]
108-rc-e	160	[79, 172]	[22, **152**]	[70, 160]	[57, 160]	[79, **152**]	128	[58, 184]	[23, **124**]	[37, 160]	[54, **124**]	[58, **124**]
114-rc-c	120	[58, 124]	[27, **104**]	[49, 140]	[47, **104**]	[58, **104**]	88	[44, 80]	[26, **72**]	[32, 112]	[40, 80]	[44, **72**]
121-rc-e	144	[70, 156]	[25, 140]	[56, 152]	[57, **132**]	[70, **132**]	96	[52, 124]	[28, **96**]	[40, 152]	[44, **96**]	[52, **96**]
126-rc-e	160	[87, 220]	[18, **152**]	[67, 184]	[63, **152**]	[87, **152**]	116	[63, 136]	[20, 108]	[44, 164]	[48, **100**]	[63, **100**]
126-r-c	172	[78, 160]	[26, **136**]	[56, 156]	[58, **136**]	[78, **136**]	108	[56, 140]	[24, **116**]	[38, 148]	[48, **116**]	[56, **116**]
144-rc-c	132	[67, 272]	[21, 128]	[47, 168]	[43, **116**]	[67, **116**]	120	[50, 132]	[23, **116**]	[35, 160]	[41, 120]	[50, **116**]
154-c-c	40	[35, -]	[14, **40**]	[8, 72]	[12, **40**]	[35, **40**]	36	[24, 36]	[14, 36]	[8, 68]	[8, 36]	[24, 36]
165-r-c	200	[88, -]	[16, 192]	[67, 224]	[68, **184**]	[88, **184**]	140	[68, -]	[15, 136]	[50, 212]	[50, **132**]	[68, **132**]
167-r-e	196	[100, -]	[16, 188]	[74, 256]	[75, **180**]	[100, **180**]	140	[73, -]	[16, **132**]	[54, 204]	[56, **132**]	[73, **132**]
173-rc-c	196	[85, 204]	[16, 188]	[59, 240]	[60, **176**]	[85, **176**]	136	[65, -]	[16, **132**]	[45, 212]	[47, **128**]	[65, **128**]
173-rc-c	152	[79, -]	[21, 148]	[48, 180]	[50, **144**]	[79, **144**]	120	[58, 172]	[20, **116**]	[37, 168]	[40, 120]	[58, **116**]
181-r-e	192	[112, -]	[18, 188]	[78, 252]	[79, **180**]	[112, **180**]	152	[82, -]	[18, **152**]	[55, 216]	[62, **152**]	[82, **152**]
185-c-c	60	[48, -]	[20, **60**]	[24, 96]	[24, **60**]	[48, **60**]	48	[32, -]	[22, 48]	[14, 96]	[24, 48]	[32, 48]
187-c-c	176	[100, 308]	[27, 172]	[65, 212]	[67, **160**]	[100, **160**]	132	[74, -]	[26, **124**]	[46, 212]	[50, 128]	[74, **124**]
198-c-c	36	[32, -]	[16, 36]	[12, 64]	[12, 36]	[32, 36]	36	[22, 36]	[16, 36]	[8, 68]	[13, 36]	[22, 36]
200-r-c	224	[105, -]	[16, **216**]	[68, 324]	[69, 220]	[105, **216**]	152	[77, -]	[16, 152]	[48, 252]	[50, **148**]	[77, **148**]
Average	131.5	[68.1, 138.0]	[26.0, 124.0]	[52.8, 150.0]	[50.2, 120.3]	[68.6, 117.2]	99.7	[49.9, 97.2]	[26.2, 94.7]	[34.4, 137.9]	[40.9, 95.9]	[51.1, 92.7]

	4 Trucks						5 Trucks					
	VRP	Model M2		Model M3		New	VRP	Model M2		Model M3		New
		[4]	Hint-start	[4]	Hint-start	bounds		[4]	Hint-start	[4]	Hint-start	bounds
Instances	UB	[LB, UB]	[LB, UB]	[LB, UB]	[LB, UB]	[LB, UB]	UB	[LB, UB]	[LB, UB]	[LB, UB]	[LB, UB]	[LB, UB]
50-r-e	116	[46, 104]	**[62, 100]**	[35, 112]	[37, 100]	**[62, 100]**	116	[47, 100]	**[61, 100]**	[30, 112]	[34, 112]	**[61, 100]**
53-r-e	112	[50, 96]	**[64, 96]**	[38, 112]	[39, 100]	**[64, 96]**	112	[50, 92]	**[64, 92]**	[32, 112]	[36, 112]	**[64, 92]**
66-rc-e	108	[41, 104]	**[44, 100]**	[34, 108]	[35, 104]	**[44, 100]**	100	[35, 100]	**[46, 100]**	[24, 120]	[32, 100]	**[46, 100]**
67-c-c	56	[21, 48]	[20, 52]	[8, 52]	[11, 52]	[21, 48]	56	[18, 52]	**[21, 52]**	[8, 52]	[11, 52]	**[21, 52]**
68-rc-c	64	[32, 52]	**[36, 60]**	[29, 88]	[30, 60]	**[36, 52]**	60	[28, 44]	**[35, 56]**	[23, 80]	[27, 56]	**[35, 44]**
76-c-c	28	[14, 24]	[12, 24]	[12, 56]	**[16, 24]**	**[16, 24]**	28	[12, 24]	[12, 24]	[12, 40]	[14, 24]	[14, 24]
82-c-e	64	[18, 64]	**[26, 64]**	[8, 64]	[10, 64]	**[26, 64]**	64	[15, 64]	**[26, 64]**	[6, 64]	[9, 64]	**[26, 64]**
82-c-c	72	[28, 68]	[32, 68]	[31, 124]	[32, **60**]	[38, **60**]	64	[32, 64]	**[33, 60]**	[24, 112]	[28, 60]	**[33, 60]**
88-c-e	76	[28, 76]	[18, **72**]	[32, 108]	[**39**, 76]	[**39**, **72**]	76	[23, 72]	[18, 72]	[32, 108]	[**38**, 76]	[**38**, 72]
91-r-c	92	[45, 96]	[33, 88]	[32, 156]	[35, **76**]	[45, **76**]	72	[38, 88]	[32, **68**]	[28, 124]	[32, **68**]	[38, **68**]
99-rc-c	76	[37, 68]	[26, 72]	[24, 120]	[26, **64**]	[37, **64**]	60	[32, 64]	[27, 60]	[20, 108]	[26, **64**]	[32, 60]
101-r-c	84	[42, 76]	[21, 80]	[30, 144]	[31, 80]	[42, 76]	72	[36, 112]	[22, **68**]	[26, 144]	[28, **68**]	[36, **68**]
103-rc-c	76	[39, 80]	[35, **76**]	[26, 140]	[30, **76**]	[39, **76**]	72	[32, 80]	[37, **68**]	[22, 120]	[25, **68**]	[37, **68**]
105-rc-e	116	[39, 116]	[27, **104**]	[26, 132]	[26, 108]	[39, **104**]	104	[33, 112]	[26, **104**]	[21, 124]	[24, **104**]	[33, **104**]
108-rc-e	120	[44, 104]	[22, **120**]	[28, 152]	[36, 120]	[46, 120]	128	[39, 120]	[23, 120]	[24, 136]	[31, 124]	[39, 120]
114-rc-c	72	[35, 88]	[28, **72**]	[26, 120]	[26, **72**]	[35, **72**]	80	[30, 64]	[28, 64]	[22, 96]	[22, 72]	[30, 64]
121-rc-e	96	[42, 104]	[34, **92**]	[29, 144]	[31, 96]	[42, **92**]	96	[34, 116]	[30, **92**]	[24, 128]	[24, 96]	[34, **92**]
126-rc-e	112	[38, 132]	[19, **84**]	[35, 164]	[34, **84**]	[50, **84**]	124	[41, 120]	[17, **76**]	[29, 148]	[36, **76**]	[41, **76**]
126-r-c	88	[45, 116]	[23, **112**]	[28, 140]	[30, **112**]	[45, **112**]	80	[37, 116]	[26, **112**]	[24, 144]	[24, **112**]	[37, **112**]
144-rc-c	80	[40, 128]	[27, 76]	[25, 144]	[26, **72**]	[40, **72**]	172	[34, 104]	[29, **92**]	[22, 136]	[23, 148]	[34, **92**]
154-c-c	36	[18, 40]	[14, **36**]	[8, 72]	[8, **36**]	[18, **36**]	36	[15, 36]	[14, 36]	[6, 68]	[12, 36]	[15, 36]
165-r-c	108	[54, 192]	[14, **108**]	[40, 192]	[40, **108**]	[54, **108**]	88	[47, 220]	[14, 88]	[34, 212]	[34, **84**]	[47, **84**]
167-r-e	124	[58, 176]	[16, **120**]	[42, 196]	[43, 124]	[58, **120**]	120	[49, 204]	[16, **116**]	[34, 204]	[35, 120]	[49, **116**]
173-r-e	104	[54, 352]	[16, **96**]	[36, 192]	[37, 104]	[54, **96**]	100	[43, -]	[16, **92**]	[32, 196]	[31, 96]	[43, **92**]
173-rc-c	88	[46, 116]	[21, 88]	[29, 164]	[29, **84**]	[46, **84**]	88	[39, 116]	[21, **80**]	[24, 164]	[24, 88]	[39, **80**]
181-r-e	128	[65, 268]	[19, **124**]	[42, 208]	[42, 128]	[65, **124**]	132	[54, 204]	[20, **120**]	[35, 204]	[36, 132]	[54, **120**]
185-c-c	48	[24, 48]	[20, **44**]	[14, 100]	[24, 48]	[24, **44**]	48	[20, 48]	[20, **44**]	[12, 60]	[**24, 48**]	[24, **44**]
187-c-c	248	[58, 216]	[27, **148**]	[37, 204]	[37, 180]	[58, **148**]	248	[48, 128]	[28, 136]	[32, 192]	[32, 176]	[48, **128**]
198-c-c	36	[16, -]	[16, 36]	[8, 68]	[13, 36]	[16, 36]	36	[16, 36]	[16, 36]	[8, 68]	[13, 36]	[16, 36]
200-r-e	124	[60, 308]	[16, **120**]	[38, 228]	[39, 124]	[60, **120**]	120	[52, 288]	[17, **120**]	[32, 216]	[34, **120**]	[52, **120**]
Average	91.7	[40.0, 120.0]	[26.3, 84.4]	[27.7, 133.5]	[29.7, 85.7]	[42.0, 82.7]	91.7	[34.3, 103.2]	[26.5, 80.4]	[23.4, 126.4]	[26.4, 86.4]	[37.2, 79.6]

performance if the information received is valuable. In the experiments reported in [4], it emerges that both the model have scalability issues on large instances, likely due to the difficulties of the *Circuit* and *Multicircuit* commands of CP-SAT of dealing with VRP problems with more than a few tens of customers. In this paper, we evaluate whether hinting a solution can make the models more effective.

In the solution considered, we will ignore the drones and solve each instance as a min-max VRP problem. The solution, using only trucks is then passed to the solver, that might benefit from this because solutions using drones can in principle be obtained by taking away some customers from the tours of the truck.

4   Experimental Results

All the models presented in previous sections have been coded in Python 3.11.2. The Constraint Programming models discussed in Sect. 3 have been solved via the CP-SAT solver of Google OR-Tools 9.6 [2], while the heuristic method adopted for retrieving min-max VRP solutions was the *Route* solver, again from OR-Tools. All the experiment reported have been carried out on a computer equipped with A CPU Intel Core i7 12700F, and 32 GB of RAM and with a maximum computation time of 1 h. The instances originally introduced in [6] for the PDSTSP-c, and available at http://orlab.com.vn/home/download are considered. The number n of customers varies from 50 to 200, the number m of drones available is between 5 and 10 and the number s of trucks is between 2 and 5. The interested reader can find all the details of the instances in [6].

The models M2 and M3, without hint-start (from [4], state-of-the-art at the time of writing) and with hint-start, are considered in Table 1. The upper and lower bounds (when available) found in the given time by each method are reported.

The experiments suggest that hint-starting the solver with a solution is beneficial when considering both lower bounds and (especially) heuristic solutions. Given that the hinted solution is only based on trucks, this was not obvious. Passing an initial solution optimized externally – even without drones – shapes up the truck tours. The CP-SAT solver appears to benefit from such information and seems more effective in taking customers out of the truck tours to assign them to drones, then to design tours from scratch. Finally, a consideration about the use of (collaborative) drones is that they allow an average time-saving in the order of 10% (comparison against the column VRP).

References

1. Bertazzi, L., Golden, B., Wang, X.: Min-max vs. min-sum vehicle routing: a worst-case analysis. Eur. J. Oper. Res. **240**(2), 372–381 (2015)
2. Google: OR-Tools (2023). https://developers.google.com/optimization/. Accessed 03 Mar 2023
3. Montemanni, R., Dell'Amico, M.: Solving the parallel drone scheduling traveling salesman problem via constraint programming. Algorithms **16**(1), 40 (2023)

4. Montemanni, R., Dell'Amico, M., Corsini, A.: Parallel drone scheduling vehicle routing problems with collective drones. Comput. Oper. Res. **163**, 106514 (2024)
5. Murray, C.C., Chu, A.G.: The flying sidekick traveling salesman problem: Optimization of drone-assisted parcel delivery. Transp. Res. Part C: Emerg. Technol. **54**, 86–109 (2015)
6. Nguyen, M.A., Hà, M.H.: The parallel drone scheduling traveling salesman problem with collective drones. Transp. Sci. **4**(57), 866–888 (2023)
7. Nguyen, M.A., Luong, II.L., Hà, M.H., Ban, H.B.: An efficient branch-and-cut algorithm for the parallel drone scheduling traveling salesman problem. 4OR **21**, 609–637 (2023)
8. Paczan, N.M., Elzinga, M.J., Hsieh, R., Nguyen, L.K.: Collective unmanned aerial vehicle configurations (2022). Patent US 11,480,958 B2
9. Raj, R., Lee, D., Lee, S., Walteros, J., Murray, C.: A branch-and-price approach for the parallel drone scheduling vehicle routing problem. SSRN Electron. J. 1–47 (2021)
10. Toth, P., Vigo, D.: The Vehicle Routing Problem. SIAM (2002)

Operations Research for Health Care

A Re-optimization Heuristic
for a Dial-a-Ride Problem
in the Transportation of Patients

Ruan Myller Magalhães de Oliveira[1,3] , Manuel Iori[2] , Arthur Kramer[3] ,
and Thiago Alves de Queiroz[2,4(✉)]

[1] Institute of Mathematics and Technology, Federal University of Catalão,
Catalão, GO 75708-560, Brazil
`ruanmyller@discente.ufcat.edu.br`
[2] Department of Sciences and Methods for Engineering,
University of Modena and Reggio Emilia, 42122 Reggio Emilia, Italy
`manuel.iori@unimore.it`
[3] Mines Saint-Étienne, Univ. Clermont Auvergne, CNRS, UMR 6158 LIMOS,
Institut Henri Fayol, F-42023 Saint-Étienne, France
`arthur.kramer@emse.fr`
[4] Institute of Mathematics and Technology, Federal University of Catalão,
Catalão, GO 75704-020, Brazil
`taq@ufcat.edu.br`

Abstract. In this paper, we handle the problem of picking and delivering patients among the distinct units of a hospital. This problem is found in hospitals with several (specialized) units covering a large area, and it emerges from a real situation faced by a hospital in northern Italy. Patient transportation requests arrive dynamically during the day, and the hospital transportation department must service them all using capacitated and homogeneous vehicles. Each request is associated with a patient urgency level (weight) and a time window. The objective is to design vehicle routes to serve all requests and minimize the total weighted tardiness. To solve the problem, we propose a re-optimization heuristic based on two policies that mimic the patients' and hospital's decision-making processes. We then improve the solutions obtained with the policies using a tabu search. Computational results show that we can obtain high-quality solutions using the tabu search compared with the policies and a simulated annealing-based heuristic from the literature.

Keywords: dynamic transportation of patients · dial-a-ride problem · re-optimization heuristic · tabu search

1 Introduction

The number of people requesting hospital services has grown significantly in the last few years, especially during the COVID-19 pandemic. At the same time,

M. Sevaux et al. (Eds.): MIC 2024, LNCS 14753, pp. 145–157, 2024.
https://doi.org/10.1007/978-3-031-62912-9_14

some hospitals offer specialized services in many care units (pavilions) occupying large areas. Pavilions, floors, and buildings are part of the same environment in these institutions. Thus, the distance between them cannot be ignored, especially considering the patients' well-being and satisfaction.

The problem addressed in this work concerns the dynamic transportation of patients between the care units of a hospital. According to Beaudry et al. [3], medical diagnosis, care, and treatment units are among the most visited by patients. Transporting patients between care units and service areas is usually under the responsibility of the hospital transportation department. In smaller hospitals, patients are commonly transported on stretchers and wheelchairs. Instead, in larger hospitals, patients can be transported by ambulances or specific vehicles of different capacities. This transport may affect hospital services, as delivering a patient after her planned time window can delay other services, cause dissatisfaction, and worsen the patient's situation.

As requests arrive dynamically, decisions should be made accordingly, considering the current routes and vehicle availability. Decisions related to requests are not simple, as they depend on the vehicle's current position, its capacity, and its *status* (e.g., whether it is responding to a request or it is waiting), as well as on complicating factors related to the patient urgency and time window [11]. Therefore, the transportation department defines which vehicle will service each request under a rolling time horizon approach.

Problems that consider the transportation of people and objects can be regarded as variants of the vehicle routing problem (VRP), as in the case of the pickup and delivery problem (PDP) [5]. Doerner and Salazar-González [6] surveyed PDPs related to the transportation of people. The authors detailed the literature's contributions to transporting elderly and disabled people, aiming to minimize costs and improve the quality of the service. A PDP variant concerning the transportation of people is the *Dial-a-Ride Problem* (DARP). In the DARP, the human perspective is considered, and hence, costs are optimized, as well as factors related to the passengers' satisfaction and well-being [14].

In this paper, we handle a Dynamic DARP that arises in the transportation of patients between hospital units. We solve the problem using a re-optimization heuristic that uses two policies that mimic the patients' and hospital's decision-making processes. To improve the decisions, we also propose a tabu search-based heuristic. The good performance of the tabu search is confirmed by an extensive computational comparison with a re-optimization heuristic based on a simulated annealing from the recent literature [8].

The remained of this paper is organized as follows. Section 2 provides a brief literature review. Section 3 describes the problem under consideration. Section 4 presents the idea behind the re-optimization heuristic, providing details of the two policies and the tabu search. Section 5 contains the experimental results, comparisons, and discussions on the solutions. Finally, Sect. 6 presents the concluding remarks and perspectives for future works.

2   Literature Review

Part of the VRP (and its variants) literature has assumed problems whose information is all known and available in advance (i.e., static problems. On the other hand, the number of contributions considering problems of dynamic nature has grown over the years [15]. In a static problem, we can plan all routes as all information is precisely known. In a dynamic problem, new information arrives under a rolling time horizon in addition to having some known initial information. According to Pillac et al. [10], dynamic problems are often solved with re-optimization-based approaches, that is, as sequences of static subproblems. Optimization starts with the set of already known requests, producing an initial set of routes to be executed by the vehicles. With new information arriving over time, routes are re-optimized accordingly.

Battarra et al. [2] discussed about the different PDPs variants. The first is related to "many-to-many" (M-M) problems, where objects/people can be transported among multiple delivery and pickup nodes. The second involves "one-to-many-to-one" (1-M-1) problems, in which deliveries and pickups involve two distinct sets: objects/people are picked from a depot/hospital and delivered to many different nodes, or they can be picked from these nodes and then delivered back to the depot. The third involves the "one-to-one" problems (1-1), in which each object/person has a specific pickup and delivery node. This is the case for mail operations, door-to-door transport services, and the problem under consideration in this paper.

Problems concerning the transportation of people (e.g., patients) involve requests containing the person's information, such as the pickup location, the delivery location, when the person will be ready to receive transportation, and the time window, if any. Respecting time windows is essential in limiting the person's waiting time. In a hospital environment, additional information from each person may be necessary, such as urgency level (weight), which can impact the time window size.

Beaudry et al. [3] handled the problem of transporting patients between distinct units of the same hospital by using a heterogeneous fleet of vehicles. Transport requests arrive dynamically, meaning they are unknown until the patient is released for transport. The authors provided a detailed problem description and then proposed a two-phase heuristic. In the first phase, a simple insertion heuristic generates a feasible solution. Next, this solution is improved with a tabu search with only two neighborhood structure types. The proposed heuristic solved instances provided by a German hospital. Kergosien et al. [9] considered the transportation of patients inside a large complex hospital in France. They also assumed a heterogeneous fleet of vehicles, besides handling constraints related to disinfection operations for contagious patients. The problem objective concerns minimizing transportation costs and patients' tardiness. The problem is solved with a tabu search heuristic that uses an adaptive memory to save routes and a cross-exchange operator to generate new solutions.

In Schmid and Doerner [12], the routing of vehicles is integrated with the scheduling of rooms for patients who undergo different examinations.

Patients are transported on stretchers, beds, wheelchairs, or simply by foot. The authors were concerned with optimizing the patients' inconvenience and hospital resources, such as porters and rooms. They proposed a hybrid metaheuristic: a shifting bottleneck heuristic was used to optimize the patients' waiting times and idle times in the rooms, while a large neighborhood search solved a multi-depot VRP with time windows. Computational experiments were performed on a randomly generated set of 130 instances based on realistic assumptions.

Elmbach et al. [7] handled the problem of transporting patients without using motorized vehicles. They considered three groups: hospital managers, who take care of the transport system with sufficient porters to avoid tardiness in surgeries or exams; patients, where the waiting time for picking and delivering should not be long; porters, who take care of the physical transportation of patients within the hospital. The authors integrated the needs of these three groups and proposed a tabu search as a solution method.

Côté et al. [4] solved the dynamic patient transportation problem from a hospital in northern Italy. The objective is to service all requests to minimize the total weighted tardiness using a heterogeneous vehicle fleet. The authors studied the impact of positioning vehicles after servicing one request: wait in the unit floor, wait in the unit parking, or return and wait in the main building. They proposed a large-adaptive neighborhood search and embedded it into a re-optimization framework. Results showed that positioning vehicles at the unit parking to wait for the next decisions is the best strategy overall.

Following the problem handled in [4], Fonseca [8] implemented a simulated annealing and five decision-making policies. The policies are based on the patient's time window, release time, urgency level, and vehicle arrival time to service the request. Recently, Aziez et al. [1] extended the work in [4] by solving a dynamic multi-trip pickup and delivery problem with time windows and a heterogeneous fleet. The authors proposed a branch-and-regret, aiming at minimizing the total weighted lateness of the delivery and the travel time of all vehicles. The simulated annealing generates new solutions by four operators that swap or insert requests in the same or different vehicles.

This paper is motivated by the problem tackled in [4] and [8], aiming to minimize the total weighted tardiness related to patient transportation. This is an important key performance indicator to assess patient satisfaction and well-being, especially when servicing urgent patients. It also directly impacts the staff schedule (e.g., doctors, nurses, and technicians) and material resources (e.g., rooms and specialized equipment). Differently from the cited literature, we assume a homogeneous fleet of vehicles and develop two policies and a tabu search within a re-optimization framework. Our tabu search comprises 12 operators based on swap and insertion movements, returning high-quality solutions in low computing times, as one decision-maker could expect for dynamic problems.

3 Problem Description

The DARP variant handled in this work considers a hospital containing several distinct care units. Requests for transportation arrive during the day when the patient is ready to be transported. A complete directed graph $G = (N, A)$ models the transport network of the hospital and its units. The set N represents the hospital units, which are the nodes for picking up and delivering patients by a fleet of homogeneous vehicles. The set A contains the arcs that connect the hospital units. Each arc is characterized by information on the travel time required to traverse it. In our work, this travel time directly corresponds with the travel distance, so large distances imply large travel times.

The problem considers a set with k homogeneous vehicles. Initially, vehicles are positioned at the main hospital unit (i.e., at node 0) and move between units to service requests. Each request is associated with a patient, providing the following information: identification (id), release time (rd), the time window in which the patient should receive care, starting at rd and having due date et, the urgency level or weight (w), the pickup node (c) and the delivery node (e). If a patient is delivered after her due date, it will incur a tardiness. We aim to minimize the total weighted tardiness by transporting all patients and thus servicing all requests. This objective follows the previous literature on the problem in which attention is given to the patient's satisfaction and well-being (see, e.g., [3,4]). Moreover, traveling times impact, to some extent, the moment when a request is serviced, which in turn may impact the patient's tardiness.

The requests are serviced on a rolling time horizon, which starts at time zero and ends at time T. This horizon is discretized in minutes. Additionally, the transportation department only becomes aware of a request when the patient is released for transportation. A feasible solution for the problem, as illustrated in Fig. 1, respects the following constraints:

- The number of patients in a vehicle cannot exceed its capacity;
  The vehicle can only operate within the time horizon;
- Each patient must be first picked up and then delivered;
- The picking of a patient can only be done after the patient is released for transport;
- The patient may be delivered after her due date; however, this will result in a weighted tardiness;
- The service time for picking up or delivering a patient is assumed to be negligible.

4   Re-optimization Heuristic

Proposing a heuristic approach is one way to deal with hard combinatorial optimization problems, such as the dynamic DARP, in a practical way and quickly produce satisfactory solutions. The heuristics that we implemented work with a vector of vehicles representing the solution. Each vehicle is then associated with another vector containing the service requests and the performed actions (pick up or delivery). Requests are sent dynamically to the transportation department according to released patients. These requests are added to a pickup list; thus, patients remain awaiting service. Once a patient is picked up, the request is

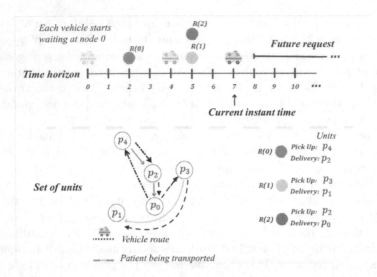

Fig. 1. Example of a solution. The vehicle starts at unit p_0 and waits until time $t = 2$ when request R(0) is revealed. Then, it departs to pick up the patient of R(0) at unit p_4, and next decides to deliver her at unit p_2 where the patient of request R(2) has already been released. It continues to deliver the patient of R(2) at unit p_0 and to, finally, service request R(1).

removed from the pickup list and added to a delivery list associated with the same vehicle.

The vector associated with each vehicle contains all its pickup and delivery actions, from those already executed or in progress to those already assigned but to be executed in the future. Vehicles with an ongoing action cannot be notified to start another action immediately. Indeed, a vehicle must finish its current action before starting another action (we assume it keeps on hold at the last visited node). Executed or ongoing actions cannot be modified or rearranged in the solution vector. However, the same does not apply to future actions. In this sense, as new requests emerge, it is said that an event occurs. Thus, to optimize and define the routes of each vehicle, we propose a re-optimization heuristic in which decisions can be made using two policies or a tabu search-based heuristic.

The re-optimization heuristic with tabu search (TS-D) is detailed in Algorithm 1. Initially, the pickup list L_p is created (line 1) and remains empty until the loop representing the time horizon starts. Over time, new requests are revealed, and patients are released for transportation. Thus, the pickup list is updated (line 3) as patients wait for pickup, raising an event. Attention now turns to the vehicles and their routes. In line 4, re-optimization may occur if a vehicle is available (i.e., on hold). In this case, a policy (i.e., a greedy heuristic) is applied (line 5) to provide a prior feasible solution before applying the tabu search to improve it (line 6). Other two versions of the re-optimization heuristic are proposed, where the tabu search is not used (line 6 is disregarded) but only policies $R1$ or $R2$ (line 5).

Algorithm 1. Re-optimization heuristic with tabu search

1: Let L_p and L_d be two initially empty lists of requests awaiting pickup and delivery, respectively
2: **for** t from 0 to the end time T **do**
3:     Update L_p with the available requests if $rd \leq t$
4:     **if** there is at least one vehicle v on hold **then**
5:         Use a policy to update the solution with the requests in L_p and L_d
6:         Apply the tabu search to optimize the solution
7:     **end if**
8:     Perform the actions to each vehicle and update lists L_p and L_d
9: **end for**

We develop two policies to define the sequence of requests to be served with a pickup or delivery action. For each available request j in L_p and L_d, among all possible requests available in the pickup (L_p) and delivery (L_d) lists, each policy selects the most promising id to be performed next as follows:

- R1: we obtain the traveling time tt_j from the node where the vehicle is to the node where j is; we obtain the due date et_j; now, we calculate the product: $tt_j \times et_j$, and select the request $id = j$ with the minimum product value to be performed next. Figure 2 illustrates an example of this policy.
- R2: we obtain the urgency level w_j; we calculate the difference between the current moment in the time horizon t and release time rd_j; we calculate the difference between the due date et_j and the release time rd_j; we divide these two differences ($\frac{t-rd_j}{et_j-rd_j}$) and multiply the result by w_j; the minimum resulting value indicates the request with $id = j$ to be performed next.

The tabu search metaheuristic behaves similarly to the local search heuristics when exploring the search space. However, one of its main features is using (short/long-term) memories. This characteristic is an attempt to escape from local optima solutions since repeated movements/solutions are avoided for a while. In this way, the metaheuristic can obtain a different solution, improving the probability of getting better solutions. Our tabu search implementation follows the framework in [13].

Different operators can be used to explore the neighborhood of a solution and then create new solutions. We can modify a vehicle route by applying two operators: N_1 and N_2. We consider the application of swaps (N_1) and insertions (N_2) in the vector of requests associated with the vehicles. Performing these operations can result in an infeasible solution (e.g., not respecting the pickup and delivery order). Therefore, the resulting solution must remain valid at the end of an operation. We only accept an operation that results in a feasible solution.

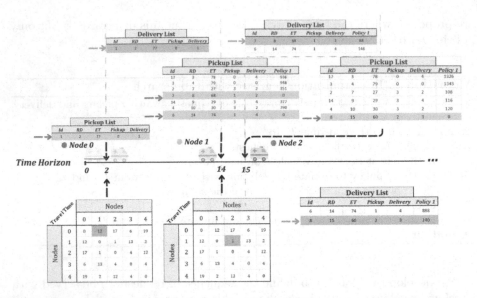

Fig. 2. Example of policy R1. The vehicle starts on hold until time $t = 2$ when the request $id = 1$ is known. The pickup list L_p is updated. As $id = 1$ is the only available option, this patient is picked up at node 0 (the same node where the vehicle is on hold), the delivery list L_d is updated with her, and then the vehicle departs to deliver her (the only action available after the pick up). The vehicle finishes the delivery action at time $t = 14$ when the pickup list L_p contains seven requests ($id = 17, 3, 2, 7, 14$, 4, and 6) and L_d is empty. Policy R1 is applied to calculate the product $tt_j \times et_j$ for each $id = j$ in L_p (the resulting values are in the last column of the pickup list table). The minimum product is 0 for $id = 7$ and $id = 6$ since these patients are at the same node where the vehicle is on hold (node 1). Both patients are picked up, and lists L_p and L_d are updated. Next, applying policy $R1$, the most promising id is 7 due to the list L_d, with the action of delivery her to node 2 (the minimum product is 68). After completing the delivery at $t = 15$, the most promising id is 8 due to the list L_p, with the action of picking up this id.

Operator N_1 randomly selects a vehicle and then randomly chooses two positions, represented by i and j, in the vector of requests. The contents of these positions are swapped. Different cases may arise from the choices of i and j. Pairs (I) pickup-pickup, (II) pickup-delivery, (III) delivery-pickup, and (IV) delivery-delivery may be selected. For example, an infeasible solution may emerge by not respecting the order of pickup and delivery for a patient in cases I, II, and IV.

Operator N_2, in turn, is responsible for randomly selecting a vehicle and then choosing two random positions i and j in the vector of requests. Next, it inserts the content of position i immediately before position j and after position $j - 1$. Similarly to operator N_1, depending on the content of each position, some cases naturally arise, and the resulting solution may be infeasible. For example, suppose we have $i > j$, the vehicle reaches full capacity when performing the action at j (a pickup), and i is associated with another pickup action. Inserting

i immediately before j will make the vehicle reach its full capacity when performing the action at i. In this way, it cannot perform the action at j since there is no capacity for another pickup action.

Operators N_1 and N_2 are used as the basis to create ten others, which are presented next. These operators perform in a best-improvement strategy (i.e., each tries all options and then performs the one leading to the best improvement, if any):

- N_3: similar to N_1 but assuming two different vehicles selected at random;
- N_4: similar to N_2 but assuming two different vehicles selected at random;
- N_5: uses N_1 where all positions of a given vehicle are tested.
- N_6: uses N_2 where all positions of a given vehicle are tested.
- N_7: uses N_3 where all positions of two vehicles are tested.
- N_8: uses N_4 where all positions of two vehicles are tested.
- N_9: similar to N_5 but it is performed for all vehicles.
- N_{10}: similar to N_6 but it is performed for all vehicles.
- N_{11}: similar to N_7, but it is performed for all combinations of two vehicles.
- N_{12}: similar to N_8, but it is performed for all combinations of two vehicles.

5   Numerical Experiments

This section details the experiments and results obtained with the re-optimization heuristic. The objective is to show the competitiveness of our proposed heuristic, comparing its results with the simulated annealing heuristic proposed by [8]. Our algorithms were implemented in the Python programming language, and the numerical tests were conducted on a computer equipped with an Intel Xeon E3-1245 3.50 GHz processor, 32 GB of RAM, and Ubuntu 18.04 LTS as the operating system.

The comparison is performed on 50 instances randomly generated and made available by [8]. These instances assume a hospital (H) with five units, while the number of requests (O) can be 20, 40, 60, 80, and 100. They also consider an arrival rate (R), which represents the frequency at which a patient is released: 0.2 (closer to each other); 0.4; 0.6; 0.8; 1.0; 1.25; 1.50; 1.75; 2.0; 3.0 (farther from each other). For each number of requests, ten distinct instances were generated, one for each arrival rate. The traveling time between any pair of hospital units was uniformly generated in the interval [1; 20] minutes. More information can be obtained from the original publication [8].

To solve these instances, a fleet of two vehicles, each with a maximum capacity to transport up to two patients simultaneously, is considered. Other combinations tested in Fonseca [8] were not considered in this work and are part of future research. We use a short-term memory in the tabu search. This way, each time a neighborhood operator is applied, the chosen positions i and j and their requests are stored in the tabu list. Then, these moves are forbidden from repeating twice until they are removed from the tabu list and can occur again. We use a maximum number of iterations as the stopping criterion of the tabu search.

After preliminary experiments using a trial-and-error methodology, we set this value to 150, recalling the tabu search is invoked by the re-optimization heuristic on each event occurrence (e.g., the release of a new request). We obtained high-quality solutions with this value, as seen in Table 1. Moreover, the preliminary experiments indicated better solutions when setting the probability of 20% to select any of the operators N_1, N_2, N_3, and N_4, and 2.5% to the remaining ones (N_5 to N_{12}).

Table 1 presents the best (b-SOL) and average (m-SOL) solutions obtained with the tabu search for five runs with different seeds (this is the same configuration considered in [8]). As the two policies, R1 and R2, are greedy heuristics and do not depend on random choices, they were run only a single time. The table rows have the solutions (SOL) and computing time in seconds (Time) obtained after solving each instance. These results are compared with the re-optimization heuristic with simulated annealing (SA-D) proposed in [8], where this author reported the best solution among five runs of SA-D. At the same time, TS-D represents the re-optimization heuristic with the proposed tabu search using policy R_1 to generate the initial/first solution.

The results in Table 1 show that policy $R1$ has an average computing time of 0.007 s and an average solution value equal to 608.36. Instead, policy $R2$ has an average time of 0.011 s and 1283.22 as the average total weighted tardiness. These values for the re-optimization heuristic with tabu search, TS-D, are 56.08 s and 130.8 (column b-SOL), respectively. If column m-SOL (average solution for each instance) of TS-D is observed, the overall average solution increases to 281.8. It is worth noting that the best solution value for 12 out of 50 instances equals 0, meaning no tardiness, with the proposed TS-D. On the other hand, there are also relatively low-quality solutions, e.g., for instance $O_{100} - H_5 - R_1$ obtained with policy $R2$.

The values in the last row of Table 1 show that the best overall performance in solution quality is due to the proposed TS-D, with an overall average value of 130.8 as the best total weighted tardiness. On the other hand, this value equals 593.66 with the re-optimization heuristic with simulated annealing, SA-D, in [8]. The latter is slightly inferior to the value obtained with policy $R1$, which is 608.36, showing this policy is somehow competitive with the literature.

Comparing the performance of the proposed TS-D with the one of SA-D, we notice TS-D obtains equal solutions to a single instance (see $O_{40} - H_5 - R_7$), worse solutions to a single instance (see $O_{60} - H_5 - R_3$), and better solutions to all other 48 instances. This difference can be explained by the number and quality of the operators TS-D has. While SA-D limits itself to operators N_1, N_2, N_3, and N_4, TS-D covers a much larger set of movements, with operators N_5 to N_{12} performing a more accurate local search.

The re-optimization heuristic with tabu search requires much more computing time than the two policies. We recall that the tabu search is applied each time an event occurs. On average, it is invoked the following number of times: 16.7, 27, 35.1, 35.6, and 42.3 to the instances with 20, 40, 60, 80, and 100 patients, respectively. Besides that, the tabu search has some operators that may be very

Table 1. Results obtained for 50 instances with the hospital having 5 units.

Instance	SA-D		R1		R2		TS-D		
	Time	b-SOL	Time	SOL	Time	SOL	Time	b-SOL	m-SOL
$O_{20} - H_5 - R_1$	22.54	33	0.001	160	0.001	180	4.67	0	43.10
$O_{20} - H_5 - R_2$	27.5	112	0.001	561	0.003	469	7.34	32	49.60
$O_{20} - H_5 - R_3$	34.1	104	0.005	399	0.001	12	1.75	0	7.80
$O_{20} - H_5 - R_4$	43.46	18	0.005	196	0.006	36	2.02	16	24.40
$O_{20} - H_5 - R_5$	37.06	80	0.003	80	0.005	76	1.46	8	24.00
$O_{20} - H_5 - R_6$	41.94	53	0.003	38	0.008	76	1.42	32	32.00
$O_{20} - H_5 - R_7$	38.73	40	0.002	4	0.003	8	1.35	4	4.00
$O_{20} - H_5 - R_8$	43.94	84	0.003	28	0.005	28	1.46	4	4.00
$O_{20} - H_5 - R_9$	43.7	36	0.002	0	0.006	0	1.46	0	0.00
$O_{20} - H_5 - R_{10}$	42.29	76	0.003	0	0.002	0	1.04	0	0.00
$O_{40} - H_5 - R_1$	133.74	985	0.006	1123	0.006	1615	52.41	130	256.40
$O_{40} - H_5 - R_2$	156.14	435	0.008	595	0.006	971	26.97	104	179.40
$O_{40} - H_5 - R_3$	162.27	103	0.005	322	0.009	130	10.75	32	73.90
$O_{40} - H_5 - R_4$	165.37	104	0.008	295	0.008	196	9.27	12	125.70
$O_{40} - H_5 - R_5$	166.14	132	0.008	84	0.014	104	5.99	48	62.40
$O_{40} - H_5 - R_6$	163.09	372	0.006	220	0.016	321	5.41	44	101.60
$O_{40} - H_5 - R_7$	171.5	120	0.009	144	0.017	124	4.28	120	120.00
$O_{40} - H_5 - R_8$	168.05	348	0.002	44	0.006	48	2.23	32	32.00
$O_{40} - H_5 - R_9$	174.13	8	0.005	8	0.008	12	2.70	4	4.00
$O_{40} - H_5 - R_{10}$	176.94	156	0.001	0	0.008	0	0.77	0	0.00
$O_{60} - H_5 - R_1$	325.40	1010	0.014	1030	0.008	4794	228.83	280	667.10
$O_{60} - H_5 - R_2$	401.68	682	0.011	798	0.013	2949	99.88	211	363.30
$O_{60} - H_5 - R_3$	396.57	33	0.011	297	0.011	304	23.69	64	89.20
$O_{60} - H_5 - R_4$	333.04	448	0.025	163	0.016	150	18.72	32	77.80
$O_{60} - H_5 - R_5$	292.18	280	0.011	480	0.016	256	7.55	80	181.77
$O_{60} - H_5 - R_6$	322.09	164	0.002	76	0.008	68	5.34	32	43.60
$O_{60} - H_5 - R_7$	370.12	84	0.005	16	0.013	16	3.24	0	0.00
$O_{60} - H_5 - R_8$	342.9	156	0.006	48	0.006	88	3.13	8	16.00
$O_{60} - H_5 - R_9$	360.01	280	0.003	40	0.009	28	2.31	0	0.00
$O_{60} - H_5 - R_{10}$	384.96	272	0.001	44	0.003	48	0.86	44	48.90
$O_{80} - H_5 - R_1$	625.69	6114	0.022	6233	0.016	12319	662.51	1302	3015.70
$O_{80} - H_5 - R_2$	704.15	712	0.019	1180	0.023	5405	99.68	382	581.30
$O_{80} - H_5 - R_3$	731.57	479	0.006	931	0.028	1410	51.10	316	425.22
$O_{80} - H_5 - R_4$	736.29	286	0.011	211	0.009	336	14.09	52	98.78
$O_{80} - H_5 - R_5$	758.52	228	0.008	131	0.017	346	7.91	4	17.80
$O_{80} - H_5 - R_6$	754.17	496	0.008	160	0.013	213	4.52	112	146.20
$O_{80} - H_5 - R_7$	773.88	144	0.005	0	0.011	0	3.07	0	2.67
$O_{80} - H_5 - R_8$	748.04	124	0.003	28	0.011	28	2.18	4	4.40
$O_{80} - H_5 - R_9$	658.13	264	0.003	4	0.005	4	2.23	0	0.00
$O_{80} - H_5 - R_{10}$	686.42	260	0.003	0	0.009	0	1.05	0	0.00
$O_{100} - H_5 - R_1$	930.49	8720	0.014	10562	0.022	25531	1158.02	1990	5792.00
$O_{100} - H_5 - R_2$	1097.17	1046	0.017	1174	0.030	2852	160.56	258	359.22
$O_{100} - H_5 - R_3$	1308.06	1344	0.022	1083	0.028	1421	55.30	361	529.77
$O_{100} - H_5 - R_4$	1251.41	638	0.008	72	0.020	370	13.25	36	79.10
$O_{100} - H_5 - R_5$	1361.84	416	0.009	223	0.013	308	15.26	56	105.77
$O_{100} - H_5 - R_6$	1242.74	536	0.009	112	0.027	104	6.36	56	82.22
$O_{100} - H_5 - R_7$	1097.14	208	0.006	121	0.009	107	2.90	0	20.55
$O_{100} - H_5 - R_8$	1083.77	496	0.003	56	0.014	60	3.22	56	62.22
$O_{100} - H_5 - R_9$	1107.65	164	0.003	184	0.011	188	2.74	56	72.80
$O_{100} - H_5 - R_{10}$	1107.38	192	0.001	52	0.003	52	1.84	0	48.88
Average	486.12	593.66	0.007	608.36	0.0112	1283.22	56.08	128.60	281.83

time-consuming, such as N_9 to N_{12} that try all possible swap and insertion movements.

6 Conclusions and Future Works

This paper develops a re-optimization heuristic that uses a tabu search-based heuristic to optimize decisions regarding a dynamic Dial-a-Ride Problem. This problem emerges from a real situation involving the picking and delivering of patients between units of the same hospital. The problem objective requires minimizing the total weighted tardiness. In addition to the tabu search, we propose two greedy heuristics (policies $R1$ and $R2$) and evaluate their performance on instances of different sizes.

Computational experiments on 50 instances from the literature show that the proposed re-optimization heuristic with tabu search is much superior in solution quality. It allows decreasing the average total weighted tardiness value by 78.34% compared to the value obtained by [8]. Its result is even better than the two heuristic policies, with an average improvement of around 78.86% over R1 and 89.98% over R2. It is worth noting that its computing time may increase when solving large instances, especially those with many requests and a small value of arrival rate.

The work in this paper can continue in different directions. The first could concern the consideration of a heterogeneous vehicle fleet. This opens the possibility of solving realistic instances from a hospital in Italy and comparing the results with the solution methods in [1]. Second, the tabu search can be improved by defining new operators (e.g., selecting sequences of positions) and new types of memories (e.g., long-term ones). Investigating each operator's influence on the final solution could also help reduce computing time. Another direction for future research could consider multiple objectives besides the weighted total tardiness, such as minimizing travel time or distance.

Acknowledgement. The authors thank the financial support provided by the National Council for Scientific and Technological Development (CNPq) [grant numbers 405369/2021-2, 408722/2023-1, and 315555/2023-8], the State of Goiás Research Foundation (FAPEG), the National Recovery and Resilience Plan (NRRP), Mission 04 Component 2 Investment 1.5-NextGenerationEU, Call for tender n. 3277 dated 30/12/2021, Award Number: 0001052 dated 23/06/2022.

Disclosure Statement. No potential conflict of interest was reported by the authors.

References

1. Aziez, I., Côté, J.F., Torkhani, M.Z., Landa, P., Coelho, L.C.: Healthcare dynamic and stochastic transportation (2023). https://doi.org/10.2139/ssrn.4622857

2. Battarra, M., Cordeau, J.F., Iori, M.: Chapter 6: pickup-and-delivery problems for goods transportation. In: Vehicle Routing: Problems, Methods, and Applications, Second Edition, pp. 161–191. SIAM (2014). https://doi.org/10.1137/1.9781611973594.ch6
3. Beaudry, A., Laporte, G., Melo, T., Nickel, S.: Dynamic transportation of patients in hospitals. OR Spectrum 32(1), 77–107 (2010)
4. Côté, J.F., Queiroz, T.A., Iori, M., Vignoli, M.: Tranporte dinâmico de pacientes dentro de um hospital. Anais do LII SBPO Simpósio Brasileiro de Pesquisa Operacional, 1–12 (2020)
5. Côté, J.F., Queiroz, T.A., Gallesi, F., Iori, M.: A branch-and-regret algorithm for the same-day delivery problem. Transportation Research Part E: Logistics and Transportation Review 177, 103226 (2023)
6. Doerner, K.F., Salazar-González, J.J.: Chapter 7: Pickup-and-delivery problems for people transportation. In: Vehicle Routing: Problems, Methods, and Applications, Second Edition, pp. 193–212. SIAM (2014). https://doi.org/10.1137/1.9781611973594.ch7
7. von Elmbach, A.F., Scholl, A., Walter, R.: Minimizing the maximal ergonomic burden in intra-hospital patient transportation. European Journal of Operational Research 276(3), 840–854 (2019)
8. Fonseca, G.S.: Heurísticas para o transporte dinâmico de pacientes dentro de hospitais (2023). Master thesis in Modelling and Optimization. Federal University of Catalão, Catalão, Brazil
9. Kergosien, Y., Lente, C., Piton, D., Billaut, J.C.: A tabu search heuristic for the dynamic transportation of patients between care units. European Journal of Operational Research 214(2), 442–452 (2011)
10. Pillac, V., Gendreau, M., Guéret, C., Medaglia, A.L.: A review of dynamic vehicle routing problems. European Journal of Operational Research 225(1), 1–11 (2013)
11. Queiroz, T.A., Iori, M., Kramer, A., Kuo, Y.H.: Dynamic scheduling of patients in emergency departments. European Journal of Operational Research 310(1), 100–116 (2023)
12. Schmid, V., Doerner, K.F.: Examination and operating room scheduling including optimization of intrahospital routing. Transportation Science 48(1), 59–77 (2014)
13. Talbi, E.G.: Metaheuristics: From Design to Implementation. John Wiley & Sons, New Jersey (2009)
14. Toth, P., Vigo, D.: Vehicle Routing: Problems, methods, and applications, 2nd edn. Siam, Philadelphia (2014)
15. Vidal, T., Laporte, G., Matl, P.: A concise guide to existing and emerging vehicle routing problem variants. European Journal of Operational Research 286(2), 401–416 (2020)

Solving the Integrated Patient-to-Room and Nurse-to-Patient Assignment by Simulated Annealing

Eugenia Zanazzo$^{(\boxtimes)}$, Sara Ceschia, and Andrea Schaerf

Polytechnic Department of Engineering and Architecture, University of Udine,
Via delle Scienze 206, I-33100 Udine, Italy
{eugenia.zanazzo,sara.ceschia,andrea.schaerf}@uniud.it

Abstract. We consider a recently-proposed integrated healthcare problem that deals with the assignment of patients to suitable rooms in wards (Patient-to-Room) and the assignment of nurses to patients to balance their workload (Nurse-to-Patient), in one single stage.

For this problem, we designed a local search approach that uses the union of two distinct neighborhoods and is guided by a Simulated Annealing metaheuristic.

We tuned our search method, ran it on the available dataset, and validated it using the available solution checker. Finally, we report our results for different running times, to show how the scores evolve based on the granted time.

Keywords: Healthcare · Problem integration · Simulated Annealing

1 Introduction

Optimization in healthcare aims to enhance efficiency, reduce costs, and ultimately improve patient outcomes and staff well-being, and it has been studied in the optimization literature for decades [3]. The integration of healthcare optimization problems involves applying optimization techniques to simultaneously improve different aspects of healthcare delivery, resource allocation, and decision-making processes. This is of paramount importance for healthcare institutions given that the flow of patients involves multiple resources and decisions at different levels [8].

The integrated approach is quite recent, and the contributions to its application have been recently surveyed by Rachuba et al [6]. In particular, the authors identify three levels of increasing integration, which go from solving one problem while incorporating the constraints coming from the others (level 1), to the sequential solution of two or more problems using the output of one problem as input for the next one (level 2), and finally to the simultaneous solution in one single stage of two or more problems (level 3).

A recent proposal for a level 3 integration comes from Brandt et al [1] and aims at the concurrent solution of the Patient-to-Room Assignment (PRA)

and the Nurse-to-Patient Assignment (NPA) problems. The resultant integrated problem, called IPRNPA, consists of assigning patients to rooms on each day of their stay and nurses to patients during each shift of their stay, spanning a given planning period of several weeks. In this problem, both the patient's admission and discharge dates and the nurse's working shifts are assumed as fixed and known at the beginning of the planning period. The objective function of the IPRNPA comprises objectives coming from both single problems as well as objectives that are expressed and evaluated at the level of the integrated problem. In particular, they regard transfers of patients between rooms, room heterogeneity concerning the age and gender of roommates, and missing room equipment for the PRA problem; minimal number of distinct nurses that take care of a patient during his/her stay (continuity of care), required nurse skill level, workload distribution for the NPA problem; and minimal nurses per room and walking distance of nurses for the interaction of the two problems.

Brandt et al. [1] provide both a mixed integer programming model and an efficient heuristic, obtained by extending the heuristic designed for PRA alone by Schäfer et al. [7]. Furthermore, they provide a dataset, written in JSON, that comprises both real-world and artificial instances. Finally, they make available a solution checker, which allows other researchers to validate their solutions and compare the obtained scores.[1] The solution checker is a Python program that receives as parameters the instance and the solution and delivers all the costs and possible violations.

In this work, we propose an alternative solution technique based on local search for the IPRNPA problem. Specifically, we design two neighborhood operators that work on the variables of each atomic problem, and we combine them in a multi-neighborhood setting. The search is guided by Simulated Annealing, which draws moves at each iteration from the union of the two neighborhoods.

This is an ongoing work and the preliminary results show that we have been able to solve to feasibility all instances of all sizes, with different levels of quality depending on the running time.

2   Search Method

In this section, we describe our solution approach. For the sake of brevity, we leave out the problem definition and we refer to Brandt et al [1] for the precise formulation and the mathematical model.

The search space is defined by two distinct data structures. The first is an integer-valued matrix, of size patients per days, that stores the room in which the patient is hosted on that day; for the days before the admittance and after the discharge, the cell contains the dummy value -1. The second structure is an integer-valued matrix, of size patients per shifts (there are three shifts per day), that stores the nurse who takes care of the patient during that shift. Similarly to

---

[1] The instance generator and the solution checker are available online at https://github.com/TLKT0M/IPRNPA_instance_generator and https://github.com/TLKT0M/IPRNPA_solution_check, respectively.

the previous matrix, when the patient is not present, the cell contains the dummy value -1. The solver also stores many additional redundant data structures, such as room occupancy and nurses' workload, that are used to accelerate delta evaluations of moves.

The initial solution is built by assigning to each patient a random room with available capacity for the admission day, and then keeping it for the subsequent days of the stay as long as its availability holds. Whenever a transfer becomes necessary, a new room is drawn, repeating this process until the discharge day. Afterward, for each shift within the patient's stay, a random nurse is selected, among those working on that shift.

We consider two neighborhood relations that operate on the two distinct data structures:

- ChangeRoom: Change the room assigned to a patient on a specific day, selecting among those that have available residual capacity.
- ChangeNurse: Change the nurse assigned to a patient in a specific shift, selecting among those working in that shift.

We consider the union of these two neighborhoods, and the selection of a random move is done in two stages: first, we select the atomic neighborhood (ChangeRoom or ChangeNurse) and then the specific move inside the neighborhood. The first selection is based on a parameter ρ, in such a way that at each step the neighborhoods ChangeRoom and ChangeNurse are drawn with probabilities ρ and $1 - \rho$, respectively. Within the selected atomic neighborhood, the specific move is drawn uniformly.

The construction of the initial solution and the preconditions of the moves guarantee that the hard constraints (capacity of rooms and availability of nurses) are always satisfied. Therefore, the cost function that guides the local search coincides with the objective function of the problem itself. This includes twelve components that are thoroughly described by Brandt et al. [1] and that we do not discuss here for the sake of brevity.

The metaheuristic that guides the search is the classic Simulated Annealing [4], with Metropolis acceptance criterion and geometric cooling scheme. As a termination criterion, we use the reaching of the minimum temperature. To have approximately the same running time for all parameter settings, we fix the total number of iterations and compute the number of iterations at each temperature based on the initial and final temperature and the cooling rate.

To speed up the early stage of the search, we include a cut-off mechanism that moves to the new (lower) temperature when either a given number of iterations has been performed or a fixed number of moves have been accepted. Iterations saved by the cut-off mechanisms are distributed among all subsequent temperatures. For a comprehensive introduction to Simulated Annealing see the work by Franzin and Stützle [2].

3 Preliminary Results

We tested our solution method on the dataset of artificial instances, because, as stated by Brandt et al. [1], they are more challenging than the real-world ones.

Our search method has several parameters, which are the typical ones of Simulated Annealing plus ρ, which represents the rate of the ChangeRoom neighborhood, with respect to the ChangeNurse one. We tuned our parameters using `irace` [5] and the winning configuration is: start temperature = 680.116, final temperature = 0.855, cooling rate = 0.991, cut-off threshold = 0.889, $\rho = 0.115$.

The dataset of artificial instances consists of 12 different groups, each one composed of 10 instances. The distinctive features of each group are reported in Table 1, in terms of number of weeks, rooms, beds, and nurses. In addition, Brandt et al [1] designed three different room configurations: variation 1 corresponds to only double rooms; variation 2 corresponds to only triple rooms, and variation 3 corresponds to a room balance of 23% single rooms, 38% double rooms, 23% triple rooms and 16% quadruple rooms (e.g., for 13 rooms, 3 single ones, 5 double ones, 3 triple ones, and 2 quadruple ones).

Table 1. Features of the groups of instances.

Group	# weeks	# rooms	variation	# beds	# nurses
G1	2	10	2	30	21
G2	2	13	3	30	21
G3	2	15	1	30	21
G4	2	20	2	60	31
G5	2	26	3	60	31
G6	2	30	1	60	31
G7	4	10	2	30	21
G8	4	13	3	30	21
G9	4	15	1	30	21
G10	4	20	2	60	31
G11	4	26	3	60	31
G12	4	30	1	60	31

Table 2 shows our results aggregated by instance group for 30 repetitions on each instance with the winning configuration with 40M iterations. It reports the average total cost, the percentage value of each separate objective with respect to it and the average running time in seconds.

Given that no detailed numerical results but only average loss w.r.t. the optimal solution are provided by Brandt et al. [1], as their paper is still a preliminary version, we could not compare our results with theirs. Nonetheless, our solutions have been validated using the solution checker made available by Brandt et al. [1], and the scores can be used for future comparisons.

The results demonstrate that the objective with the largest impact on the total cost (about 25%) is the one related to the number of different nurses that are assigned to the same room. The other objectives with a high contribution

Table 2. Impact (%) of objectives, average total cost and running time.

Inst. group	transfers	age diff.	gender mix	missing equip.	cont. of care	missing skill	work load	nurses × room	walking distance	total cost	time [s]
G1	1.03	5.74	1.04	2.32	20.82	7.60	26.37	24.67	10.41	3922.44	91.27
G2	1.65	5.07	1.18	3.62	19.53	6.97	24.78	24.39	12.81	4174.19	88.00
G3	1.98	3.94	0.70	3.80	18.43	6.41	21.81	27.01	15.92	4450.56	90.63
G4	3.10	5.40	1.26	2.65	20.08	8.83	17.02	25.05	16.61	8845.65	97.81
G5	4.16	4.69	1.38	3.32	18.01	7.99	17.20	23.07	20.18	9951.02	96.04
G6	5.72	3.89	0.90	3.25	17.10	7.62	15.12	24.25	22.15	10469.18	99.18
G7	1.51	5.98	1.24	3.13	21.36	8.31	20.94	26.23	11.30	7571.14	92.97
G8	2.35	5.52	1.48	3.95	20.14	7.58	20.03	25.47	13.48	8061.36	92.92
G9	2.75	4.22	0.89	3.99	18.82	7.35	17.64	27.84	16.50	8686.72	94.70
G10	4.58	5.52	1.46	2.63	19.68	9.03	13.52	25.77	17.81	18062.50	103.54
G11	5.98	5.01	1.71	3.08	18.08	8.26	11.96	24.18	21.74	19812.79	102.40
G12	7.51	4.10	1.05	3.63	17.13	7.84	10.84	24.79	23.11	20870.22	103.75

are related to the continuity of care, the work balance, and the walking of the nurses. This reveals that the most critical part in terms of penalty regards the assignment of the nurses to patients rather than the assignment of patients to rooms.

We then investigated the performance of SA with different running times, with 4M, 40M, and 400M iterations, corresponding to approximately 10, 100, and 1000 s in our machine. The outcome is illustrated in Fig. 1, where we further gathered instance groups with similar behavior.

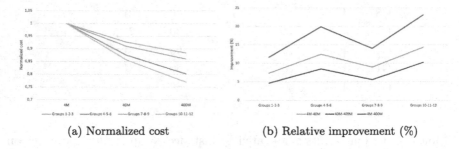

(a) Normalized cost   (b) Relative improvement (%)

Fig. 1. Comparative results for 4M, 40M, and 400M iterations.

From Fig. 1a, we can see that for all groups the scores continue to improve as the number of iterations increases, without any flattening effect. Figure 1b plots the percentage improvement between 4M and 40M, 40M and 400M, and 4M and 400M. As expected, the improvement between 4M and 40M (green line) is greater than for 40M and 400M (purple line), with average values of about 10% and 7% respectively. We also noticed that the improvement is more significant for instances with 60 beds (groups 4–5–6, 10–11–12), probably because the difficulty level of an instance depends more on the number of beds than on other features.

4   Conclusions

We designed a metaheuristic approach to solve a novel integrated healthcare problem and we tested it on the publicly available dataset. Even though at present a fair comparison with previous results is not viable, our solutions have been validated with the official solution checker of the problem, and can be used as a baseline for future comparisons.

For the future, we plan to try to improve our results by designing additional neighborhoods and adding them to our search method. For example, we are currently working on two operators that swap assignments, rather than changing one single assignment.

In addition, we plan to work on the design and solution of a new formulation of the problem that could also include the management of the operating rooms, which are generally the most critical resource in hospitals.

Acknowledgements. We thank Fabian Schäfer and Tom Klein for kindly answering all our questions about their work.

This work has been funded by the project "Models and algorithms for the optimization of integrated healthcare management" (no. 2020LNEZYC) supported by the Italian Ministry of University and Research (MUR) under the PRIN-2020 program.

References

1. Brandt, T., et al.: Integrated patient-to-room and nurse-to-patient assignment in hospital wards (2023). arXiv preprint 2309.10739
2. Franzin, A., Stützle, T.: Revisiting simulated annealing: a component-based analysis. Comput. Oper. Res. **104**, 191–206 (2019)
3. Hulshof, P.J., Kortbeek, N., Boucherie, R.J., Hans, E.W., Bakker, P.J.: Taxonomic classification of planning decisions in health care: a structured review of the state of the art in OR/MS. Health Syst. **1**, 129–175 (2012)
4. Kirkpatrick, S., Gelatt, D., Vecchi, M.: Optimization by simulated annealing. Science **220**, 671–680 (1983)
5. López-Ibáñez, M., Dubois-Lacoste, J., Cáceres, L.P., Birattari, M., Stützle, T.: The irace package: Iterated racing for automatic algorithm configuration. Oper. Res. Perspect. **3**, 43–58 (2016)
6. Rachuba, S., Reuter-Oppermann, M., Thielen, C.: Integrated planning in hospitals: a review (2023). arXiv preprint 2307.05258
7. Schäfer, F., Walther, M., Hübner, A., Kuhn, H.: Operational patient-bed assignment problem in large hospital settings including overflow and uncertainty management. Flex. Serv. Manuf. J. **31**, 1012–1041 (2019)
8. Vanberkel, P., Boucherie, R., Hans, E., Hurink, J., Litvak, N.: A survey of health care models that encompass multiple departments. Int. J. Health Manag. Inf. (IJHMI) **1**(1), 37–69 (2010)

Enhancing Real-World Applicability in Home Healthcare: A Metaheuristic Approach for Advanced Routing and Scheduling

Sara Ceschia[1] , Luca Di Gaspero[1] , Simona Mancini[2] ,
Vittorio Maniezzo[3] , Roberto Montemanni[4] , Roberto Maria Rosati[1] ,
and Andrea Schaerf[1(✉)]

[1] Polytechnic Department of Engineering and Architecture,
University of Udine, Via delle Scienze 206, I-33100 Udine, Italy
{sara.ceschia,luca.digaspero,robertomaria.rosati,andrea.schaerf}@uniud.it
[2] Department of Engineering, University of Palermo,
Viale delle Scienze, I-90128 Palermo, Italy
simona.mancini@unipa.it
[3] Department of Computer Science, University of Bologna,
Via dell'Università 50, I-47521 Cesena, Italy
vittorio.maniezzo@unibo.it
[4] Department of Sciences and Methods for Engineering,
University of Modena and Reggio Emilia,
Via Amendola, 2, I-42122 Reggio Emilia, Italy
roberto.montemanni@unimore.it

Abstract. We consider the home healthcare scheduling and routing problem, and we extend the classic formulation introduced by Mankowska et al, by adding several real-world features. For this novel problem, we created a new realistic dataset, and we developed a metaheuristic approach based on a combination of neighborhoods guided by a Simulated Annealing procedure. Our solver, properly engineered and tuned, is able to solve all instances in a short time. Our experimental results highlight the relative importance of the various (original and new) cost components.

Keywords: Homecare · Routing with time windows · Route synchronization

1 Introduction

Home healthcare (or simply homecare) refers to providing healthcare services and assistance to individuals in their homes rather than in a hospital or other healthcare facilities. Homecare offers a range of benefits, including personalized care, cost-effectiveness, comfort, and the promotion of independence. Depending on the individual's health needs and preferences, it can be a valuable alternative or complement to institutional care.

M. Sevaux et al. (Eds.): MIC 2024, LNCS 14753, pp. 164–170, 2024.
https://doi.org/10.1007/978-3-031-62912-9_16

Providing homecare services is an optimization problem involving scheduling and routing issues. Many problem formulations have been proposed in the literature, depending on the different settings and horizons. For an overview of the available formulations and solution techniques, we refer to the following surveys [3,4,9].

We consider here the well-known formulation proposed by Mankowska et al. [8], which comes along with a large and challenging dataset that has attracted the attention of many researchers, which dealt with it mainly using metaheuristics (see [1,2,5–7]). Although this formulation is interesting and challenging from a computational point of view, it lacks some specific features that would make the problem more attractive in reality. For this reason, we introduce hereby an extended formulation that comprises additional real-world features, i.e., multiple departure points for caregivers, incompatibilities between patients and caregivers, and working shifts for caregivers. In addition, the objective function penalizes caregiver waiting times and overtime and unbalanced workload distribution. For this novel formulation, we propose a new dataset and a search method based on Simulated Annealing, obtained by extending our previous work on the original problem [1,2]. For the formulation obtained, we created a new artificial dataset by using real road distances and sampling the relevant locations, considering the area's actual population distribution. Our search method, properly tuned, has been tested on the new dataset, highlighting which are the most impactful components of the objective function.

2 Problem Formulation

We introduce the formulation in two steps. First, we recall the basic one by Mankowska et al. [8], and then we illustrate the extensions we introduced.

2.1 Basic Formulation

The most relevant elements of the Home Health Care Routing and Scheduling Problem (HHCRSP) are:

Patients: Patients are categorized as *single service* (requiring one service) or *double service* (requiring two services, either *simultaneous* or *sequential*). Sequential double service patients require a specific minimum and maximum time gap between services. Additionally, each patient has a designated time window for starting the (first) service.

Services: Each service duration, in minutes, varies by patient.

Caregivers: Each caregiver is qualified for a specific subset of services. They begin and end their workday at the central office.

The planning horizon consists of a single day. Distances represent the travel time (in minutes) from one location to another (either a patient's home or the central office). The hard constraints of the problem are:

– Each patient must be visited during the planning horizon (either by one or two caregivers).
– A service cannot be provided by a caregiver who is not qualified for it.
– For each double service patient, the minimum and maximum time separations between the first and the second service have to be respected. In the case of simultaneous services, the separation is strictly equal to 0.
– Each double service patient needs two separate caregivers.
– A service cannot start before the beginning of the patient time window. In case of early arrival, the caregiver has to wait until the time window starts.

Conversely, it is permissible for a patient to be served late (after the end of the time window), but this tardiness contributes to the objective function.

The objective function to be minimized includes three components: *i*) the total travel time, *ii*) the total tardiness encompassing all services, and *iii*) the highest tardiness. In cases of double service patients, each service contributes separately to the total tardiness.

2.2   Extended Formulation

We now discuss the extensions that we introduced, along with their motivations from a practical point of view.

Multidepot: In some cases, it is rather unrealistic to assume that all caregivers move to the central office at the beginning of their shift. For this reason, we assume that a caregiver departs either from the central office or from their home and returns to the same place at the end of their shift. This decision is fixed in the input data and cannot be changed based on the route. In this situation, the distance matrix is extended to include the locations of all caregivers who depart from home.

Compatibility: It may happen that, for various reasons, some caregivers are not acceptable to some patients. To deal with this limitation, some patient/caregiver pairs are fixed as *incompatible* so that the given caregiver cannot serve that patient.

Waiting times: When a caregiver arrives early at a patient's home, she/he waits until the time window of the patient starts. This situation is rather inconvenient for the caregiver, but since it receives no penalty in the basic formulation, it actually occurs quite often in the solutions. For this reason, we introduce a cost component for the total *waiting time* spent by all caregivers in this specific situation.

Work shift and overtime: Caregivers are assumed to be available within the full horizon. This is rather unrealistic, as they normally work in specific shifts, which can span over the entire day (full-time), or be set either on the morning or the afternoon (part-time). Therefore, we introduce the working shift of the caregivers, so that each caregiver leaves their location at the beginning of the shift (or later) and should return by the end of the shift. If the return time is after the end of the shift, this accounts for *overtime*, which should be minimized and contribute to the objective function.

Work balance and fairness: In the basic formulation, there is no notion of work balance, causing situations in which one caregiver visits very few patients (even zero in some cases), while others visit up to ten patients. To fix this unfair situation, we introduce a measure of balance in the objective function. To this aim, we introduce the *idle time* of a caregiver, which is defined as the length of the caregiver's shift minus their working time, which in turn is measured as the service time plus the traveling time. In other words, the idle time is the waiting time defined above, plus the time before going out to the first patient, plus the time between the return to the starting point and the end of the shift (the latter only if bigger than zero). We count as fairness cost the *highest idle time* among all caregivers.

According to these extensions, we move from the three-component objective function of the original formulation to a six-component one for the new formulation, by adding waiting times, overtime and highest idle time (fairness).

These objectives might have different impacts on the quality of the solution, determining whether we give more importance to the point of view of patients or the one of caregivers and the company. In the original formulations, in order to keep the objective function simple, all components were given identical weights, thus assuming that one minute of traveling time costs as much as one minute of tardiness. We maintain this approach, applying the same weight to additional components, and defer a detailed cost analysis to future work.

3   Solution Technique

For the solution of this problem, we extend the multi-neighborhood Simulated Annealing approach proposed for the original formulation in our previous work [1,2]. This approach works on an indirect search space composed of the permutations of the patients and the assignments of the caregivers to the patients. The actual schedule is obtained by a forward greedy procedure that processes the patients one at a time according to the permutation and adds the patient at the end of the route(s) of their caregiver(s) at the earliest time.

The neighborhood relation is the combination of three atomic neighborhoods:

MovePatient: Reposition one patient in the global ordering and assign new caregiver(s) to the patient.

SwapPatients: Swap both the positions of two patients in the global ordering and the caregiver(s) assigned to them. A swap is possible only between patients with the same number of services and with current caregivers with the required abilities for the other patient.

InRouteSwap: Swap the positions of two patients within the route of a given caregiver. If one or both patients are double-service ones, the route of the *side* caregiver(s) serving the patient(s) are modified accordingly.

In order to draw a random move, first we perform a *biased* random selection to establish which of the three atomic neighborhoods should be sampled, and then a *uniform* selection within the chosen neighborhood. For the first selection, we use two parameters called σ_{SP} and σ_{IRS}, so moves of the three types are drawn with probability $1 - \sigma_{SP} - \sigma_{IRS}$, σ_{SP} and σ_{IRS}, respectively.

As the metaheuristic that guides the search, we make use of the classic Simulated Annealing (SA). The SA procedure starts from a random initial solution and then, at each iteration, draws a random move. This is always accepted if it is improving or sideways, whereas worsening moves are accepted based on the time-decreasing exponential distribution (known as *Metropolis Acceptance*).

SA starts with an initial high temperature T_0, which is decreased after a fixed number of samples are drawn according to the geometric cooling scheme with rate α. The search is stopped when the final temperature T_f is reached. In order to speed up the early stages of the search, we add the customary *cut-off* mechanism, such that the temperature also decreases if a fraction ρ of the moves has been accepted. The iterations saved by the cut-off are redistributed uniformly to all the remaining temperatures.

4   Experimental Results

We adapted the generator developed for the basic problem [1] and we created 500 training instances for the tuning phase plus 10 validation ones. They are available at https://github.com/iolab-uniud/hhcrsp, along with their best solutions. The tuning procedure on the training instances has been done using RACE in two stages: first the parameters of SA and then the two rates σ_*. The winning configuration turned out to be: $T_0 = 28.77$, $T_f = 0.94$, $\alpha = 0.987$, $\rho = 0.138$, $\sigma_{SP} = 0.2$ and $\sigma_{IRS} = 0.08$. Table 1 reports average and minimum results of 30 runs on the validation instances with the above configuration and with 100M total iterations. The table also reports the average percentage cost for each component: total distance (TD), total tardiness (TT), highest tardiness (HT), total waiting time (TWT), total overtime (TOT), and highest idle time (HIT).

Table 1. Results on the validation instances

Inst	Patients	Caregivers	avg	min	time(s)	TD	TT	HT	TWT	TOT	HIT
0	220	42	29829.8	27763	532.3	34.31	48.33	1.61	2.14	12.74	0.87
1	68	13	13776.3	13550	181.3	13.19	63.70	4.23	4.39	11.58	2.90
2	261	50	35113.6	32535	652.5	33.34	41.91	1.26	1.76	21.42	0.30
3	304	54	11119.3	10547	723.1	62.38	16.55	1.37	2.85	13.81	3.03
4	493	96	23628.1	21999	1349.6	51.95	36.34	1.39	2.85	6.78	0.70
5	233	36	25278.6	24265	518.6	22.80	59.69	1.60	1.39	14.13	0.40
6	490	87	45648.6	41273	1277.3	34.86	50.41	0.98	1.47	12.14	0.14
7	217	43	6196.4	6022	500.4	84.20	1.18	0.23	3.33	5.22	5.84
8	136	22	26517.3	25475	326.6	16.02	61.12	3.42	4.21	13.73	1.50
9	159	30	14816.3	13827	389.0	34.35	35.45	2.79	2.29	22.82	2.30

The results show that there is big variability among different instances, in terms of total cost and distribution of the cost among the various components. In particular, in some cases, the traveling cost is dominant (instances 3 and 7); in others, the tardiness component is dominant (instances 1, 5, 6, and 8). Unsurprisingly, when the tardiness is high, also overtime is relevant because some services are postponed after both the time window of the patient and the working shift of the caregiver. This reveals the presence of either a significant understaffing or a bad matching between patient needs and caregiver skills.

5 Conclusions and Future Work

We have extended a classic formulation of the homecare routing and scheduling problem, creating a novel, more realistic problem, for which we created a new dataset, properly split into training and validation instances, and a metaheuristic method based on our previous work on the original formulation [1,2].

This is a preliminary work and for the future we plan to further refine the general formulation, the cost components, and their weights, in order to capture real-world situations. In parallel, we plan to improve our metaheuristic and to hybridize it with exact methods, bringing forth a matheuristic approach.

Acknowledgements. This research has been partly funded by the European Union - *NextGenerationEU*, under the project "Modeling and solving a real-world home health-care routing and scheduling problem".

References

1. Ceschia, S., Di Gaspero, L., Rosati, R.M., Schaerf, A.: Multi-neighborhood simulated annealing for the home healthcare routing and scheduling problem (2023). https://doi.org/10.21203/rs.3.rs-4086164/v1. preprint available at Research Square
2. Ceschia, S., Di Gaspero, L., Schaerf, A.: Simulated annealing for the home healthcare routing and scheduling problem. In: Dovier, A., Montanari, A., Orlandini, A. (eds.) AIxIA 2022. LNCS, vol. 13976, pp. 402–412. Springer, Cham (2022). https://doi.org/10.1007/978-3-031-27181-6_28
3. Cissé, M., Yalçındağ, S., Kergosien, Y., Şahin, E., Lenté, C., Matta, A.: OR problems related to home health care: a review of relevant routing and scheduling problems. Oper. Res. Health Care **13**, 1–22 (2017)
4. Fikar, C., Hirsch, P.: Home health care routing and scheduling: a review. Comput. Oper. Res. **77**, 86–95 (2017)
5. Kummer, A.: A study on the home care routing and scheduling problem. Ph.D. thesis, Universidade Federal do Rio Grande do Sul (2021)
6. Kummer, A., de Araújo, O., Buriol, L., Resende, M.: A biased random-key genetic algorithm for the home health care problem. Int. Trans. Oper. Res. **31**(3), 1859–1889 (2024)

7. Lasfargeas, S., Gagné, C., Sioud, A.: Solving the home health care problem with temporal precedence and synchronization. In: Talbi, E.-G., Nakib, A. (eds.) Bioinspired Heuristics for Optimization. SCI, vol. 774, pp. 251–267. Springer, Cham (2019). https://doi.org/10.1007/978-3-319-95104-1_16
8. Mankowska, D., Meisel, F., Bierwirth, C.: The home health care routing and scheduling problem with interdependent services. Health Care Manag. Sci. **17**(1), 15–30 (2014)
9. Soares, R., Marques, A., Amorim, P., Parragh, S.N.: Synchronisation in vehicle routing: classification schema, modelling framework and literature review. Eur. J. Oper. Res. **313**(3), 817–840 (2024)

Solving the Two-Stage Robust Elective Patient Surgery Planning Under Uncertainties with Intensive Care Unit Beds Availability

Salma Makboul[✉]

LIST3N, Université de Technologie de Troyes, Troyes, France
salma.makboul@utt.fr

Abstract. This paper explores the intricate challenges of the elective surgery scheduling problem, considering uncertainties in both surgery duration and length of stay in the intensive care unit. We present a novel two-stage robust approach employing the column-and-constraint generation algorithm to address the master surgical schedule and surgery case assignment problems under these uncertainties. Our approach differs from traditional methods by incorporating a specific modeling of uncertainty using independent uncertainty sets and accounts for surgical teams and resource availability. Comparative analysis with the cutting-plane method demonstrates the effectiveness of our approach, offering valuable insights for the enhanced management of uncertainties in elective surgery planning.

Keywords: Downstream Resource Constraint · Operating Rooms Planning · Robust Optimization · Column-and-Constraint Generation

1 Introduction and Related Works

Surgical suites, comprising Operating Rooms (ORs), Post-Anesthesia Care Units (PACU), and Intensive Care Units (ICU), play a crucial role in determining the costs and revenues of a hospital facility [4,15]. The scheduling of elective patients (those whose surgeries are anticipated in advance) poses a challenge due to various factors, including resource availability, limited capacity, and the stochastic nature of surgeries and patients' length of stay (LOS) in the ICU. Most literature papers tend to concentrate on upstream resources, often overlooking the capacity of post-surgery units. Surgery duration and LOS in the ICU, if post-surgery units are considered, are typically treated as deterministic or, when considered stochastic, follow a lognormal distribution. Our emphasis is on the limited literature that addresses approaches utilizing free probability distributions [3]. To tackle the uncertainty, there are three primary frameworks: Stochastic Programming (SP), Robust Optimization (RO) [12], and Distributionally Robust Optimization (DRO) [13]. Recently, the two-stage modeling has gained particular attention due to its efficiency, where the first-stage decision must be finalized before the uncertainty is revealed. In contrast, the second stage, often referred

M. Sevaux et al. (Eds.): MIC 2024, LNCS 14753, pp. 171–177, 2024.
https://doi.org/10.1007/978-3-031-62912-9_17

to as recourse decisions, is decided after the first stage. Unlike a static approach, the two-stage framework allows adjustments based on the information received about uncertain data. Some papers in the literature have addressed the advance scheduling using SP such as [6,7,17]. While other authors used static RO, as referenced in [1,9,10]. [11] introduced the first and only two-stage RO model for the Surgical Case Assignment Problem (SCAP), specifically focusing on uncertainty in surgery duration and LOS in the ICU. The proposed approach enables Operating Theater (OT) managers to adjust risk levels. They employed an adapted Column-and-Constraints Generation (C&CG) algorithm to obtain exact solutions. Due to the uncertainty set defined for LOS, which depends on the first-stage decisions, the Cutting-Plane (CP) [8] and the standard C&CG algorithm cannot be applied. Therefore, they have used an adapted C&CG to solve the problem. Regarding DRO for SCAP, [14] proposed a Distributionally Robust Elective Surgery Scheduling (DRESS) model to optimize elective surgery assignments under uncertainties. They minimized costs in worst-case scenarios, addressing issues like surgery delays, overtime, and ICU capacity limitations. Numerical experiments demonstrated the effectiveness of employing DRO over traditional stochastic programming, offering insights into managing uncertainties with ambiguous probability distributions. The C&CG algorithm was introduced by [16], inspired by Benders' decomposition. It consists of identifying all the scenarios and formulating the Master Problem (MP) by adding deterministic constraints for every scenario, instead of using dual variables for the recourse as proposed by [8]. C&CG has proven its efficiency compared to CP in solving robust two-stage problems [16]. In this paper, we propose a two-stage robust approach to address the elective OR planning under uncertainty associated with surgery duration and LOS. We use a modeling of uncertainty that generates optimal valid constraints. Subsequently, we tackle the problem using the C&CG algorithm [16] and compare the results obtained with those from the CP algorithm, using the symmetry breaking inequalities proposed by [14].

2   Solving the Two-Stage Robust Elective Surgery Planning

To handle uncertainties, the OT manager can flexibly decide the number of surgeries with deviations in duration and LOS. This approach, employing polyhedral uncertainty sets [2] and constrained by robustness budgets Γ_d and Γ_l. The uncertain surgery duration for each patient i of $s \in \mathcal{S}$ (set of surgical specialties) belongs to the range $[\bar{d}_{is}, \bar{d}_{is} + \hat{d}_{is}]$ ($\bar{d}_{is}$ is the nominal duration and $\hat{d}_{is}$ is the maximum deviation), and the uncertain LOS to the range $[\overline{l_{is}^{ICU}}, \overline{l_{is}^{ICU}} + \widehat{l_{is}^{ICU}}]$ ($\overline{l_{is}^{ICU}}$ is the nominal LOS and $\widehat{l_{is}^{ICU}}$ is the maximum deviation), such as λ_{is} and η_{is} are the normalized deviations for surgery duration and LOS, respectively. The uncertainty sets are explicitly defined as follows:

$$\Xi_{rj}^d = \left\{ d_{is} \in \mathbb{R}^n \mid \quad d_{is} = \bar{d}_{is} + \hat{d}_{is}\lambda_{is}, \sum_{s \in \mathcal{S}} \sum_{i \in \mathcal{I}_s} \lambda_{is} \leq \Gamma_d, 0 \leq \lambda_{is} \leq 1 \right\} \quad (1)$$

$$\Xi_j^{ICU} = \left\{ l_{is}^{ICU} \in \mathbb{R}^n \mid \; l_{is}^{ICU} = \overline{l_{is}^{ICU}} + \widehat{l_{is}^{ICU}} \eta_{is}, \sum_{s \in \mathcal{S}} \sum_{i \in \mathcal{I}_s} \eta_{is} \leq \Gamma_l, 0 \leq \eta_{is} \leq 1 \right\}$$
$$(2)$$

Algorithm 1: C&CG algorithm for the robust elective planning

Initialization:
$LB = -\infty$, $UB = +\infty$, $K = 0$, $O = \varnothing$
Master: Solve the Master Surgical Schedule (MSS) (Master Problem)

min $\displaystyle\sum_{s \in \mathcal{S}} \sum_{i \in \mathcal{I}_s} \sum_{r \in \mathcal{R} \cup \{r'\}} \sum_{j \in \mathcal{J}} \chi_{isrj} \phi_{isr} + \eta$

s.t. $\eta \geq \displaystyle\sum_{r \in \mathcal{R}} \sum_{j \in \mathcal{J}} c_{rj} o_{rj}^k + \sum_{j \in \mathcal{J}} p_j z_j^k \quad \forall k \in O$; /* c: overtime cost, p: cost for denied ICU

bed */

$\displaystyle\sum_{r \in \mathcal{R} \cup \{r'\}} \sum_{j \in \mathcal{J}} \chi_{isrj} = 1 \quad \forall s \in \mathcal{S} \quad \forall i \in \mathcal{I}_s$

Add specialty-to-OR restrictions constraints
Add Limits on specialty parallelism constraints
Add OR sessions-per-specialty restrictions constraints

$\displaystyle\sum_{s \in \mathcal{S}} \sum_{i \in \mathcal{I}_s} d_{is}^k \chi_{isrj} \leq O^{\max} + o_{rj}^k \quad \forall r \in \mathcal{R} \quad \forall j \in \mathcal{J} \quad \forall k \leq K$;   /* OR capacity */

$\displaystyle\sum_{s \in \mathcal{S}} \sum_{i \in \mathcal{I}_s} \sum_{r \in \mathcal{R}} \sum_{\substack{j' \in \mathcal{J} \\ j' > j - l_{is}^{ICU} k}}^{j} r_{is} \chi_{isrj'} \leq \nu_j + z_j^k \quad \forall j \in \mathcal{J} \quad \forall k \leq K$;   /* ICU capacity */

$\chi_{isrj} \in \{0,1\} \quad \forall s \in \mathcal{S} \quad \forall i \in \mathcal{I}_s \quad \forall r \in \mathcal{R} \quad \forall j \in \mathcal{J} \, \forall k \leq K$;   /* +other MSS variables */
Obtain the optimal solution $(\chi_{K+1}^*, \eta_{K+1}^*, o^{1*}, ..., o^{K*}, z^{1*}, ..., z^{K*})$
Set $LB = \phi_{isr} \chi_{K+1}^* + \eta_{K+1}^*$
Recourse:
Solve the sub-problems that tackle uncertainty and get objective values $\mathcal{O}_{K+1}^*$ and $\mathcal{D}_{K+1}^*$
Update $UB = min\{UB, \phi_{isr} \chi_{K+1}^* + \mathcal{O}_{K+1}^* + \mathcal{D}_{K+1}^*\}$
if $UB - LB \leq \epsilon$ **then**
| The optimal solution is found
else
| **Add-Cut:**
| Add variables o_{rj}^{K+1} and z_j^{K+1} and the following constraints to the MP
| $\eta \geq \displaystyle\sum_{r \in \mathcal{R}} \sum_{j \in \mathcal{J}} c_{rj} o_{rj}^{K+1} + \sum_{j \in \mathcal{J}} p_j z_j^{K+1}$
| $\displaystyle\sum_{s \in \mathcal{S}} \sum_{i \in \mathcal{I}_s} d_{is}^{K+1} \chi_{isrj} \leq O^{\max} + o_{rj}^{K+1} \quad \forall r \in \mathcal{R} \quad \forall j \in \mathcal{J}$
| $\displaystyle\sum_{s \in \mathcal{S}} \sum_{i \in \mathcal{I}_s} \sum_{r \in \mathcal{R}} \sum_{\substack{j' \in \mathcal{J} \\ j' > j - l_{is}^{ICU \, K+1}}}^{j} r_{is} \chi_{isrj'} \leq \nu_j + z_j^{K+1} \quad \forall j \in \mathcal{J}$
end
where d_{is}^{K+1} and $l_{is}^{ICU \, K+1}$ are the optimal scenarios solving the $\mathcal{O}_{K+1}^*$ and $\mathcal{D}_{K+1}^*$
Update $K \leftarrow K+1, O \leftarrow O \cup \{K+1\}$ and go to **Master.**

Algorithm 1 provides the pseudo-code of C&CG for solving the two-stage robust elective surgery planning under uncertainty. χ_{isrj} is a decision variable set to 1 if surgery $i \in \mathcal{I}_s$ is assigned to day $j \in \mathcal{J}$ (set of days) in room $r \in \mathcal{R}$ (set of ORs) and 0 otherwise, r' is a dummy OR for postponed patients. ϕ_{isr} is the assignment cost of patient i to OR r. $O^{\max}$ is the capacity of the OR session, and ν_j represents the number of available ICU beds on day j. r_{is} is 1 if patient $i \in \mathcal{I}_s$ requires an ICU bed and 0 otherwise. o_{rj} is a decision variable capturing the overtime in the OR session on day $j \in \mathcal{J}$ and OR $r \in \mathcal{R}$, and z_j is a decision variable capturing the extra beds required in the ICU on day j. In each iteration, the algorithm solves the MP, minimizing the cost assignment of patients to the

OR while accounting for penalties related to uncertain parameters. The first-stage decisions are obtained and used to derive the worst-case realization of uncertain parameters through recourse formulations. This information is then fed back into the MP by introducing new variables and constraints, leading to an updated solution. The process continues until convergence, where the gap between upper and lower bounds is within a specified value.

We solve the recourse problem $\mathcal{O}(\Gamma_d)$ to capture the worst-case cost under surgery duration uncertainty. The model is a bilinear bi-level optimization problem. The outer problem maximizes over the uncertainty set (1) to find the worst-case overtime scenario for surgery duration, while the inner level minimizes overtime costs based on actual surgery duration. LOS recourse problem $\mathcal{D}(\Gamma_l)$ is resolved with the same strategy as $\mathcal{O}(\Gamma_d)$.

$$\max_{\sum_{s\in\mathcal{S}}\sum_{i\in\mathcal{I}_s}\lambda_{is}\leq\Gamma_d, 0\leq\lambda_{is}\leq 1} \quad \min \quad \sum_{r\in\mathcal{R}}\sum_{j\in\mathcal{J}}c_{rj}o_{rj}^k \tag{3}$$

$$\sum_{s\in\mathcal{S}}\sum_{i\in\mathcal{I}_s}(\bar{d}_{is}+\widehat{d}_{is}\lambda_{is})\chi_{isrj}\leq O^{\max}+o_{rj} \quad \forall r\in\mathcal{R} \quad \forall j\in\mathcal{J} \tag{4}$$

$$o_{rj}\geq 0 \quad \forall r\in\mathcal{R} \quad \forall j\in\mathcal{J} \tag{5}$$

We reformulate $\mathcal{O}(\Gamma_d)$ as a MILP. Using strong duality, the linear inner-problem (since χ is known in the second stage) can be written as a maximization problem. Let u be the dual variable associated to constraint (4). (3–5) can be written as a maximization problem as follows:

$$\max\sum_{r\in\mathcal{R}}\sum_{j\in\mathcal{J}}\left[\sum_{s\in\mathcal{S}}\sum_{i\in\mathcal{I}_s}(\bar{d}_{is}+\widehat{d}_{is}\lambda_{is})\chi_{isrj}-O^{max}\right]u_{rj} \tag{6}$$

$$\sum_{s\in\mathcal{S}}\sum_{i\in\mathcal{I}_s}\lambda_{is}\leq\Gamma_d \tag{7}$$

$$0\leq u_{rj}\leq c_{rj} \quad \forall r\in\mathcal{R} \quad \forall j\in\mathcal{J} \tag{8}$$

$$0\leq\lambda_{is}\leq 1 \quad \forall s\in\mathcal{S} \quad \forall i\in\mathcal{I}_s \tag{9}$$

If Γ_d is an integer, then λ^* is binary (proof in [5]). Consequently, using the big-M method, we define $T_{isrj}=\lambda_{is}u_{rj}$. The sub-problem $\mathcal{O}(\Gamma_d)$ can be reformulated as follows:

$$\max\sum_{r\in\mathcal{R}}\sum_{j\in\mathcal{J}}\sum_{s\in\mathcal{S}}\sum_{i\in\mathcal{I}_s}(\bar{d}_{is}\chi_{isrj}u_{rj}+\widehat{d}_{is}\chi_{isrj}T_{isrj}-O^{max}u_{rj}) \tag{10}$$

$$\sum_{s\in\mathcal{S}}\sum_{i\in\mathcal{I}_s}\lambda_{is}\leq\Gamma_d \tag{11}$$

$$u_{rj}\leq c_{rj} \quad \forall r\in\mathcal{R} \quad \forall j\in\mathcal{J} \tag{12}$$

$$T_{isrj}\leq M\lambda_{is} \quad \forall s\in\mathcal{S} \quad \forall i\in\mathcal{I}_s \quad \forall r\in\mathcal{R} \quad \forall j\in\mathcal{J} \tag{13}$$

$$T_{isrj} \leq u_{rj} \qquad \forall s \in \mathcal{S} \quad \forall i \in \mathcal{I}_s \quad \forall r \in \mathcal{R} \quad \forall j \in \mathcal{J} \tag{14}$$

$$T_{isrj} \geq u_{rj} - (1 - \lambda_{is})M \qquad \forall s \in \mathcal{S} \quad \forall i \in \mathcal{I}_s \quad \forall r \in \mathcal{R} \quad \forall j \in \mathcal{J} \tag{15}$$

$$u_{rj} \geq 0 \qquad \forall r \in \mathcal{R} \quad \forall j \in \mathcal{J} \tag{16}$$

$$\lambda_{is} \in \{0, 1\} \qquad \forall s \in \mathcal{S} \quad \forall i \in \mathcal{I}_s \tag{17}$$

$$T_{isrj} \geq 0 \qquad \forall s \in \mathcal{S} \quad \forall i \in \mathcal{I}_s \quad \forall r \in \mathcal{R} \quad \forall j \in \mathcal{J} \tag{18}$$

3   Computational Experience

In this section, we presents the results obtained using C&CG and CP with two instances with 99, 124 patients and 10 ORs.

Table 1. Comparison between C&CG and CP algorithms for solving the two-stage robust elective surgery planning problem

instance	Γ_d	Γ_l	C&CG Obj value	# of Iter	Run time (s)	Gap (%)	# Sess	# of Cancel	CP Obj value	# of Iter	Run time (s)	Gap (%)	# of Sess	# of Cancel
(P.99, R.10)	0	0	7866	1	370	0	22	3.99	7866	3	430	0	22	3.95
	2	2	8044	5	990	0	21	3.37	8044	8	1202	0	21	3.45
	4	4	8506	30	-	0.66	21	3.74	8654	583	-	22.33	20	3.56
	6	6	8567	101	-	1.41	22	1.74	8693	2171	-	16.30	20	3.43
	8	8	8612	75	-	0.70	22	1.56	8837	2069	-	13.36	21	2.96
	10	10	8781	87	-	0.07	22	1.36	9055	2311	-	10.53	22	2.14
(P.124, R.10)	0	0	10536	3	305	0	28	5.42	10536	4	450	0	28	5.38
	2	2	10598	105	-	0.09	30	4.21	10654	2072	-	0.29	26	4.89
	4	4	10694	74	-	0.94	28	3.60	10796	1354	-	1.13	26	4.02
	6	6	10767	71	-	1.48	29	2.80	10988	1420	-	2.18	30	3.42
	8	8	10843	60	-	1.53	30	2.63	11099	2722	-	2.71	26	3.52
	10	10	10945	65	-	1.70	30	2.31	11212	2345	-	2.13	26	2.96

We use real-world data from five surgical specialties, focusing on instances with 99 and 124 patients and 10 ORs. Tests are limited to 1 h. The C&CG and CP algorithms are implemented in the Julia programming language, using the CPLEX solver for solving the MILPs. The dataset is available at https:// github.com/SMAKBOUL/RMSS. Results, presented in Table 1, highlight algorithm performance metrics. (-) means that the run time reached one hour. Objective values increase with the Budget of Robustness (BOR), and both algorithms maintain consistently low optimal gaps across different Γ_d and Γ_l values. Cancellations are calculated following a Monte Carlo simulation, providing an idea of planning risk and how much additional resources may be required. The risk is lower as the BOR increases because of providing a more robust planning, and expect a continuous decrease until reaching 0 cancellations with a very high budgets in the worst case (in the extended results). The imposed Gap is ($\epsilon = 1.e - 5$). C&CG demonstrates efficiency with minimal objective values, number of iterations and a lower gap, emphasizing its effectiveness and ability to provide near-optimal solutions. Sensitivity to BOR parameter variations is evident, impacting objective values. These findings contribute insights into OR planning algorithmic performance and sensitivity to parameter tuning for OT managers.

4 Conclusion and Perspectives

Our paper introduces a robust two-stage approach using the C&CG algorithm for elective surgery scheduling under uncertainties in surgery duration and LOS. The method, emphasizing downstream resources, outperforms CP algorithm, providing valuable insights for enhanced risk management. The real-world experiments, validate the effectiveness of our C&CG algorithm, providing near-optimal solutions, with a maximum gap of (1.70%). Our work highlights the robustness of the approach across different levels of BOR and sensitivity to parameter variations. In future work, we plan to extend our analysis to more instances and budgets, evaluating various metrics such as overtime, OT utilization rate, and additional ICU using the C&CG algorithm. This expanded investigation aims to offer OT managers more comprehensive information about risks, further improving elective surgery operational planning.

Acknowledgments. The author acknowledges the financial support for the exploratory project "Research on robust optimization and its application FRORA" provided by the Département de l'Aube, Troyes Champagne Métropole, and Université de Technologie de Troyes under the grant [OPE-2024-0045].

References

1. Addis, B., Carello, G., Tànfani, E.: A robust optimization approach for the advanced scheduling problem with uncertain surgery duration in operating room planning - an extended analysis (2014)
2. Bertsimas, D., Sim, M.: The price of robustness. Oper. Res. **52**(1), 35–53 (2004)
3. Denton, B., Viapiano, J., Vogl, A.: Optimization of surgery sequencing and scheduling decisions under uncertainty. Health Care Manag. Sci. **10**(1), 13–24 (2007)
4. Denton, B., Miller, A., Balasubramanian, H., Huschka, T.: Optimal allocation of surgery blocks to operating rooms under uncertainty. Oper. Res. **58**(4-part-1), 802–816 (2010)
5. Horst, R., Tuy, H.: Special Problems of Concave Minimization. In: Global Optimization: Deterministic Approaches, pp. 447–515. Springer, Heidelberg (1996). https://doi.org/10.1007/978-3-662-03199-5_9
6. Jebali, A., Diabat, A.: A stochastic model for operating room planning under capacity constraints. Int. J. Prod. Res. **53**(24), 7252–7270 (2015)
7. Jebali, A., Diabat, A.: A chance-constrained operating room planning with elective and emergency cases under downstream capacity constraints. Comput. Ind. Eng. **114**, 329–344 (2017)
8. Kelley, J., James, E.: The cutting-plane method for solving convex programs. J. Soc. Ind. Appl. Math. **8**(4), 703–712 (1960)
9. Lalmazloumian, M., Baki, M., Ahmadi, M.: A robust multiobjective integrated master surgery schedule and surgical case assignment model at a publicly funded hospital. Comput. Ind. Eng. **163**, 107826 (2022)
10. Makboul, S., Kharraja, S., Abbassi, A., El Hilali Alaoui, A.: A two-stage robust optimization approach for the master surgical schedule problem under uncertainty considering downstream resources. Health Care Manag. Sci. **25**, 63–88 (2022)

11. Neyshabouri, S., Berg, B.: Two-stage robust optimization approach to elective surgery and downstream capacity planning. European J. Oper. Res. **260**(1), 21–40 (2017)
12. Poss, M.: Robust combinatorial optimization with variable cost uncertainty. European J. Oper. Res. **237**(3), 836–845 (2014)
13. Shehadeh, K.: Data-driven distributionally robust surgery planning in flexible operating rooms over a wasserstein ambiguity. Comput. Oper. Res. **146**, 105927 (2022)
14. Shehadeh, K., Padman, R.: A distributionally robust optimization approach for stochastic elective surgery scheduling with limited intensive care unit capacity. European J. Oper. Res. **290**(3), 901–913 (2020)
15. Shehadeh, K., Padman, R.: Stochastic optimization approaches for elective surgery scheduling with downstream capacity constraints: models, challenges, and opportunities. Comput. Oper. Res. **137**, 105523 (2022)
16. Zeng, B., Zhao, L.: Solving two-stage robust optimization problems using a column-and-constraint generation method. Oper. Res. Lett. **41**(5), 457–461 (2013)
17. Zhang, J., Dridi, M., El Moudni, A.: A two-level optimization model for elective surgery scheduling with downstream capacity constraints. European J. Oper. Res. **276**(2), 602–613 (2019)

Extracting White-Box Knowledge from Word Embedding: Modeling as an Optimization Problem

Julie Jacques[(⊠)][iD] and Alexander Bassett

Univ. Lille, CNRS, Centrale Lille, UMR 9189 CRIStAL, 59000 Lille, France
{julie.jacques,alexander.bassett.etu}@univ-lille.fr

Abstract. Explainability is crucial to building the confidence of the medical team to adopt natural language processing (NLP) techniques. In the majority of recent studies in medical informatics, Deep Learning performed better than other machine learning (ML) techniques for natural language processing (NLP) on medical documents. However, the generated models are black-box models difficult to explain. One of these models is word embedding which allows a representation of text and words as vectors, which makes them more exploitable by machines. This paper proposes a new method to add explainability to word embedding. We propose a modelization as an optimization problem. The first results on the text8 dataset and 5 target words show the local search can obtain explanations with an improvement of cosine similarity by 11% to 30%.

Keywords: Word Embedding · Optimization Problem · Local Search · Explainable Artificial Intelligence

1 Introduction

Deep neural networks have led to many advances in computer science in recent years. In the majority of recent studies in medical informatics, Deep Learning performed better than machine learning (ML) for natural language processing (NLP) on medical documents [8]. In recent years, different methods have been proposed to perform word embedding, including neural networks. Word embedding allows words and text documents to be represented more richly, by vectorizing them. This vector representation makes the texts more exploitable by machines. Several word embedding techniques have been proposed such as Word2Vec [5], BERT [2] which works in word sub-units, or GPT-3. Despite a growing interest in XAI (explainable AI) in medicine [1], models generated by a majority of ML methods - including word embedding - are black-box models that are difficult to explain. However, explainability is fundamental to building the confidence of the medical team to adopt NLP techniques [7].

Meta-heuristics have often been used to solve NP-hard problems, such as big data problems [3]. They allow high flexibility in the modeling of the representation of a solution and its evaluation, which is particularly adapted to

M. Sevaux et al. (Eds.): MIC 2024, LNCS 14753, pp. 178–183, 2024.
https://doi.org/10.1007/978-3-031-62912-9_18

generate explainable solutions. One strength of word embedding is the ability to express semantic analogies using simple arithmetic on vectors, such as $king - man + woman \sim queen$. This capacity has been verified on real data, such as medical data [4]. We propose to add explainability by exploring the vector representation using combinatorial optimization. This is preliminary work on non-medical data, before an extension on clinical documents.

The main contribution of this paper is to propose a novel method to extract white-box knowledge from word embedding. We approach the problem as an optimization problem. The paper is organized as follows: first, we present some background on word embedding and the associated vector representation. Secondly, we detail our proposed modeling and the associated components for a resolution with a local search. Finally, we present several first results obtained by our proposed method on *text8* dataset.

2   Background on Word Embedding

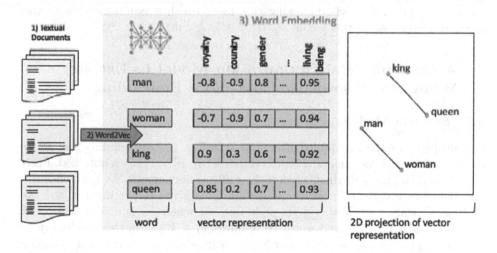

Fig. 1. Representation of words, their corresponding word vector, and projection in 2D

In 2013, Mikolov et al. introduced Word2Vec: a new method to learn high-quality word vectors for datasets composed of billions of words [5]. Each word is represented by a vector of a fixed dimension (from 50 to 600 dimensions, depending on their experiments). One of the aims of this representation is that two words with a similar meaning should be represented in a close way. It turns out that the resulting representation is even richer, as it allows algebraic operations to be performed on vectors. As an illustration, $\overrightarrow{King} - \overrightarrow{Man} + \overrightarrow{Woman}$ is the closest representation of $\overrightarrow{Queen}$. Moreover, this representation preserves semantic relationships between words: the relation between France to Paris is the same as Germany is to Berlin. Figure 1 shows an illustration of the process. Word2Vec is trained on a set of documents (steps 1) and 2)) which generates the word vectors (step 3). Step 3 illustrates the obtained text embedding: some words and their associated vector representation (here 4 dimensions are illustrated). Each dimension can be associated with meaning, for example *queen* and *king* have a strong positive association to the *royalty* dimension. Each dimension has been labeled for illustration purposes, but the word embedding discovery process does not give any clue to the interpretation of the dimensions. On the 2D projection on the right, we can observe that the relations between the words are preserved: it is the same between *man* and *woman* than between *king* and *queen*.

Şenel et al. proposed an approach called BiImp to train word embedding with interpretable dimensions [10]. Since word embedding can be costly to train, we want to propose a method that can be plugged into the pre-trained word embedding such as Word2Vec (or BERT [2] in the future). Zhang and Ogasawara proposed an interesting approach based on Grad-CAM that highlights the words the model pays attention to when making predictions [9]. However, it needs to be plugged into a supervised classification task, which is not the case with the method we propose in this paper.

3   A Combinatorial Optimization Model to Extract White-Box Knowledge from Word Embedding

3.1   Solution Modeling

Our proposed method is applied after step 3) in Fig. 1. After training Word2Vec on a set of textual documents, a set of vectors $V(\overrightarrow{v_1}, \overrightarrow{v_2}, \ldots, \overrightarrow{v_n})$ is obtained. Each $\overrightarrow{v_i} \in V$ is matching a distinct word of the textual documents. We formulate the knowledge extraction from word embedding for a given target word $\overrightarrow{t} \in V$ as finding $\overrightarrow{s}$ defined as $\overrightarrow{s} = \sum_{\overrightarrow{v_s} \in V_s, \overrightarrow{v_a} \in V_a} \overrightarrow{v_a} - \overrightarrow{v_s}$ where $\overrightarrow{s}$ is a combination of words to add ($V_a \subset V$) and words to remove ($V_s \subset V$) from the vocabulary V that maximize cosine_similary($\overrightarrow{s}, \overrightarrow{t}$). As an illustration, on Rouen University Hospital data with 641 279 documents, Word2Vec generates a vocabulary V of 355 597 words [4]. This represents $1,51 \times 10^{27}$ candidate solutions $\overrightarrow{s}$ of 5 words.

Solution Representation. A solution S can be seen as variable-length sets of tokens to add ($token_{ai}$) or remove ($token_{ri}$) to approximate a target token T:

$$+\{token_{a1}, token_{a2}, ..., token_{an}\} - \{token_{r1}, token_{r2}, ..., token_{rn}\} \sim target$$

As an illustration, the analogy presented previously can be represented as:

$$+\{king, woman\} - \{man\} \sim queen$$

Evaluation Function. In text mining, cosine similarity is frequently used to determine the proximity between two text documents [6]. It can be used on text vectors as well. A value close to 1 indicates that the two vectors are collinear and therefore have a similar meaning. A value close to -1 indicates the texts are opposite. In our case, we want to maximize the similarity between the solution S and the target T:

$$\text{cosine similarity}(S, T) = \frac{S \cdot T}{\|S\|\|T\|}$$

3.2 Resolution with a Local Search

Neighborhood. The vocabulary V is the set of all tokens for which we have a vector in Word2Vec representation. The neighborhood ensures that every token contained in V can be introduced into a solution S. The neighborhood of S is the set of all solutions with a 1-token difference from S. To avoid bloat we ordered the neighbors to prioritize token suppression and token replacement:

- suppression: remove a token from the set of tokens to add, or tokens to subtract. ie : $+\{king, woman\} \sim queen$ or $+\{king\} - \{man\} \sim queen$
- replacement: replace a token from the solution S by another token from the vocabulary V. i.e.: $+\{king, woman\} - \{cat\} \sim queen$
- addition: add a token from the vocabulary V to solution S. i.e.: $|\{king, woman\} - \{man, cat\} \sim queen$

Local Search. We choose a simple Hill Climbing approach with a first improve strategy: the first encountered enhancing neighbor is chosen as the next solution. The associated algorithm is given in Algorithm 1. The algorithm is initialized with a random solution of 6 tokens, distributed in the two sets of tokens to add or subtract. At each iteration, the neighborhood of the current solution is generated. It is scanned in order until an improving neighbor is found. The stopping criterion is when all the neighborhood of the current solution have been visited without an improvement of the cosine similarity.

Algorithm 1. Local Search for Word Embedding Explanations

Require: vocabulary V, target token T
 $S \leftarrow$ selects 6 random tokens from V
 $N \leftarrow$ generate neighbors of S
 while $len(N) > 0$ **do**
  if $cosine\_similarity(N[0], T) > cosine\_similarity((S, T)$ **then**
   $S \leftarrow N[0]$
   $N \leftarrow$ generate neighbors of S
  else {next neighbor}
   remove $N[0]$ from N
  end if
 end while

---

4   Experiments and Results

This experiment aims to investigate if the proposed algorithm could find better semantic analogies than those suggested in Word2Vec paper. We create 5 semantic analogies inspired by relations presented in the Word2Vec paper. They are in Tab 1 with the cosine similarity obtained with their associated target. For each target, the local search is executed 50 runs. The training of the Word2Vec is done using Python Gensim Library, on the dataset *text8* which is an extraction of the first 10^8 bytes of Wikipedia. The default parameters are used: vector_size = 100, epoch = 5, window = 5. After training, the vocabulary is composed of 71290 tokens. The experiments are carried out on an Apple M1 Pro with 32 GB of ram.

Table 1. Average and standard deviation of cosine similarity (CS), iterations, and seconds obtained by Local Search over 50 runs for each of the 5 targets

Target	Example Analogy	CS	Local Search CS	iterations	seconds
queen	king-man+woman	0.731	0.827 ±0.017	9.560 ±1.643	53.228 ±74.508
berlin	paris-france+germany	0.797	0.811 ±0.019	9.980 ±1.363	46.543 ±11.025
brother	sister-woman+man	0.706	0.847 ±0.011	9.680 ±1.911	39.697 ±8.912
euro	dollar-usa+europe	0.556	0.838 ±0.019	10.240 ±1.813	46.560 ±14.015
athens	oslo-norway+greece	0.635	0.831 ±0.015	10.420 ±1.401	50.697 ±43.576

Table 2. Best solutions obtained by Local Search (with maximum cosine similarity with target)

Target	CS	Best Local Search Solution
queen	0.887	king +lord +elizabeth +princess -freiherr
berlin	0.867	german +city +works +germany +boulder +hobhouse -tortilla
brother	0.884	son +brothers +sister +nephew
euro	0.891	currency +european +standard +dollar -detective
athens	0.871	city +age +greek +bethlehem -cd -gene

Table 1 gives for each target the average and standard deviation of cosine similarity, number of iterations, and elapsed time in seconds obtained by the local search. Local search obtains solutions with better cosine similarity than the initial solutions, in a reasonable time and number of iterations, with gains from 11% to 33% (except for target berlin where the proposed solution does not improve cosine similarity more than 1%). Table 2 gives the best solution in terms of cosine similarity obtained by the local search over the 50 runs. For reasons of space we will not detail all the solutions obtained, but among all the solutions obtained by LS, the ones with the best cosine similarity sometimes introduce strange associations, like "-detective" for "euro". It could be caused by the cosine similarity which may not be the best-adapted measure for this task. Another hypothesis is that the dataset used for training Word2Vec is too small: some words are not very frequent, which can lead to irrelevant associations.

5   Conclusion and Further Research

In this paper, we proposed a novel method to extract white-box knowledge from word embedding, with modeling as an optimization problem. The first results on the text8 dataset and 5 target words show that the proposed method can improve the cosine similarity by 11% to 33% by finding better explanations. One perspective of this work would be to apply it to larger datasets, particularly in healthcare, to assess whether it can generate new hypotheses or explanations for specific patient profiles or diseases. Evaluation criteria other than cosine similarity can also be considered, to see if a multi-objective approach can find interesting solutions. Since Word2Vec, more complex methods like BERT [2] and GPT have been proposed, in which a word can have multiple vectors, depending on the context: it could be interesting to study how to adapt the modeling.

References

1. Combi, C., et al.: A manifesto on explainability for artificial intelligence in medicine. Artif. Intell. Med. **133**, 102423 (2022)
2. Devlin, J., Chang, M.W., Lee, K., Toutanova, K.: BERT: pre-training of deep bidirectional transformers for language understanding (2019)
3. Dhaenens, C., Jourdan, L.: Metaheuristics for data mining: survey and opportunities for big data. Ann. Oper. Res. **314**(1), 117–140 (2022)
4. Dynomant, E., et al.: Word embedding for the French natural language in health care: comparative study. JMIR Med. Inform. **7**(3), e12310 (2019)
5. Mikolov, T., Chen, K., Corrado, G., Dean, J.: Efficient Estimation of Word Representations in Vector Space (2013)
6. Singhal, A., et al.: Modern information retrieval: a brief overview. IEEE Data Eng. Bull. **24**(4), 35–43 (2001)
7. Sobanski, V., Lescoat, A., Launay, D.: Novel classifications for systemic sclerosis: challenging historical subsets to unlock new doors. Curr. Opin. Rheumatol. **32**(6), 463–471 (2020)
8. Wu, S., et al.: Deep learning in clinical natural language processing: a methodical review. J. Am. Med. Inform. Assoc.: JAMIA **27**(3), 457–470 (2020)
9. Zhang, H., Ogasawara, K.: Grad-CAM-based explainable artificial intelligence related to medical text processing. Bioengineering **10**(9), 1070 (2023)
10. Şenel, L.K., Şahinuç, F., Yücesoy, V., Schütze, H., Çukur, T., Koç, A.: Learning interpretable word embeddings via bidirectional alignment of dimensions with semantic concepts. Inf. Process. Manag. **59**(3), 102925 (2022)

A Hybrid Biased-Randomized Heuristic for a Home Care Problem with Team Scheme Selection

Ana Raquel de Aguiar[1]([✉]) [iD], Maria Isabel Gomes[2] [iD], Tânia Ramos[1] [iD], and Helena Ramalhinho[3] [iD]

[1] Centre of Management Studies for Instituto Superior Técnico (CEGIST), Universidade de Lisboa, Lisbon, Portugal
a.raquel.aguiar@tecnico.ulisboa.pt
[2] Center for Mathematics and Applications (NOVA Math) and Department of Mathematics, NOVA SST, Caparica, Portugal
[3] Department of Economics and Business, Universitat Pompeu Fabra, Barcelona, Spain

Abstract. The increasing demand for home care services imposes effective human resource management. The problem concerns the creation of teams of one or two caregivers, serving patient requiring one or two caregivers. The number of teams of each type makes up the team scheme. A single-caregiver team can synchronize for tasks requiring two caregivers. Introducing a novel methodology, we employ a biased-randomized greedy constructive algorithm for route design, comparing it with its hybridization with a local search algorithm. Then, the performance of the hybrid method is compared with that of a mixed integer linear program model and a biased random-key genetic algorithm implementation.

Keywords: Heuristic · home care · VRPTW · synchronization

1 Introduction

This study is focused on the home care routing and scheduling, specifically how synchronization can be used to better assign caregivers to teams and serve more patients [1]. Previous approaches, such as biased random-key genetic algorithms (BRKGA) [2], failed to balance solution quality and computational time when extended into multi-period. This led to a search for a better solution methodology and improved characterization of component performance.

A novel biased-randomized greedy constructive (BR-GC) algorithm is introduced, enhanced with a swap local search (LS) algorithm (hBR-GC), to address these shortcomings. It aims to improve solution quality and computational efficiency compared to existing methods. In doing so, we aim to answer the following research questions: 1) What are the solution quality and computational performance variations resulting from hybridizing the BR-GC with LS?, 2) How does hBR-GC's performance measure against a mixed integer linear program (MILP) model's solution? and 3) How does hBR-GC compare to the BRKGA?

The plan is for the hBR-GC to become the component for generating initial solutions for an implementation of an Iterated Local Search (ILS) algorithm, justifying the characterization of this component.

2 Solution Methodology

A set of patients requires one daily task provided at home, $i \in N_C$, associated to a duration, W_i, a time-window, $[e_i, l_i]$. Tasks requiring one caregiver are placed by semi-dependent (SD) patients, $i \in N_S$ whereas bedridden (BR) patients' tasks require two, $i \in N_B$. All tasks are classified as either SD or BR: $N_C = N_B \dot\cup N_S$.

The care is provided by a homogeneous set of A^C caregivers, with a maximum shift length, H, starting and finishing their routes at the day-care center. There are Q cars available. The organization assigns the caregivers to teams, $k \in V$, of either one or two caregivers, denominated single, $k \in V_S$, and double, $k \in V_D$, teams respectively. Set $V = V_S \dot\cup V_D$, where $|V_S| = \min\{Q, A^C\}$ and $|V_D| = \min\{Q, \lfloor \frac{A^C}{2} \rfloor\}$. The number of teams on the solution is at most equal to Q. Parameter M_k represents the number of caregivers in each team.

The decisions concern task allocation to teams, selecting the team scheme, when to synchronize, and visiting sequence. The choice of team scheme is influenced by factors like the ratio between SD and BR services and their geographical distribution. The objective function is given by Eq. 1. If an SD service is assigned to a double team, the objective function will incur a penalty since the service duration is multiplied by the caregivers assigned to the task. The parameter T_{ij} represents the travel time between two locations, and decision variable x_{ijk} equals 1 if team k travels from node i to node j, and 0 otherwise.

$$\min_x \sum_{k \in V} \sum_{i,j \in N} (T_{ij} + W_i) M_k x_{ijk} \tag{1}$$

Exact Approach (MILP) - One of the solution methodologies tested is the MILP implementation of the mathematically formulated model available in [1], corresponding to scenario with synchronization and without continuity of care (wSyn_woCC), using the modeling software GAMS 34.2 and CPLEX 20.1.

BRKGA - The BRKGA encodes solutions using a vector of random-key values ranging from 0 to 1, with each value corresponding to a gene's allele. A decoder translates these keys into solutions, which are categorized into Elite and Non-Elite groups. Evolutionary operations, including mutation and uniform crossover, generate offspring, with one parent selected from each group. Offspring inherit alleles from Elite parents with a higher probability. Then, the BRKGA explores the solution space by iteratively evolving the population, until meeting a stopping criterion. Its modular design distinguishes the problem-independent evolutionary routine from the problem-dependent fitness evaluation, facilitating adaptation to different problems and simplifying implementation.

The initial population and the decoder are the relevant components of the BRKGA in [2]. The initial population is homogeneous, with allele values uniformly distributed between 0 and 1, sequentially assigned based on increasing values of e_i. The decoder is a deterministic GC algorithm. It inserts tasks into routes with the least-cost insertion. For each task, it determines the insertion strategy (insertion_stg) by applying function *insertion_strategy*. It begins by checking if it's SD or BR. For SD tasks, insertion is tested in every route; for BR tasks, insertion is tested in double routes or a pair of single routes for synchronization. If feasible insertions are found, the solution is updated with the least-cost insertion. The *insertion_strategy* function ensures compliance with time window and shift length constraints, returning ∞ if constraints are not respected. *update_sol* function updates the partial solution by inserting the new task and updating the solution attributes. The algorithm iteratively checks team availability during insertions until all caregivers are assigned (see line 3), then fixes the team scheme. Function *final_update* adds the costs between the last task in each route and the day-care center.

Algorithm 1 : The Greedy Constructor

Require: O, δ
 1: current_solution = *initialize_solution*(A^C, Q)
 2: **for** task $\in O$ **do**
 3:    **if** *check_teamScheme_feasibility*(current_solution) **then**
 4:       insertion_stg= *insertion_strategy*(task, current_solution)
 5:       **if** *feasible*(inserion_stg) **then**
 6:          current_solution $\leftarrow$ *update_sol*(current_solution, insertion_stg, task)
 7:       **else return** None
 8:       **end if**
 9:    **end if**
10: **end for**
11: current_solution $\leftarrow$ *final_update*(current_solution)
12: **return** Least-cost Solution

BR-GC and hBR-GC - The BR-GC method extends the GC of the BRKGA with a new routine to enhance solution diversity. Its framework is displayed in Algorithm 2. The routine adds biased randomness when constructing the task list O^R fed to the GC. Tasks are randomly selected from the δ remaining tasks with earliest e_i. This step promotes diversity compared to BRKGA's homogeneous initial population, exploring a broader range of team schemes and accelerating feasible solution discovery. After the CG application a feasible solution is added to the *sol_list*, until reaching σ applications, then sorting the feasible solutions and returning the best. The hBR-GC applies LS after the GC phase (between lines 8 and 9). The LS algorithm involves swapping two tasks within the same route. A feasible move must secure dependencies resulting from synchronization.

Algorithm 2 : The framework of the BR-GC
Require: σ, δ , $task\_list$
1: sol_list ←Initialize solutions list; $k = 1$
2: **while** $k \leq \sigma$ **do**
3:    Initialize list O^R
4:    **while** $task\_list$ not empty **do**
5:       i ← Randomly select one of the δ tasks with the lowest e_i from $task\_list$
6:       Append i to O^R; Remove i from $task\_list$
7:    **end while**
8:    new_sol ← $greedyConstructor(O^R)$
9:    Append new_sol to sol_list; $k+ = 1$
10: **end while**
11: Sort sol_list by cost **return** Least-cost Solution

3   Results

The results presented in this section for the exact method were obtained in a workstation with an Intel(R) Core(TM) i9-10850K CPU @ 3.60 GHz 3.60 GHz and with a RAM of 128 GB, whereas the results for the heuristic method were obtained on a computer with Intel(R) Core(TM) i5 8500 CPU @ 3.00 GHz 3.00 GHz processor and 8 GB of RAM. The instances used are the same as in [2]. For each of the instances, the methodology is implemented from scratch in Python and run for 5 seeds.

1) What are the solution quality and computational performance variations resulting from hybridizing the BR-GC with LS? The parameter settings for size-25 instances are $\sigma = 3000$ and $\delta = 4$, and for size-50 instances, $\sigma = 6000$ and $\delta = 8$. The average OF performance variation between BR-GC and hBR-GC for size-25 instances is -1.1%, and for size-50 instances, it's -1% Table 1. Regarding runtime, both algorithms take 2 s for size-25 and 18 s for size-50 instances, with hBR-GC showing a negligible increase. The application of hBR-GC increases runtime by 1.5% for size-25 and 0.7% for size-50. Including LS after BR-GC yields an average OF improvement of around 1%, with negligible impact on runtime, which diminishes relatively with instance size. Consequently, subsequent analysis focuses on hBR-GC.

2) How does hBR-GC's performance measure against a traditional MILP solution? The comparison between the MILP model and hBR-GC focuses on size-25 instances due to the inability to obtain solutions for size-50 instances within two hours using the exact method. For size-25 instances, hBR-GC found optimal solutions in 5 instances and reached the memory limit in the rest Table 2. On average, hBR-GC solutions had a 6.9% higher objective function (OF) value than MILP. However, hBR-GC required only 2 s on average, while MILP took 9.4k times longer (18 772 s). The best solution from multiple hBR-GC runs was around 5.3% worse than the MILP method. Running hBR-GC five times to initiate an ILS would yield good starting points with an average runtime of 10 s. More time would be required for larger instances.

Table 1. Objective Function results for size-25 and size-50 instances. avg-average, min-minimal OF value within 5-seed runs. Avg - average performance over all instances.

Id	Size-25						Size-50					
	BR-GC		hBR-GC		variation(%)		BR-GC		hBR-GC		variation(%)	
	min	avg	min	avg	min	avg	min	avg	min	avg	min	avg
1	762	770.9	755.5	763.8	−0.9%	−0.9%	1414.9	1423.8	1407.0	1415.3	−0.6%	−0.6%
2	740.2	749.64	725.5	740.2	−2.0%	−1.3%	1232.5	1245.1	1212.7	1230.1	−1.6%	−1.2%
3	822.1	830.02	802.3	818.0	−2.4%	−1.4%	1509.6	1524.5	1502.7	1514.7	−0.5%	−0.6%
4	877.9	885.92	866.7	879.0	−1.3%	−0.8%	1561.7	1568.1	1544.2	1557.4	−1.1%	−0.7%
5	657	662.96	638.7	653.1	−2.8%	−1.5%	1198.2	1208.1	1161.5	1187.9	−3.1%	−1.7%
6	919.3	935.04	918.6	928.8	−0.1%	−0.7%	1730.4	1748.1	1709.7	1730.6	−1.2%	−1.0%
7	693.3	699.28	677.2	683.1	−2.3%	−2.3%	1269.6	1282.8	1242.3	1261.9	−2.2%	−1.6%
8	714	729	707.9	722.8	−0.9%	−0.8%	1323.4	1333.0	1295.3	1313.0	−2.1%	−1.5%
9	888	902.62	880.3	898.7	−0.9%	−0.4%	1637.2	1650.5	1626.5	1633.3	−0.7%	−1.0%
10	845	859.62	845.0	851.4	0.0%	−1.0%	1475.1	1485.0	1461.2	1478.4	−0.9%	−0.4%
Avg	791.9	802.5	781.8	793.9	−1.3%	−1.1%	1435.3	1446.9	1411.3	1427.1	−1.4%	−1.0%

Table 2. Comparison between hBR-GC and Exact method (MILP). avg-average, min-minimal OF value within 5-seed runs. Avg - average performance over all instances. Bolt OF for MILP represents optimality.

		hBR-GC				MILP		Variation(%)		
		OF		RT						
Size	Id	min	avg	min	avg	OF	RT	min_OF	avg_OF	avg_RT
25	1	755.5	763.8	2.2	2.2	**742.7**	13094	1.7%	3%	−100.0%
	2	725.5	740.2	2.1	2.1	651.2	11301	11.4%	14%	−100.0%
	3	802.3	818.0	1.5	1.6	**766.1**	918	4.7%	7%	−99.8%
	4	866.7	879.0	1.8	1.8	**825.2**	21346	5.0%	7%	−100.0%
	5	638.7	653.1	2.7	2.7	604.9	7546	5.6%	8%	−100.0%
	6	918.6	928.8	2.4	2.4	888.7	20887	3.4%	5%	−100.0%
	7	677.2	683.1	1.4	1.4	**620.1**	2407	9.2%	10%	−99.9%
	8	707.9	722.8	1.7	1.7	**688.6**	20516	2.8%	5%	−100.0%
	9	880.3	898.7	2.3	2.3	852.9	48415	3.2%	5%	−100.0%
	10	845.0	851.4	2.2	2.2	798.1	41292	5.9%	7%	−100.0%
Avg		781.8	793.9	2.0	2.0	743.9	18772.2	5.3%	6.9%	−100.0%

3) How does hBR-GC compare to the BRKGA? For comparison between hBR-GC and BRKGA, see Table 3 for summarized results. Due to space constraints, the full table isn't provided, but [2] contains BRKGA results. Objective function values are similar on average, with a negligible variation of about 0% for size-25 and 2% for size-50. This difference between instance sizes may relate to parameter settings; $\delta = \lfloor 0.15|N_C| \rfloor$ and σ proportional to the reasonable value of 3000 found for size-25. Increasing σ could improve solution quality for size-50 but also affect runtime. Despite this, hBR-GC achieves solutions close to BRKGA quality with about 90% less runtime on average. A slightly higher σ could significantly improve solution quality with a small impact on RT.

Table 3. Summary of results comparing the hBR-GC and BRKGA

	Average				Variation(%)			
	Objective Funtion		Runtime		OF		RT	
size	hBR-GC	BRKGA	hBR-GC	BRKGA	min	avg	min	avg
25	793.9	795.0	2.0	23.0	0%	0%	−91%	−91%
50	1427.1	1409.0	18.3	187.2	1%	2%	−90%	−90%

4   Conclusions and Future Work

In conclusion, including LS after BR-GC yields an average OF improvement of around 1%, with negligible impact on runtime. Notably, the hBR-GC achieves solutions close to the MILP method with significantly less runtime. Comparison with BRKGA shows similar objective function values, with hBR-GC requiring about 90% less runtime on average. Adjusting parameter σ could further improve solution quality with minimal impact on runtime. Overall, hBR-GC presents a promising solution for solving the routing and scheduling optimization, warranting its exploration as a component of an ILS metaheuristic within a methodology to solve a rich VRPTW with synchronization and team-scheme selection.

Acknowledgments. This study was funded by: Ph.D. Grant SFRH/BD/148773/ 2019; UIDB/00297/2020; UIDP/00297/2020; UIDB/00097/2020.
Disclosure of Interests. The authors have no competing interests to declare that are relevant to the content of this article.

References

1. de Aguiar, A.R.P., Ramos, T., Gomes, M.I.: Home care routing and scheduling problem with teams' synchronization. Socioecon. Plann. Sci. **86**, 101503 (2023). https://doi.org/10.1016/j.seps.2022.101503
2. Aguiar, A.R., Ramos, T., Gomes, M.I.: A biased random-key genetic algorithm for the home care routing and scheduling problem: exploring the algorithm's configuration process. In: Almeida, J.P., Geraldes, C.S., Lopes, I.C., Moniz, S., Oliveira, J.F., Pinto, A.A. (eds.) IO 2021. Springer Proceedings in Mathematics & Statistics, vol. 411, pp. 1–21. Springer, Cham (2023). https://doi.org/10.1007/978-3-031-20788-4_1

Optimization for Forecasting

Extended Set Covering for Time Series Segmentation

Vittorio Maniezzo(✉) [ID]

Department of Computer Science, University of Bologna, Via dell'Università 50,
47521 Cesena, Italy
vittorio.maniezzo@unibo.it

Abstract. Time series analysis plays a critical role in data analytics, an
effective modeling of nonlinear trends is essential for obtaining action-
able results, notably for forecasting and missing values imputation. The
segmentation of time series and the corresponding detection of change
points stand out for their practical implications. This paper presents
preliminary results of a study on the applicability of mathematical programming, and in particular matheuristics, to time series segmentation.

Keywords: Time series · Set covering · matheuristics · segmentation ·
change point detection

1 Introduction

Time series analysis plays a critical role in the field of data analytics, revealing
patterns, trends, and anomalies in a variety of fields ranging from finance and
healthcare to environmental monitoring and industrial processes. The inherent
sequential nature of time series data requires specialized techniques for effective
analysis and interpretation. Time series segmentation and the associated change
point detection are key tools in this context, providing a means to decompose
complex temporal data into meaningful segments for in-depth analysis.

The process of time series segmentation [5] involves dividing a continuous
time series into distinct, non-overlapping segments, each characterized by homogeneous patterns or behaviors. This decomposition not only facilitates a clearer
understanding of the underlying structure within the data, but also enables the
application of targeted analytical methods to each segment, potentially increasing the accuracy of predictions and insights.

Segmenting a time series involves identifying points of change in the series
[1,8]. Time series often exhibit temporal shifts, transitions, or abrupt changes
that carry critical information. The process of change point detection serves as
a central tool for uncovering these moments, enabling a deeper understanding of
the underlying dynamics and facilitating timely responses to emerging patterns.
Change points identify events in a time series where the statistical properties of
the data undergo a significant change. These changes can manifest themselves

as shifts in mean, variance, or other structural characteristics, and thus indicate transitions in the underlying processes. Effective change-point detection methods therefore contribute not only to the understanding of evolving trends, but also to the prediction of future behavior and the identification of anomalous events.

The identified segments can correspond to any pattern of interest, but because of its computational efficiency and immediacy of understanding, linear regression plays a predominant role. Segmentation into linear intervals is known as *piecewise linear regression* [4]. Unlike traditional linear regression models, piecewise linear regression recognizes the presence of distinct segments in the data, each characterized by its own linear trend. This approach is particularly valuable because it can capture nonlinear components, such as shifts, abrupt changes, or varying slopes, that may exist within the evolution of a phenomenon.

The importance of time series segmentation becomes apparent when dealing with real-world applications, where the complexity of temporal patterns often requires nonlinear models. Because of its importance, several approaches have been proposed for time series segmentation and change point detection, including classical statistical techniques, Bayesian approaches, machine learning algorithms, and hybrid models. This paper presents a new possibility, using Mixed-Integer Programming (MIP) to derive a model for optimal segmentation. Given the non-polynomial nature of the method, a Lagrangian matheuristic is also derived.

Through the presentation of real-world applications and case studies, the paper illustrates the versatility of MIP-based change point detection in addressing domain-specific challenges. Applications ranging from financial markets to epidemiology to environmental modeling are presented, and challenges and common obstacles such as noise, irregularities, and the curse of dimensionality are briefly addressed. This research is still in progress, so its strengths and weaknesses cannot yet be fully determined, but current results already testify to the viability of the approach.

The paper introduces the MIP model and its Lagrangian relaxation in Sect. 2, some obtained computational results in Sect. 3 and provisional conclusions in Sect. 4.

2   An Extended Set Covering Model

The core idea of the model presented here is to list all subsequences of series values that satisfy specific structural constraints, in the following denoted as *feasible runs* of values, quantify a quality measure of each run, and select the subset of runs that collectively cover the entire series while minimizing a cost of the difference between actual and model data (the model *residuals*) or maximizing a model fitness measure. This is the core of the model, further constraints can be added to accommodate specific desiderata on the model, taking advantage of the modeling flexibility of mixed-integer programming and the effectiveness of state-of-the-art general MIP solvers [2].

A Set Partitioning Problem (SPP) is at the core of the model, where the objective is to minimize the cost of the residuals and the constraints ensure that each point of the series has a corresponding one in the model. Note that no assumption is made about the nature of the model of the segments, it can be linear regression, giving rise to piecewise linear regression, but also any other nonlinear model. It is also possible to use different models for different runs at no cost to the optimization process. The SPP is defined over a set $X = [x_j]$, $j = 1, \ldots, nruns$ of binary decision variables, each associated with one of the runs, i.e., of the segments. The generation of feasible runs can already implement some dominance, e.g. avoiding the generation of runs with too few or too many values. The cost c_j of each run can be computed according to any of the fitness measures proposed in the literature, including R^2, SER, χ^2, RMSE, simple variance etc. The constraints ensure that each point of the series is covered by a selected run.

Unfortunately, the number of feasible runs for a real-world application can be very large, and the SPP can take an unacceptable amount of time to solve. Therefore, we relax the equality constraints into inequalities, transforming the problem into a Set Covering Problem (SCP), which can typically be solved in much higher dimensions, at the cost of having multiple runs covering some of the series points, therefore requiring a postprocessing to get a feasible segmentation. The resulting TSSC (standing for Time Series Set Covering) model is the following, where coefficients a_{ij} are $0/1$ coefficients taking the value of 1 if and only if run j covers the i-th series point, $i \in \{1, \ldots, npoints\}$ and $j \in \{1, \ldots nruns\}$.

$$(TSSC) \quad z_{TSSC} = \min \sum_{j=1}^{nruns} c_j x_j \qquad (1)$$

$$s.t. \sum_{j=1}^{nruns} a_{ij} x_j \geq 1, \qquad i = 1, \ldots, npoints \qquad (2)$$

$$\sum_{j=1}^{nruns} x_j \leq maxruns \qquad (3)$$

$$x_j \in \{0, 1\}, \qquad j = 1, \ldots, nruns \qquad (4)$$

State-of-the-art MIP solvers are very effective on SCP, but very large instances can still require high computational times. We have therefore also implemented a Lagrangian matheuristic [6] on the TSSC problem, relaxing all constraints (2) by associating a Lagrangian penalty λ_i to each i-th constraint, $1 \leq i \leq npoints$, and keeping only constraint (3) in the model. This results in model LTSSC (standing for Lagrangian TSSC) and we solved it by a subgradient algorithm, where the subproblem is very easy to solve, requiring to select at most the $maxruns$ most negative variables at each subgradient iteration. A simple fixing heuristic takes the incumbent subproblem solution at each iteration, which can be infeasible because it does not cover all relaxed constraints, and adds selected variables until it becomes feasible.

3 Computational Experience

We implemented models TSSC and LTSSC in C++ and ran them on a Windows 11 Intel i7 machine equipped with 32 Gb of RAM. The MIP solver used was CPLEX 22.11. All codes and data are available from the project repository.

The computational results reported here, still to be considered as preliminary, are mainly obtained on environmental monitoring data series, which initially prompted this research. The series were produced in the context of the SMARTLAGOON project, an EU H2020 project, born with the primary objective of developing a tool to allow real-time monitoring, analysis, to predict socio-environmental evolution of the vulnerable area *Mar Menor*, which is the largest saltwater coastal lagoon in Europe. [7]. Within the framework of the project, smart buoys were deployed in the lagoon and their sensors generated series on 12 attributes.

The sensed variables used here are: the steam pressure (Vapor-Pressure-Avg), the average wind speed (WS-ms-Avg), water temperature measured by a thermistor at the depth of 0.5 m (ThermTemp1-Avg), water temperature measured by oximeter at different depths (1 m - Wtemp-C1-Avg, 3 m - Wtemp-C2-Avg), water temperature measured by conductimeter at the depth of 1 m (SDI-Temp-1m), minimum current in the battery (IBatt-Min), average and maximum temperature obtained by a datalogger panel temp thermocouple (PTemp-C-Avg, PTemp-C-Max) and average charging voltage (V-in-chg-Avg). The overall dataset was recorded from August 2022 to April 2023. We complemented these datasets with two others from different domains: economics (bitcoin-US dollar exchange rate, BTC-USD) and healthcare (COVID infections in Italy, Covid-Italia-22) to get a first indication of the generality of the approach across domains.

Table 1. Data series results

Dataseries	npoints	nruns	nsegm	truns	tsolve	QMSE	QMSE1	QMSE2
Vapor-Pressure-Avg	194	15400	6	0	1	0.90	1.14	1.03
Vapor-Pressure-Avg-2	1139	618828	7	30	832	2.85	3.26	3.32
WS-ms-Avg	1139	618828	2	30	1136	162	193.03	210.80
ThermTemp1-Avg	194	15400	8	0	1	5.12	5.98	na*
Wtemp-C1-Avg	1139	618828	19	30	1212	13.89	22.63	na*
Wtemp-C2-Avg	194	15400	8	0	1	50.75	68.15	na*
SDI-Temp-1m	1139	618828	21	30	1354	15.06	25.06	na*
IBatt-Min	194	15400	3	0	1	3.19e-3	5.29e-3	3.27e-3
PTemp-C-Avg	194	15400	5	0	1	46.40	68.56	58.12
PTemp-C-Max	194	15400	4	0	1	114.24	139.61	125.09
V-in-chg-Avg	194	15400	2	0	1	9.20	9.38	9.38
BTC-USD	366	60378	12	3	4	1.88e7	7.51e7	2.42e7
Covid-Italia-22	507	109746	5	–	11	5070.95	na	na

The results are summarized in Table 1. The columns show the name of the series (*Dataseries*), the number of data points (*npoints*), the number of generated runs (*nruns*), the number of chosen segments (*nsegm*), the CPU time in seconds to generate all runs (*truns*), the CPU time in seconds to generate all runs (*truns*), the CPU time to solve the extended SCP model (*tsolve*), and a *quasi RMSE* quality measure of the solutions obtained by the algorithm described in Sect. 2, *QMSE*, and, for comparison, the same measure for the solutions obtained by the codes of [3], *QMSE1*, and of [4], *QMSE2*. A comparison with the results from *Ruptures* [8] is being finalized and will be presented at the conference. Data with asterisks require special comments, which are incompatible with the page limit of this note, but will be presented at the conference and in the full version of the paper, *na*'s mean that no feasible solution was produced.

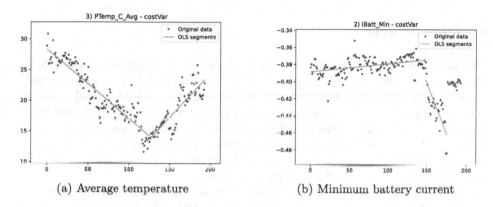

(a) Average temperature   (b) Minimum battery current

Fig. 1. Environmental dataseries

Not surprisingly, the set partitioning model could always produce the better solution, as it was the only algorithm that explicitly used the proposed cost function for optimization. More interesting is the limited computational time needed to solve instances with up to a few hundred variables, while the "curse of dimensionality" becomes apparent when solving instances with more than 1000 variables. The healthcare instance, here only a validation case, proves that the approach is also effective for nonlinear segmentation, but the search for optimal fitting parameters requires a very high CPU time. The Figs. 1 show the solution for two environmental data series, while the Figs. 2 show the two non environmental data series.

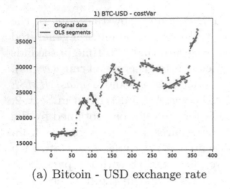

(a) Bitcoin - USD exchange rate

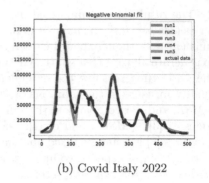

(b) Covid Italy 2022

Fig. 2. Economics and healthcare dataseries

4   Conclusions

The paper presents an MIP-based approach, both a Lagrangian matheuristic and an exact model, to time series segmentation. The underlying mathematical model is able to adapt to different requirements on the resulting model, making it more tolerant to residuals or more representative of short trends. The analysis imposes no constraints on the model associated with each segment, which can be linear, polynomial, or defined on any distribution function of interest. It can even be a combination of different models, allowing for linearity on some segments and, for example, exponential increases on others.

This flexibility comes at the cost of having to solve a NP problem on large size instances. The increase in effectiveness of general MIP solvers allows to consider the solution of real-world instances to optimality, and it also provides the basis for the design of mathematically grounded heuristics. We report preliminary results on sensor data series obtained in an environmental monitoring, but also two experiments on financial and healthcare use cases.

The results so far provide only a first indication of the possibilities offered by mathematical models, but also of the difficulties to be overcome. Besides the obvious limit imposed by the NP-hardiness of the problem (the "curse of dimensionality"), which is only partially alleviated by heuristic solving, there are problem-specific issues to be faced, such as which cost function to use or which constraint to impose on the runs. In fact, the final result is only indirectly determined by our model, and there are still cases where results are unsatisfactory to the eye, even though they are numerically satisfying. Current results have already proven useful in the motivating application domain, both for missing value imputation and for short-term forecasting, obtained by decomposing the series down to the last segment and using the identified components to extend it into the near future. Moreover, these results do not seem to be affected by the application domain.

Acknowledgements. This work has been supported by the European Union's Horizon 2020 research and innovation programme under grant agreement No 101017861, SMARTLAGOON project (smartlagoon.eu).

References

1. Aminikhanghahi, S., Cook, D.J.: A survey of methods for time series change point detection. Knowl. Inf. Syst. **51**, 339–367 (2017)
2. Fischetti, M., Lodi, A., Salvagnin, D.: Just MIP it!. In: Maniezzo, V., Stützle, T., Voß, S. (eds.) Matheuristics. Annals of Information Systems, vol. 10, pp. 39–70. Springer, Boston, MA (2009). https://doi.org/10.1007/978-1-4419-1306-7_2
3. Keogh E.J., Chu S., Hart D.M., Pazzani M.J.: An online algorithm for segmenting time series. In: Proceedings of the 2001 IEEE International Conference on Data Mining, San Jose, CA, USA, 2001, pp. 289–296 (2001)
4. Keogh E.J., Chu S., Hart D.M., Pazzani M.J.: Segmenting Time Series: A Survey and Novel Approach (2002). https://api.semanticscholar.org/CorpusID:8365617
5. Lovrić, M., Milanović, M., Stamenković, M.: Algorithmic methods for segmentation of time series: an overview. J. Contemp. Econ. Bus. Issues **1**(1), 31–53 (2014). ISSN 1857–9108. Skopje
6. Maniezzo, V., Boschetti, M.A., Stützle, T.: Matheuristics, Algorithms and Implementations. EATOR, Springer, Cham (2021). https://doi.org/10.1007/978-3-030-70277-9
7. The Smartlagoon project. https://www.smartlagoon.eu/. Accessed December 2023
8. Truong, C., Oudre, L., Vayatis, N.: Selective review of offline change point detection methods. Signal Process. **167**, 1–20 (2020)

Quantum Meta-Heuristic for Operations Research

Indirect Flow-Shop Coding Using Rank: Application to Indirect QAOA

Gérard Fleury[1], Philippe Lacomme[1(✉)], and Caroline Prodhon[2]

[1] Université Clermont Auvergne, Clermont Auvergne INP, UMR 6158 LIMOS, 1 rue de la Chebarde, 63178 Aubière, France
{gerard.fleury,philippe.lacomme}@isima.fr
[2] Université de Technologie de Troyes, 12 rue Marie Curie, CS 42060, 10004 Troyes Cedex, France
caroline.prodhon@utt.fr

Abstract. The Flow-Shop Scheduling Problem (FSSP) is one of the most famous scheduling problems. The Flow-Shop scheduling problem is a disjunctive problem, meaning that a solution is fully described by an oriented disjunctive graph where the earliest starting times are computed with a longest path algorithm. We propose a new approach based on Quantum Approximate Optimization Algorithm (QAOA) to find high quality solutions to FSSP instances using a vector representation. This approach permits to solve the well-known Carlier's instances with 64 operations to schedule. All the experiments have been achieved using the Qiskit library and carried on the IBM simulator. Presently, quantum methods cannot compete with classical ones because we lack quantum computers capable of solving large instances, and we have yet to figure out how to integrate the vast body of research results accumulated in flow-shop resolution over the last few decades into quantum algorithms. The ability of quantum approaches to effectively solve optimization problems in the future depends both on technical advancements in quantum machines and on the capacity to incorporate theoretical findings from scheduling into quantum optimization strategies.

Keywords: QAOA · IQAOA · Flow-Shop · Indirect representation · rank

1 Introduction

The Flow-Shop Scheduling Problem (FSSP) stands as a well-known optimization challenge extensively applied in manufacturing scheduling scenarios. It involves a collection of n jobs ($i = 1, n$) to be processed across m machines ($j = 1, m$). Each job comprises a sequence of tasks associated with specific machines, defining the problem's dimensions as typically represented by $n \times m$. Additionally, the FSSP adheres to several constraints: (i) prohibiting concurrent execution of multiple tasks within a job; (ii) restricting each machine to handle only one operation simultaneously; (iii) mandating job operations to follow a predetermined sequence without interruptions once initiated. Every operation O_{ij}, associated with job i and rank j, has a designated duration, p_{ij}. The primary objective revolves around scheduling these operations, considering precedence constraints,

M. Sevaux et al. (Eds.): MIC 2024, LNCS 14753, pp. 203–218, 2024.
https://doi.org/10.1007/978-3-031-62912-9_21

to minimize the total makespan (C_{max}). Notably, the problem complexity escalates to NP-hard status when $m \geq 3$, as verified in [1]. While the sequence of operations on machines remains independent of jobs, the durations p_{ij} vary based on the jobs. An effective problem representation is the disjunctive graph model introduced by Roy and Sussmann [2] and applied to scheduling in 1978 [16].

The utilization of the disjunctive graph model enables the visualization of any job shop problem instance via a directed graph denoted as $G = (V, A, E)$. Here, V signifies the set of nodes, A represents conjunctive arcs, and E denotes pairs of disjunctive arcs. The node set encompasses elements for each operation O_{ij}, a source node (0) linked to the initial operation of each job, and a sink node ($*$) connected to the final operation of each job. Conjunctive arcs depict the scheduling of operations within jobs, linking consecutive operations in the same job. Disjunctive arcs, on the other hand, connect operations from different jobs, designated to be handled on the same machine. A solution can be depicted by an acyclic subgraph encompassing all conjunctive arcs and one disjunctive arc for each pair within the set of disjunctive arcs. An optimal solution is derived from the feasible subgraph that minimizes the makespan.

The classical flob-shop scheduling problem makes the following assumptions:

- A job comprises a finite number of operations.
- The processing time for each operation on a specific machine is predefined.
- The sequence of machines used by the operations is the same for all job.
- Each machine executes each job only once.
- A machine can handle only one job at a time.
- Interruptions are not permitted until the completion of every operation within each job.
- Interruption or preemption is prohibited.
- No due date constraints are specified.
- There are no setup or tardiness costs associated.

A solution is fully defined by.

- the definition of the earliest starting times $st_{i,j}$ that meet the Job-shop constraints;
- the definition of the same sequence of jobs on all the machines.
- An optimal solution is modeled by a feasible subgraph with the minimal makespan C_{max}.

The flow-shop problem received a lot of attention for decades and recent publications focus on flow-shop extensions including for example maintenance and release time [10], delay [11], due-date [12]. Depending on both objective and constraint, numerous review exists in the literature including but not limited to [13] for distributed permutation flow-shop, [14] for multi-objective hybrid flow-shop, [15] for the multi-objective permutation flow-shop [16]. The fundamental principles of modeling and solving the flow-shop problem were introduced very early in J. Carlier's publication in 1978 [16].

The mathematical formulation of the Flow-Shop is based on x_{ij} a binary variable that value one if the job i is schedule before the job j and 0 if not and on St_{ij} variable that model the starting time of operations O_{ij} which refers to the jth operations of job i. The classical MILP model (that is a special case of the Manne's formulation [24]) is as follow with M is a large integer value that must be an upper bound of the makespan.

$$\min \ C_{max}$$

(1)

$$\forall i = 1..n, \forall j = 1..m \ \ St_{ij} \geq St_{i,j-1} + p_{i,j-1}$$

$$\forall i = 1..n, \forall k = 1..n, \ \ St_{kj} \geq St_{i,j} + p_{i,j} + M.(x_{ik} - 1)$$
$$\forall j = 1..m | k \neq i \quad\quad St_{ij} \geq St_{k,j} + p_{k,j} + M.x_{ik}$$

(2)

$$\forall i = 1..n, \forall k = 1..n, \forall j = 1..m | k \neq i \ \ x_{ik} + x_{ki} = 1$$

(3)

$$\forall i = 1..n \ \ C_{max} \geq St_{i,m} + p_{i,m}$$

(4)

Numerous alternative formulations have been introduced for the flow-shop [22, 23] and tackle numerous objective and additional constraints including for example the flow-time, batch or assembly operations. The major considerations on MILP for scheduling come from the early publication of Wagner [25]. The paper is organized into two sections. Section 2 introduces the method of resolution based on the rank (indirect representation of solution) and the Indirect QAOA approaches used to solve the Flow-Shop. Numerical experiments are presented at the end of the Sect. 2 including resolution of several instances of Carlier. Section 3 is a conclusion.

2 Indirect Flow-Shop Coding Using Rank

2.1 Graph Modeling

Let's explore a Flow-Shop instance comprising 3 jobs and 3 operations as detailed in Table 1. This table presents the operations set for each job, along with the corresponding machine (m_1, m_2, or m_3) and the processing duration on that specific machine. The order of operations establishes an identical sequence on the machine (independent of the job), yet with varying processing times. For instance, Job 1 necessitates 10 time units on Machine M1, while Job 2 requires 15 time units on Machine M1. The modelization of a flow-shop takes advantages of the disjunctive graph commonly used in flow shop and are based on the same construction rules using disjunctive arcs to model resource constraint. The job-shop modelization is described in numerous publications including for example [9].

Table 1. Example of FSSP instance data

Jobs	Operation 1	Operation 2	Operation 3
$i = 1$	$(m_1, 10)$	$(m_2, 35)$	$(m_3, 25)$
$i = 2$	$(m_1, 15)$	$(m_2, 8)$	$(m_3, 14)$
$i = 3$	$(m_1, 100)$	$(m_2, 1)$	$(m_3, 10)$

A disjunctive graph that depicts the problem is composed of solid-line arc linking two consecutive operations $(O_{i,j}; O_{i,j+1})$ that symbolizes the sequential constraint within job i. These arcs are weighted by the minimum time delay between the starting times of two consecutive operations: $st_{i,j+1} \geq st_{i,j} + p_{i,j}$. Every set of disjunctive arcs is commonly represented by a dashed line, defining the constraint between two operations planned to be processed on the same machine.

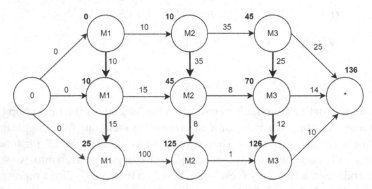

Fig. 1. Oriented disjunctive graph.

One acyclic conjunctive graph is given in Fig. 1 where all operations in disjunction (one set per machine) are reduced to disjunctives arcs modeling the sequencing of operations processed on the same machine. The left-shifted solution (semi-active) is obtained by execution of one longest-path algorithm that permits to compute the earliest starting times of operation $st_{i,j}$. The starting time of operation $*$ is 136 and the full solution is defined by the starting time of all operations.

2.2 Quasi-Direct Representation

A solution can be derived from the full order of job that defines a quasi-direct representation of a solution. The solution of Fig. 3 is derived from the order Job 1 first, Job 2 s and Job 3 last. The significance of establishing a specific and direct solution representation has long been emphasized. Notably, in the subsequent publication of [3], the authors explicitly outlined that a quasi-direct representation allows definition of: (1) a coding space and (2) a solution space. As stressed by [3], an effective mapping function should associate any element within the coding space with a solution that meet constraints. Whatever the mapping function used, a flow-shop solution is defined by a directed disjunctive graph. Computing the longest path within this graph determines the earliest starting times of operations, defining a semi-active solution.

2.3 Indirect Representation of Solutions

Resolution of scheduling problem based on quantum approaches focus on problems that are the corner stone of the scheduling theory. In 2016, the resolution of the job-shop introduced in [18] is based on a time-indexed quadratic unconstrained binary optimization problem (QUBO). The potential resolution is evaluated in the scientific report of Carughon, Dacrema and Cremonesi in [19]. The experiments have been achieved on Dwave. QUBO formulation are well-known to require a set of appropriate weighted terms for each constraint included in the objective function. The recent publication [20] focuses on Quantum Annealing and include both job-shop and flow-shop instances for the experiments taking advantages of one Hamiltonian formulation. In [21] the authors demonstrate how to efficiently apply QAOA to the Job-Shop and introduce numerical experiment proving that JSPP resolution on gate-based computer remains possible. Research related to the application of quantum approaches to scheduling problems encompasses both modeling and implementation aspects. Scheduling problems stand out as a favored domain for quantum technology application, primarily due to their highly combinatorial nature most of the time. Our proposition is a methodological proposal aiming to define a new resolution perspective based not on modeling as a Hamiltonian of the function to minimize, but on the rank of solutions. The objective is to achieve compact circuits using few gates that can be used on current machines for classical instances found in literature. This is why we will utilize the concept of rank as defined in [4] and an approach resembling IQAOA (Indirect QAOA). The indirect representation of a solution can harness the inherent one-to-one relationship between permutations and a concept known as subexceedant functions, ultimately simplifying the modeling of permutation ranks using a single integer. Various methods exist for establishing this one-to-one correspondence, with the most renowned being the Lehmer code, also referred to as the inversion table. An algorithmic description of this approach is initially presented in Knuth's work from 1981 [5].

f is the subexceedant [6, 7] function defined by:

$f(i)$ is the number of indices $j \leq i$ such that $\sigma_j < \sigma_i$

Obviously the following remarks holds (subexceedant function):

$$\forall i = 1..n, 0 \leq f(i) \leq i$$

Let a permutation σ in f. The subexceedant function f related to σ can be obtained by iteratively scanning σ and by assigning $f[i] = \sigma[i]$ at each iteration. The remaining elements of σ that occurs on the right of i, such that $\sigma[j] > f(i)$, have to be decreased of one unit, to ensure that at the position $i+1$ to n the number are in the interval $[0; n - i]$. A algorithm description of both computation of f from one permutation σ and computation of a permutation σ from f has been provided lately in [4].

```
Algorithm 1. Compute_f ()
Input parameters:
    σ : a permutation of n element
    [n] : the interval
Output parameters:
    f : the subexceedant function
Begin
  For i = n − 1 to 1 do
    f[i] = σ[i]
    For j = i − 1 to 0 do
      If (σ[j] > i) then
          σ[j] = σ[j] − 1
      Endif
          EndFor
  EndFor
  Return f
End
```

Algorithm 1. Conversion of σ into a subexceedant function f

```
Algorithm 2. Compute_Permutation()
Input parameters:
    f : a subexceedant function
    [n] : an interval
Output parameters:
    σ : a permutation of n elements
Local parameters:
    v : an ordered list of n elements beginning at 0
Begin
  v = [n − 1, n − 2, ..., 1, 0]
  σ = []
  For i = n − 1 to 0 do
    x = f(i)
    y = v(x)
    σ[i] = y
    v = v − {y}
  EndFor
  Return σ
End
```

Algorithm 2. Computation of σ_f where f is subexceedent.

The total number of solutions is $n!$ and represents the total number of job-permutations with one job-permutation that fully defined a disjunctive oriented graph that models a solution. Let us consider $x \in [0; n! − 1]$ a rank in the list of permutations. To any rank $x \in [0; n! − 1]$, it is possible to defined the subexceedant f composed of decomposition in the factorial basis and by consequence the permutation σ and finally a solution [6].

Contrary to QAOA [8] that requires a Hamiltonian that defines the function to minimize and that required potentially a large number of gates, IQAOA only required a Hamiltonian that models the ranks leading to very compact circuit with a very low number of gates and qubits. The IQAOA algorithm uses $\vec{\beta}$ and $\vec{\gamma}$ weights parametrized a quantum state $\left|\varphi\left(\vec{\beta}, \vec{\gamma}\right)\right\rangle$ that defines a solution rank x with probability $\left|\left\langle x|\varphi\left(\vec{\beta}, \vec{\gamma}\right)\right\rangle\right|^2$ and an expectation value $\left\langle\varphi\left(\vec{\beta}, \vec{\gamma}\right)|C^p|\varphi\left(\vec{\beta}, \vec{\gamma}\right)\right\rangle$ evaluated by sampling. Each shot gives a rank in the list of Flow-Shop solution list that can be evaluated by transforming the rank into the corresponding permutation and the permutation into a fully oriented disjunctive graph. For a fixed $\vec{\beta}$, $\vec{\gamma}$, the quantum computer is to defined the state $\left| \varphi\left(\vec{\beta}, \vec{\gamma}\right) \right\rangle$ and a measure in the computational basis is required to obtain a string x that permits to evaluate $\left\langle \varphi\left(\vec{\beta}, \vec{\gamma}\right) | C^p | \varphi\left(\vec{\beta}, \vec{\gamma}\right) \right\rangle$. The main interest lies in simultaneously manipulating all ranks (i.e., indirectly over all solutions) through the associated probability distribution. Traditional resolution of scheduling problems requires traversing a portion of the solution space, with common operators defined to transform one solution into another (consider, for instance, mechanisms like operation permutations situated on the critical path to generate neighbors). The challenge of exploring a subset of the solution space vanishes with quantum approaches: the difficulty lies in modeling it as a Hamiltonian and determining the angles associated with the probability distribution.

The binary representation of rank is

$$rank = \sum_{j=0}^{n} x_j . 2^j \text{ with } x_j \tag{5}$$
$$\in \{0; 1\}$$

with

$$H_P = \frac{1}{2} \sum_{j=0}^{n} (Id - Z_j) . 2^j \tag{6}$$

The algorithm principle is illustrated on Fig. 2. IQAOA efficiency relies on the following key-points:

- The capability to give a good ratio between the estimation $C^p\left(\vec{\beta}, \vec{\gamma}\right)$ as regards as the number of shots required that have to tuned carefully.
- The best found distribution $\left| \psi\left(\vec{\beta^*}, \vec{\gamma^*}\right) \right\rangle$ must be estimated on a small subset of solutions as regards as the total number of solutions to avoid a costly enumerations, i.e. the algorithm must converged to the optimal and quasi solutions meaning that a large part of probabilities are on the optimal or high quality solutions.
- The availability of a dedicated method to computed the $\left(\vec{\beta^*}, \vec{\gamma^*}\right)$.
- The number of qubits p such that $2^p \geq n!$ and the circuit required a small number of gates to encode a permutation on the quantum circuit as regards the circuit length required using the classical QAOA approach.

The best parameters $\left(\vec{\beta^*}, \vec{\gamma^*}\right)$ are computed using the C_GRASP × ELS introduced in [4] with the following parameters:

- Minimization of the expectation value of the decile plus the expectation value of the distribution meeting the remark of [4];
- The parameters $np = 10$, $ne = 10$, $nd = 5$ for the first C_GRASP × ELS execution to optimize simultaneously $\vec{\beta}$ and $\vec{\gamma}$ with np the number of GRASP iteration (restart), ne number of ELS iterations and nd number of neighborhoods.

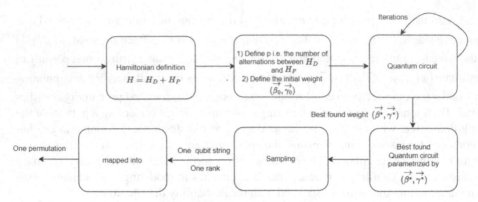

Fig. 2. IQAOA principles [4].

- The parameters $np = 10$, $ne = 10$, $nd = 5$ for the first C_GRASP $\times$ ELS execution to optimize $\vec{\gamma}$ only with np the number of GRASP iteration (restart), ne number of ELS iterations and nd number of neighborhoods.
- During the local search both $\Delta\vec{\beta}$ and $\Delta\vec{\gamma}$ vary from 0.1 to 0.001 but the value 0.001 is slowly decreased (divided by 10) at each iteration neighborhood generation.
- The quantum circuit is used with $p = 2$.
- 80 shots are used to evaluate the probability distribution.
- at the end of the optimization, 1000 shots are used to estimate the probability distribution.

The C_GRASP $\times$ ELS is an adaptation of the GRASP $\times$ ELS to non-discrete function and the GRASP $\times$ ELS is a hybridization between a GRASP (Greedy Adaptative Search Procedure) introduced early by [26] with an ELS (Evolutionary Local Search) introduced in [27]. Note that the results introduced in this paper are not strongly method dependent since similar outcomes have been achieved using a genetic algorithm, pushing us to consider the possibility that numerous methods could be employed to compute $\left(\vec{\beta^*}, \vec{\gamma^*}\right)$. Nevertheless, this study does not concentrate on the optimization method, which is a highly specialized research domain. Typically, in QAOA-based approaches, parameter optimization is accomplished using somewhat of a black-box method. We did not perform any specific optimization of the method's parameters: the parameters were set following a simple experimental study. Determining the parameters and designing a dedicated method is a separate and distinct subject.

The experiments have been achieved using the Carlier's instances that are available in the OR Library:

http://people.brunel.ac.uk/~mastjjb/jeb/info.html

2.4 Resolution of the Carlier 7 Jobs 7 Machines Instance

The Carlier 7×7 is a flow-shop (with 7 jobs and 7 machines) defining a total of 49 operations to schedule. The total number of permutations is $7! = 5040$ but there is only 1693 different costs (Table 2).

Table 2. Instance Carlier 7 jobs – 7 machines

M0	M1	M2	M3	M4	M5	M6
692	310	832	630	258	147	255
581	582	14	214	147	753	806
475	475	785	578	852	2	699
23	196	696	214	586	356	877
158	325	530	785	325	565	412
796	874	214	236	896	898	302
542	205	578	963	325	800	120

A sampling of permutations proves the high quality solutions have a very low probability (Fig. 3) and we note that, for example, the optimal solution 6590 [17] has a probability of about 0.019.

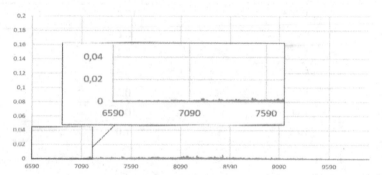

Fig. 3. Initial distribution (Car_7_7)

After the IQAOA execution, the sampling with 1000 shots gives 6550 with 194 shots (Table 3) meaning that about 19.4% of the probabilities is now on the optimal solution: the amplification is about 900 times (Table 3). These results prove that IQAOA succeeds into transforming the amplitude of an initial distribution into that of a target state (Fig. 4).

The rank 3281 correspond to the permutation

$$\sigma = [5, 4, 2, 6, 7, 3, 1]$$

with a makespan of 6590. Note that the final distribution shows that the optimal solution which values 6590 has a probability about 19% .

The partial representation of the solution introduced in Fig. 5 shows the operations scheduled at each machine, where the jobs are executed in the order defined by σ. Each starting time is defined as the maximal value between the previous operation of the job and the previously scheduled operation on the machine.

The job 4 on the machine $M1$ starts at 483 where $483 = \max(158 + 325; 158 + 23)$ since:

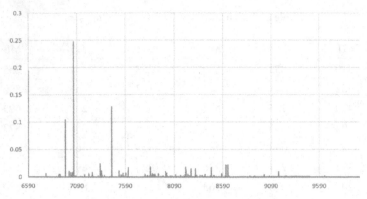

Fig. 4. Final distribution of solutions (*Car_7_7*)

Table 3. Instance Carlier 7 jobs – 7 machines

Cost	Number of shots	Probabilities (%)
6590	194	19.4
6772	6	0.6
6878	1	0.1
6905	5	0.5
6917	5	0.5
6972	104	10.4
7010	10	1.0

- the job P4 cannot start before the end of P5 on M1 i.e. not before 158 (earliest starting time of P5 on M1) plus the processing time of P5 on M1 that values 325.
- the job P4 cannot start before the end of P4 on M0 i.e. not before 158 (earliest starting time of P4 on M0) plus the processing time of P4 on M0 that values 23.

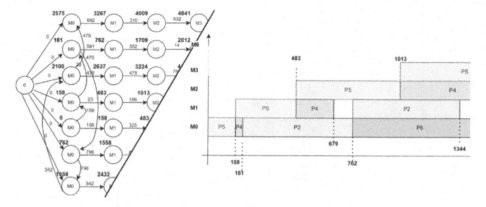

Fig. 5. Partial representation of the optimal solution

2.5 Resolution of the Carlier 8 Jobs 8 Machines Instance

The Carlier 8×8 is a flow-shop with 8 jobs and 8 machines defining a total of 64 operations to schedule. The total number of permutations is $8! = 40320$ but there is only 1996 different costs.

Similarly to the previous Carlier's instance, a sampling of permutations leads to a very similar conclusion (Fig. 6): we have only a probability of 1.47% to find a solution with a makespan lower than 8866 meaning that we have only 1.47% of chance to be at less of 6% of the optimal solution.

After the IQAOA execution, the sampling with 1000 shots gives 8366 [17] with 17 shots meaning that about 1.7% of the probabilities is now on the optimal solution: the amplification is about 685 times. These results prove that IQAOA succeeds into transforming the amplitude of an initial distribution into that of a target state (Fig. 7).

The numerical experiments push us to consider that it is possible to define a probability distribution that focuses on high-quality solutions, as emphasized, for instance, in Fig. 8.

It is important to note that the parameters have been fixed after a brief numerical study and should require a specific attention. All the experiments have been achieved using Qiskit (IBM) using the simulator.

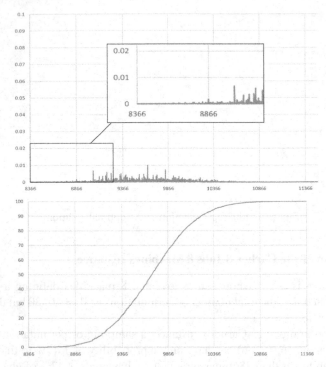

Fig. 6. Initial distribution (*Car_8_8*) and cumulative function

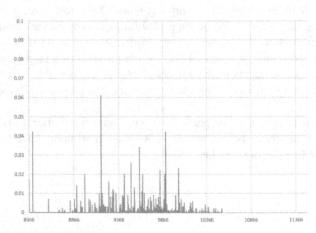

Fig. 7. Final distribution of solutions (*Car_8_8*)

2.6 Resolution of the Carlier 8 Jobs 9 Machines Instance

The Carlier 8×9 is a flow-shop with 8 jobs and 8 machines defining a total of 72 operations to schedule. The total number of permutations is $8! = 40320$ but there is only 1996 different costs.

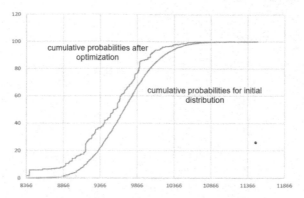

Fig. 8. Cumulative probabilities (comparison)

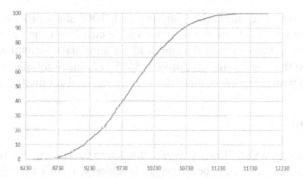

Fig. 9. Initial cumulative distribution (Car_8_9)

The cumulative distribution of Fig. 9 proves that the probability to have a solution with a cost lower that 8730 is about 0.002%.

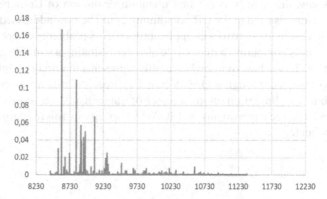

Fig. 10. Final distribution of solution (Car_8_9)

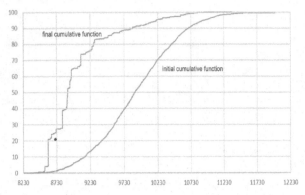

Fig. 11. Final cumulative function (Car_8_9) as regards the initial one

After the IQAOA execution, the sampling with 1000 shots gives a probability of 27% (Fig. 10 and Fig. 11) to have a solution with a cost lower than 8730 proving the IQAOA capability in transforming the amplitude of an initial distribution into that of a target state. Note that the optimal solution is 8505 [17] meaning that a probability to obtain a solution with a gap lower that 2.64% of the optimal solution, is about 27%.

3   Conclusion

We have assessed IQAOA's ability to solve the flow-shop problem, and the results demonstrate that Carlier's instances can be successfully addressed using this approach. To the best of our knowledge, this is the first quantum resolution of Carlier's instances. This approach is suitable for all problems where the indirect representation boils down to computing a rank on one hand, and having a suitable method for optimizing the angles on the other. We have evaluated the efficacy of IQAOA in tackling the flow-shop problem, and the results demonstrate its capability in effectively solving the Carlier's instances. As far as we know, this stands as the first quantum resolution of Carlier's instances, signifying that instances used in the OR community could be now addressed with quantum approaches. This method proves its applicability across problems where the indirect representation entails computing a rank while also necessitating a suitable approach for optimizing angles. To the best of knowledge, this represents the first resolution of Carlier's flow-shop instance through a QAOA-based approach that is an extension of the QAOA algorithm. The limitation on the number of qubits using the simulator does not permit to make intensive experiments and IQAOA capacity in solving larger instances has not been evaluated.

References

1. Garey, M.R., Johnson, D.S., Seth, R.: The complexity of flowshop and jobshop scheduling. Math. Oper. Res. **1**, 117–129 (1976)
2. Roy, B., Sussmann, B.: Les problèmes d'ordonnancement avec contraintes disjunctives. In: Note DS N°9 bis. SEMA, Paris (1964)

3. Cheng, A., Gen, M., Tsumjimura, Y.: A tutorial survey of job-shop scheduling problems using genetic algorithms – representations. Comput. Ind. Eng. **30**(4), 983–997 (1996)

4. Bourreau, E., Fleury, G., Lacomme, P.: Indirect quantum approximate optimization algorithms: application to the TSP (2023). arXiv:2311.03294

5. Knuth, D.: The Art of Computer Programming - Volume 3. Sorting and Searching. 2nd edn. Addison-Wesley, Reading (1981)

6. Laisant, C.A.: Sur la numération factorielle, application aux permutations. Bull. de la S.M.F. tome **16**, 176–173 (1888)

7. Mantaci, R., Rakotondrajao, F.: A permutation representation that knows what Eulerian means. Discrete Math. Theor. Comput. Sci. **4**, 101–108 (2001)

8. Hadfield, S.: Quantum algorithms for scientific computing and approximate optimization. Submitted in partial fulfillment of the requirements for the degree of doctor of Philosophy in the Graduate School of Arts and Sciences. Columbia University (2018)

9. Chassaing, M., Fontanel, J., Lacomme, P., Ren, L., Tchernev, N., Villechenon, P.: A GRASP× ELS approach for the job-shop with a web service paradigm packaging. Expert Syst. Appl. **41**(2), 544–562 (2014)

10. Anunay, F.A., Pandey, A., Kumar, S.K.: Mathematical models for multi-stage hybrid assembly flow-shop scheduling with preventive maintenance and release times. Comput. Ind. Eng. **186**, 109719 (2023)

11. Khatami, M., Salehipour, A., Cheng, T.C.E.: Flow-shop scheduling with exact delays to minimize makespan. Comput. Ind. Eng. **183**, 109456 (2023)

12. Geng, X.-N., Sun, X., Wang, J., Pan, L.: Scheduling on proportionate flow shop with job rejection and common due date assignment. Comput. Ind. Eng. **181**, 109317 (2023)

13. Mraihi, T., Driss, O.B., EL-Haouzi, H.B.: Distributed permutation flow shop scheduling problem with worker flexibility: review, trends and model proposition. Expert Syst. Appl. **238**, 121947 (2023)

14. Neufeld, J.S., Schulz, S., Buscher, U.: A systematic review of multi-objective hybrid flow shop scheduling. Eur. J. Oper. Res. **309**, 1–23 (2023)

15. Yenisey, M.M., Yagmahan, B.: Multi-objective permutation flow shop scheduling problem: literature review, classification and current trends. Omega **45**, 119–135 (2014)

16. Carlier, J.: Ordonnancements à contraintes disjonctives. RAIRO. Recherche opérationnelle. **12**(4), 333–350 (1978)

17. Ren, J., Ye, C., Yang, F.: Solving flow-shop scheduling problem with a reinforcement learning algorithm that generalizes the value function with neural network. Alex. Eng. J. **60**, 2787–2800 (2021)

18. Venturelli, D., Marchand, D.J.J., Rojo, G.: Quantum annealing implementation of job-shop scheduling (2016). arXiv:1506.08479v2

19. Carugno, C., Dacrema, M.F., Cremonesi, P.: Evaluating the job shop scheduling problem on a D-wave quantum annealer. Sci. Rep. **12**, 6539 (2022)

20. Schworm, P., Wu, X., Glatt, M., Aurich, J.C.: Solving fexible job shop scheduling problems in manufacturing with Quantum Annealing. Prod. Eng. Res. Devel. **17**, 105–115 (2023)

21. Kurowski, K., Pecynaa, T., Slysz, M., Rózycki, R., Waligóra, G., Weglarz, J.: Application of quantum approximate optimization algorithm to job shop scheduling problem. Eur. J. Oper. Res. **310**(2), 518–528 (2023)

22. Wilson, J.M.: Alternative formulations of a flow-shop scheduling problem. J. Opl. Res. Soc. **40**(4), 395–399 (1989)

23. Seda, M.: Mathematical models of flow-shop and job-shop scheduling problems. World Academy of Science, Engineering and Technology. 31 (2007)

24. Manne, A.S.: On the job-shop scheduling problem. Oper. Res. **8**, 219–223 (1960)

25. Wagner, H.M.: An integer linear-programming model for machine scheduling. Nav. Res. Logist. Q. **6**(2), 131–140 (1959)

26. Feo, T.A., Resende, M.G.C.: Greedy adaptative search procedures. J. Glob. Optim. **6**(2), 109–133 (1995)
27. Wolf, S., Merz, P.: Evolutionary local search for the super-peer selection problem and the p-hub median problem. In: Bartz-Beielstein, T., et al. (eds.) HM 2007. LNCS, vol. 4771, pp. 1–15. Springer, Heidelberg (2007). https://doi.org/10.1007/978-3-540-75514-2_1

Utilizing Graph Sparsification for Pre-processing in Max Cut QUBO Solver

Vorapong Suppakitpaisarn[1]([⊠]) [iD] and Jin-Kao Hao[2] [iD]

[1] The University of Tokyo, Tokyo, Japan
vorapong@is.s.u-tokyo.ac.jp
[2] University of Angers, Angers, France
jin-kao.hao@univ-angers.fr

Abstract. We suggest employing graph sparsification as a pre-processing step for max cut programs using the QUBO solver. Quantum(-inspired) algorithms are recognized for their potential efficiency in handling quadratic unconstrained binary optimization (QUBO). Various meta-heuristic approaches, including those based on the Quantum Approximate Optimization Algorithm, have been suggested for addressing QUBO challenges in this context. Given that max cut is an NP-hard problem and can be readily expressed using QUBO, it stands out as an exemplary case to demonstrate the effectiveness of quantum(-inspired) QUBO approaches. Here, the non-zero count in the QUBO matrix corresponds to the graph's edge count. Given that many quantum(-inspired) solvers operate through cloud services, transmitting data for dense graphs can be costly. By introducing the graph sparsification method, we aim to mitigate these communication costs. Experimental results on classical and quantum-inspired solvers indicate that this approach substantially reduces communication overheads and yields an objective value close to the optimal solution.

Keywords: quantum-inspired optimization · max cut problem · pre-processing · graph sparsification · quadratic unconstrained binary optimization (QUBO)

1 Introduction

Quantum and quantum-inspired computing are considered to have the potential to enhance the efficiency of solving various computational problems [17,22,49]. Consequently, many meta-heuristics have been proposed for solving QUBO on both quantum and quantum-inspired computers [10]. These methods include

We are grateful to the reviewers, Prof. Philippe Codognet, and Prof. Hiroshi Imai for their comments and suggestions, which helped us to improve the paper. This work was partially supported by the Japanese-French cooperation project JST SICORP Grant Number JPMJSC2208, Japan.

M. Sevaux et al. (Eds.): MIC 2024, LNCS 14753, pp. 219–233, 2024.
https://doi.org/10.1007/978-3-031-62912-9_22

those based on quantum annealing [47] and the Quantum Approximate Optimization Algorithm [39]. Given that several combinatorial and network optimization problems can be reformulated as QUBO, numerous researchers are actively exploring the most proficient methods for addressing these optimization problems with the aid of quantum(-inspired) QUBO solvers [8,9].

Researchers are particularly drawn to the maximum cut problem (max cut) [30,41] because it is an NP-hard problem [26] that can be easily expressed within the QUBO framework [11,46]. It has been observed that pre-processing the input before feeding it to QUBO solvers can yield good solutions more efficiently than using the original data directly. As a result, various studies have introduced pre-processing strategies specifically designed for the max cut problem to enhance the solution process [13,14,31].

Although minimizing computation time is important for solving the max cut problem, there is an additional challenge in addressing the problem with quantum or quantum-inspired QUBO solvers. Since quantum-inspired computers will not be commercially available for the next several decades, we are compelled to utilize these solvers through cloud services. This requires us to transmit our problems to the service providers, a step which often results in communication becoming a significant bottleneck [28,43]. Therefore, our focus in this paper is on diminishing the costs associated with this communication.

1.1 Our Contributions

The communication cost of the max cut problem is strongly related to the number of edges in the input graph. We therefore propose to use the graph sparsification technique by the effective resistance edge sampling [2,25,50] to reduce the communication cost. The effective resistance technique has been demonstrated to significantly reduce the number of edges in a graph while preserving the cut size [50].

Let the symbol $|V|$ represent the total number of nodes in our input graph. Building upon the theoretical foundations presented in [50], we demonstrate that for any chosen $\epsilon > 0$, the outcome of our sampling method can yield a solution for the max cut problem that approximates within a factor of $1 + \epsilon$. Simultaneously, this approach manages to decrease the edge count to $O\left(\frac{|V|\log|V|}{\epsilon^2}\right)$.

While a graph with $O\left(\frac{|V|\log|V|}{\epsilon^2}\right)$ edges is typically considered sparse in many applications, our experimental findings with both classical and quantum-inspired solvers demonstrate that setting the number of edges to fewer than $5|V|$ can still yield a viable approximate solution. Our study encompassed tests on 17 distinct networks, with node counts ranging from 100 to 12912 and edge numbers varying from 2124 to 807535. Moreover, these networks have a variety of topological structures. Remarkably, we have been able to reduce the number of edges – and consequently, the communication cost – by as much as 90%, while consistently achieving solutions where the cut size is at most 10% smaller than the maximum cut. Furthermore, we extracted subgraphs of varying sizes from two of the networks and verified that similar experimental outcomes are achievable in each of these subgraphs. This consistently suggests that our findings can be scaled up to larger networks, for which it is not feasible to upload all information to the cloud service.

We have also noticed a decrease in computation time when using classical QUBO solvers like Gurobi [20] on max cut instances where the edges have been sparsified using our sampling technique. For instance, while Gurobi could not complete the task on the original, dense max cut problem within two hours, it was able to finish in under two seconds after the sparsification has eliminated 90% of the edges. However, we do not consider this improvement as significant, because these solvers can still quickly find a reasonably good solution for both the original and the sparsified max cut instances. The reason Gurobi does not terminate with dense input graphs is primarily due to the extensive time required to prove the optimality of its solution.

As a pre-processing, our technique can benefit all solvers for max cut. The solvers which would have the biggest benefit from our pre-processing is the algorithm designed for addressing max cut on sparse graphs such as McSparse [7]. We believe that our pre-processing technique could improve the computation time of the McSparse algorithm, particularly when applied to dense graphs.

1.2 Related Works

The max cut problem has garnered widespread interest among researchers, leading to the development of numerous approximation and exact algorithms. Prominent among these are the well-known SDP relaxation algorithm [6,18,37] and algorithms for specific graph types [19,36,40,48]. In this paper, however, our focus is not on the algorithms for solving the max cut problem itself, but rather on its pre-processing. Consequently, our algorithm is designed to be compatible with all these various algorithms.

As outlined in [45], several pre-processing techniques for QUBO solvers have been developed. Among the most significant are those based on autarkies and persistencies, which enable the determination of some binary variable values in the optimal solution [21,42]. Additionally, there are methods that utilize the upper bound of the relaxed program to enhance solver efficiency [5,12], as well as approaches centered around variable fixing [3]. These methods have been shown to yield smaller QUBO instances that can exactly solve the original problems. In contrast, our paper introduces a pre-processing technique aimed at generating approximate QUBO instances. Importantly, our approach is designed to be compatible with these existing pre-processing methods.

The graph sparsification by edge sampling technique has been introduced to give an efficient algorithm for the maximum flow problem and the sparsest cut problem [27]. Also, it has been used as a pre-processing of the maximum cut problem in [1]. The goal of using the sparsification in that paper is not to reduce the communication cost as in this paper but to increase the precision of publishing the maximum cut results under differential privacy. Consequently, the sampling technique in [1] is different from the effective resistance sampling, which we have used in this paper.

2   Preliminaries

2.1   Max Cut Problem

Consider a weighted graph (V, E, w), where V represents the set of nodes in the graph. The set of edges is denoted as $E \subseteq \{\{u, v\} : u, v \in V, u \neq v\}$, and $w : E \to \mathbb{R}_{\geq 0}$ is the weight function assigning a non-negative weight $w(e)$ to each edge $e \in E$. A cut in graph G is defined as any subset $S \subseteq V$, with the weight of the cut S being $w_G(S) = \sum\limits_{\{u,v\} \in E : u \in S, v \notin S} w(\{u, v\})$. The max cut problem aims to find the cut in G that has the highest weight.

2.2   QUBO Formulation for the Max Cut Problem

The quadratic unconstrained binary optimization (QUBO) is the following mathematical programming problem

$$\max \sum_u \sum_v Q_{u,v} x_u x_v$$

subject to $x_i \in \{0, 1\}$ for all i.

To express the max cut problem stated in the previous section using QUBO, we let $x_u = 1$ if $u \in S$ and $x_u = 0$ otherwise. Also, let $w'(\{u, v\}) = w(\{u, v\})$ when $\{u, v\} \in E$ and $w'(\{u, v\}) = 0$ otherwise. Since $x_u^2 = x_u$ when $x_u \in \{0, 1\}$, the weight of a cut S is then

$$w(S) = \sum_{\{u,v\} \in E : x_u = 1, x_v = 0} w(\{u, v\}) = \sum_{\{u,v\} \in E} w(\{u, v\}) x_u (1 - x_v)$$

$$= \sum_{u,v} w'(\{u, v\}) x_u (1 - x_v)$$

$$= \sum_u \left[\sum_v w'(\{u, v\}) \right] x_u - \sum_{u \neq v} w'(\{u, v\}) x_u x_v$$

$$= \sum_u \left[\sum_v w'(\{u, v\}) \right] x_u^2 - \sum_{u \neq v} w'(\{u, v\}) x_u x_v.$$

By defining $Q_{u,u} = \sum\limits_v w'(\{u, v\})$ and $Q_{u,v} = -w'(\{u, v\})$ for $u \neq v$, we establish that the objective value of the QUBO corresponds to the cut size, which is also the objective value of the max cut problem. Consequently, maximizing this objective value leads to an optimal solution for the max cut problem.

In the context of solving the max cut problem with QUBO solvers available through cloud services, it becomes necessary to transmit the values of $Q_{u,v}$ for every pair of u, v. Consequently, the quantity of real numbers required to be sent is on the order of $O(|V|^2)$. This count becomes substantially large for large graphs, turning the communication cost into a critical bottleneck for the max cut solver.

By opting to submit only the non-zero entries of $Q_{u,v}$, we can significantly reduce the communication cost. This means sending the QUBO problem in the format $(u, v, Q_{u,v})$ where $Q_{u,v} \neq 0$. From our definition of $Q_{u,v}$, it is evident that for $u \neq v$, $Q_{u,v}$ is non-zero if and only if $\{u, v\} \in E$. Therefore, the communication cost with this method of submission is $O(|E|)$, which is substantially more efficient for scenarios where $|E| \ll |V|^2$, or in other words, when the input graph is sparse.

2.3 Graph Sparsification by Effective Resistances [50]

In this section, we explore the concept of graph sparsification through effective resistances. Consider the input graph denoted as $G = (V, E, w)$. Our objective is to construct a graph $\mathcal{G} = (V, \mathcal{E}, \mathbf{w})$ in such a way that for any cut $S \subseteq V$, the relationship $w_G(S) \approx w_{\mathcal{G}}(S)$ holds true. This approach aims to ensure that the weight of any given cut S in the original graph G closely approximates the weight of the same cut in the sparsified graph $\mathcal{G}$.

Given a parameter q, this method begins with an initially empty set $\mathcal{E}$. The process involves selecting edges from the graph G a total of q times to be added to $\mathcal{E}$. During each selection, every edge $e \in E$ has a chance of being chosen, with this probability denoted as p_e and to be detailed in the following paragraph. If an edge e that is not already in $\mathcal{E}$ is selected, we assign its weight in $\mathcal{G}$ as $\mathbf{w}(e) = w(e)/(q \cdot p_e)$. In cases where e is already in $\mathcal{E}$, we increase $\mathbf{w}(e)$ by $w(e)/(q \cdot p_e)$. This approach ensures that each edge's contribution to the total weight is adjusted based on its probability of selection and the number of selections, thereby maintaining the graph's structure in $\mathcal{G}$.

To establish the probability distribution $(p_e)_{e \in E}$, we start by defining the concept of effective resistance for each edge e in E, denoted as R_e. We treat the graph G as if it was an electrical circuit, where each edge e is equivalent to a resistor, the resistance of which is inversely proportional to the weight of the edge, given as $1/w_e$. In this analogy, the effective resistance R_e of an edge $e = \{u, v\}$ is understood as the electrical resistance experienced between nodes u and v.

Subsequently, the probability p_e for each edge e is defined as

$$p_e = w_e R_e \Big/ \sum_{e' \in E} (w_e R_{e'}).$$

This formulation assigns higher probabilities to edges with greater effective resistance, reflecting their relative importance in the electrical flow analogy of the graph.

The following theorem is shown in [50].

Theorem 1. *If* $q = 9|V| \cdot \log |V| / \epsilon^2$, *then, for all* $S \subseteq V$, $w_G(S) \leq w_{\mathcal{G}}(S) \leq (1 + \epsilon) w_G(S)$.

We have from the theorem that we would obtain a sparse graph with $|\mathcal{E}| = O(|V| \log |V|)$ that preserves the cut size by the sparsification technique.

3 Proposed Method

Our approach is depicted in Fig. 1. Rather than directly sending the original graph G to the QUBO solver provided by cloud services, we initially apply effective resistance sampling to sparsify the graph. The resultant sparsified graph, denoted as $\mathcal{G}$, is then submitted to the solver.

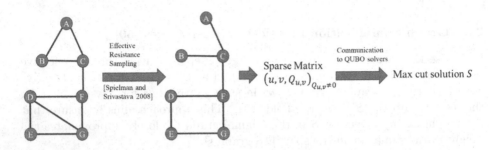

Fig. 1. Outline of our proposal

The following theorem is directly followed from Theorem 1.

Theorem 2. *Given that S' is a cut derived from the QUBO solver using our method, and S^* represents the optimal maximum cut, we can establish that:*

$$w_G(S') \leq w_G(S^*) \leq (1 + \epsilon)w_G(S').$$

Consequently, our algorithm is a $(1 + \epsilon)$-approximation algorithm for the max cut problem.

Proof. Because S' is the optimal max cut solution for the graph $\mathcal{G}$, we have that $w_{\mathcal{G}}(S') \geq w_{\mathcal{G}}(S^*)$. Applying Theorem 1, we obtain

$$w_G(S') \geq \frac{1}{1+\epsilon}w_{\mathcal{G}}(S') \geq \frac{1}{1+\epsilon}w_{\mathcal{G}}(S^*) \geq \frac{1}{1+\epsilon}w_G(S^*).$$

Hence, $w_G(S^*) \leq (1 + \epsilon)w_G(S')$.

Theorem 1 reveals that $|\mathcal{E}| = O(|V| \log |V|)$, indicating that the communication cost associated with sending the sparsified graph $\mathcal{G}$ to cloud servers is also $O(|V| \log |V|)$. Therefore, our method can achieve an asymptotic improvement in communication costs for dense input graphs where the number of edges is on the order of $O(|V|^2)$. However, when dealing with sparse input graphs, our approach does not yield a significant reduction in communication costs.

In Theorem 2, we assume that our QUBO solver is exact, meaning it always delivers the optimal solution. However, this result can be extended to scenarios where the solver is approximate. If the QUBO solver functions as an α-approximation algorithm, then the outcome produced by our method can be demonstrated to be an $\alpha(1 + \epsilon)$-approximation.

4   Experimental Results

We conduct experiments on the proposed method and give the experimental results in this section.

All experiments were carried out on a personal computer running Windows 11, equipped with an 11th Gen Intel(R) Core(TM) i7-1165G7 @2.80 GHz CPU and 16 GB of RAM. The code for these experiments was written in Python. Furthermore, we utilized publicly available datasets as provided in [38]. However, as it is assumed by the effective resistance samplings that all weights are non-negative, the weights used in our experiments are absolute values of those provided in the publicly available datasets. The values presented in this paper represent the mean of ten separate replications.

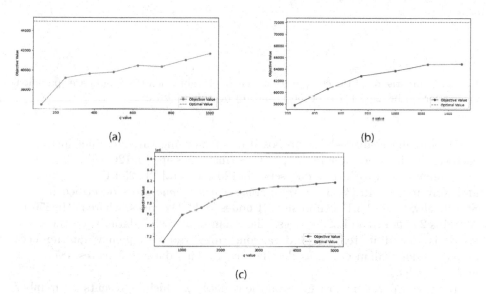

Fig. 2. Comparisons of the optimal values derived from the original max cut instances against the objective values from the sparsified graphs for (a) be120.3.1, (b) be250.1, and (c) mannino_k487.c

4.1   Gap in Solutions Due to Graph Sparsification

In this subsection, we examine the extent to which the optimal solutions are changed by effective resistance sampling. While our primary focus is on developing an algorithm suitable for quantum-inspired optimization, for these experiments, we have opted to use a classical solver, specifically Gurobi [20]. The rationale behind this choice is Gurobi's ability to guarantee the optimality of the solutions it generates. We employ the QUBO optimization feature available in Gurobi Optimods of Gurobi version 10.0.3.

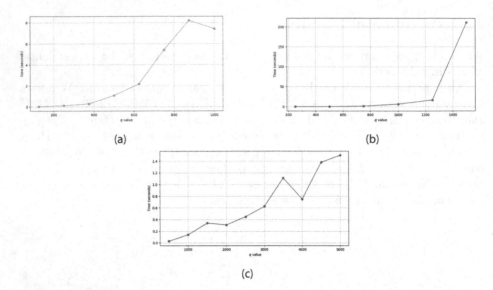

(a)   (b)

(c)

Fig. 3. Comparisons of the computation time of Gurobi when the inputs are sparsified graphs for (a) be120.3.1, (b) be250.1, and (c) mannino_k487.c

We initially evaluated our proposed methods using three distinct instances, each varying in node count and type. These instances are "be120.3.1", "be250.1", and "mannino_k487c". The datasets "be120.3.1" and "be250.1" are synthetic and were utilized in [3]. They were created using generators described in [44]. Specifically, "be120.3.1" comprises 121 nodes and 2242 edges, whereas "be250.1" contains 251 nodes and 3269 edges. The "mannino_k487c" dataset, on the other hand, is rooted in real-world data concerning radio frequency interferences among major Italian cities, as detailed in [4]. This dataset features 487 nodes and 8511 edges.

In these experiments, we focus on the variable q, which represents the number of times edges are sampled from the input graph. We experiment with varying the value of q. It is crucial to understand that q does not directly correspond to the number of edges in the sparsified graph, denoted as $|\mathcal{E}|$, since an edge can be selected multiple times during sampling. However, it is evident that $|\mathcal{E}| \leq q$, and generally, a larger q tends to result in a higher number of edges in the sparsified graph.

Theorem 1 suggests setting q to $\frac{9|V| \cdot \log |V|}{\epsilon^2}$. While this value is theoretically smaller than the edge count for dense graphs in an asymptotic sense, the sizeable constant factor $\frac{9}{\epsilon^2}$ can lead to a q that exceeds the actual number of edges, especially when the input graph G is relatively small. Take, for instance, when ϵ is 0.1, this results in approximately 522261 for "be120.3.1", 1248200 for "be250.1", and 2712316 for "mannino_k487c". Because of this, we have opted to use a reduced q value for our experimental evaluations.

Figure 2 presents the outcomes for the three specified instances. Examination of the figure reveals that the objective function improves as more edges are sampled and the value of q increases. Our method achieves a cut size that exceeds 90% of the optimal cut size for $q \geq 500$ in "be120.3.1", for $q \geq 1500$ in "be250.1", and for $q \geq 2000$ in "mannino_k487c". Correspondingly, these thresholds yield edge counts of 424 for "be120.3.1", 1092 for "be250.1", and 1231 for "mannino_k487c". These results indicate that our approach not only secures a 0.9-approximation to the solution but also facilitates a substantial reduction in communication costs-81% for "be120.3.1", 67% for "be250.1", and 86% for "mannino_k487c".

Table 1. The reduction in communication costs and the approximation ratios achieved by our algorithms across various graph types are detailed in [38]

Dataset Name	$|V|$	$|E|$	$|\mathcal{E}|$	Reduction in Comm. Cost	Optimal Value	Our Objective Value	Approx. Ratio
bqp250-1	251	3339	1163.6	0.65151	143669	129863	0.90390
gka1e	201	2124	810.7	0.61831	48263	42829	0.88741
ising2.5-150_6666	150	10722	387.3	0.96388	9067341	8502808	0.93774
g05_100.0	100	2475	452.3	0.81725	1430	1309	0.91538
w05_100.0	100	2343	432.3	0.81549	7737	7033.9	0.90912
G_1	800	19176	3509.3	0.81905	11624[a]	10412.3	0.89576

[a]Since the optimal value of G_1 is currently unknown, we instead use the best known objective value here.

Our experiment results with these three datasets yield a 0.9-approximation solution when setting q to roughly $5|V|$. Consequently, we extrapolate this finding to additional instance types in [38]. As demonstrated in Table 1, a similar approximation ratio is achieved for all tested instance types with q set at $5|V|$. Notably, there is a substantial decrease in communication cost particularly when the original graph G is dense.

4.2 Computation Time in Classical Solver

Because graph sparsification techniques are often employed to reduce computation time, it is worth investigating whether our sparsification method also reduces the computational times for classical solvers.

Figure 3 demonstrates that sparsification does indeed have a significant effect on reducing the computational time for Gurobi. For the original "be120.3.1" and "be250.1" inputs, the solver requires more than three hours to find a solution, whereas with the sparsified graphs at $q \approx 5|V|$, the computation times drop dramatically to 2.18 s and 16.6 s, respectively. There is also a clear pattern where larger values of q and increased edge counts correlate with longer computation times.

Despite this, it is noteworthy that Gurobi is able to quickly find reasonably good solutions for denser graphs. In every test conducted, solutions surpassing those of the sparsified graphs were obtained in under five seconds using the original graphs. A large part of the time that the solver spends on the dense graphs is devoted to proving that its solutions are optimal. This leads us to conclude that reduced computation time may not be a decisive advantage of sparsification techniques. Gurobi is capable of providing viable solutions for larger graphs. Nevertheless, given our focus on quantum-inspired solvers in this paper, we opt to limit our experiments with Gurobi to cases where it can assure the optimality of its results.

4.3 Experiments on Quantum-Inspired Solvers

We employed the Fixstars Amplify Annealing Engine [16] to corroborate our findings with QUBO quantum-inspired solvers. For all instances, whether original or sparsified, we imposed a solver time constraint of 10 s. Consistent with the methodology outlined in the preceding section, we set the value of q to be $5|V|$ in this experiment.

Table 2. Reduction in communication cost and changes in objective value by the graph sparsification technique when conducting experiments on QUBO quantum-inspired solver

Dataset Name	Reduction in Comm. Cost	Changes in Objective Value
be120.3.1	0.779	0.911
be250.1	0.704	0.891
mannino_k487.c	0.835	0.92
bqp250-1	0.71	0.895
gka1e	0.65	0.909
ising2.5-150_6666	0.964	0.947
g05_100.0	0.816	0.908
w05_100.0	0.812	0.902
G_1	0.812	0.897

Table 2 shows that the outcomes obtained using quantum-inspired solvers align closely with those presented in Table 1, confirming consistency across all datasets tested with the classical solver.

We also conducted experiments to test the effectiveness of our pre-processing methods on large graphs and real-world social networks. Our findings, presented in Table 3, demonstrate that our approach yields consistent results even with graphs exceeding 100,000 edges. We used graphs generated by a tool given in [24]. Specifically, rnd_graph1000_10_1 is a randomly generated graph with 1,000 nodes, a 10% connection density, and seed = 1. The leap_xx_y_z graphs represent leap graphs on y-dimensional chessboards of size xx, with the "z" parameter indicating the graph type: root graphs for z = 1 and bishop graphs for z

= 2. Additionally, "facebook", "congress", and "wiki_vote" were sourced from the Stanford large network dataset collection, representing real social networks [33]. These networks have been previously analyzed in several notable studies on social network behavior and characteristics [15,32,34]. The graph labeled "gplus_100000" represents a subgraph of the gplus network, induced by the initial 100,000 edges listed in the file. Given that this selection includes some repeated edges, the resulting graph comprises 12,912 nodes and 807,535 edges. This specific subset was chosen because the basic package of the Fixstars Amplify Annealing Engine supports a maximum of 16,000 nodes.

Table 3. Reduction in communication costs and changes in objective values by the graph sparsification technique when conducting experiments on QUBO quantum-inspired solver

Dataset Name	$|V|$	$|E|$	$|\mathcal{E}|$	Reduction in Comm. Cost	Objective Value Obtaining from Original Graph	Our Objective Value	Reduction in Obj. Value
rnd_graph1000_10_1	1000	49950	4752	0.90486	28587	26019.8	0.91020
leap_30_2_1	900	26100	401	0.98464	202500	178281.6	0.88040
leap_30_2_2	900	17110	3739.8	0.78143	101250	89984.5	0.88874
leap_10_3_1	1000	13500	3982.2	0.70502	37500	32921.9	0.87792
leap_10_3_2	1000	17100	4148.1	0.75742	31284	27711.2	0.88579
facebook	4039	88234	15874.2	0.82009	50600	46368.3	0.91637
congress	475	10209	1940.5	0.83598	38.98	34.37	0.88173
wiki_vote	7115	103689	22078.7	0.78707	73407	67675.3	0.92192
gplus_1000000	12912	807535	56175.7	0.93043	563656	400932.1	0.71131

Moreover, experiments were carried out on subgraphs of varying sizes derived from wiki_vote. These subgraphs were obtained through random walks to maintain the integrity of the network's structure. The data presented in Table 4 indicates that across all subgraph sizes tested, the cost reduction and approximation ratios achieved were consistent. This consistency leads us to posit that comparable outcomes are attainable for substantially larger graphs. Consequently, our pre-processing technique has the potential to markedly diminish communication costs in such scenarios.

Table 4. Reduction in communication costs and changes in objective values of subgraphs of the wiki_vote graph obtained from random walks on QUBO quantum-inspired solver

Random Walk Length	$|V|$	$|E|$	$|\mathcal{E}|$	Cost Reduction	Optimal Solution	Our Objective Value	Approx. Ratio
500	411	6523	1640.8	74.85%	4371	3749.4	85.78%
1000	755	18703	3208.6	82.84%	12563	10658.3	84.84%
1500	985	25812	4169.4	83.85%	17295	14861.8	85.93%
2000	1206	34352	5154	85.04%	23352	20100.4	86.08%
2500	1411	43040	6087.5	85.86%	29164	25123.8	86%
3000	1524	47535	6544.3	86.23%	32296	27871.2	86.30%
3500	1653	51371	7099.4	86.18%	35361	30478.3	86.19%
4000	1807	56964	7742.6	86.41%	39199	34066.9	86.91%
4500	1928	61425	8207.1	86.64%	42419	37179.1	87.65%
5000	2038	63534	8679.2	87.26%	43846	38260.9	87.26%

Table 5 displays the results for the Facebook network's subgraphs. It is evident that there's a variation in cost reduction among these subgraphs. This variation is likely a consequence of the network's multi-cluster structure, with each cluster possessing a distinct edge density, leading to differing levels of cost reduction. The random walk method does not uniformly sample nodes across clusters, resulting in varied cost reduction outcomes for the graphs generated by the algorithm. Although the cost reduction figures exhibit some fluctuation, they consistently fall within the range of 75% to 90%. On the other hand, despite the varied structures of each subgraph, the approximation ratio remains consistent across all experiments, lying within the 91% to 93% bracket.

Table 5. Reduction in communication costs and changes in objective values of subgraphs of the facebook graph obtained from random walks on QUBO quantum-inspired solver

Random Walk Length	$|V|$	$|E|$	$|\mathcal{E}|$	Cost Reduction	Optimal Solution	Our Objective Value	Approx. Ratio
500	312	13305	1390.1	89.55%	7298	6772.1	92.79%
1000	684	12186	2715.3	77.72%	7043	6452.4	91.61%
1500	882	25028	3658.2	85.38%	14029	12948.9	92.30%
2000	1022	16441	4045.3	75.40%	9600	8775.5	91.41%
2500	982	35349	4238.3	88.01%	19685	18181.6	92%
3000	1355	42222	5799	86.26%	23689	21843.5	92.21%
3500	1465	30107	5968.2	80.20%	17375	15889.8	91.45%
4000	1709	50934	7285.1	85.70%	28637	26409.1	92.22%
4500	1898	47302	7949.8	83.19%	26859	24618.1	91.66%
5000	1828	57824	7851.5	86.42%	32543	29918	91.93%

4.4 Discussions on Results on Classical and Quantum-Inspired Solvers

The argument could be made that similar or even superior approximation ratios to those achieved in our research might be attainable using an approximation algorithm based on semi-definite programming, as demonstrated in previous studies [18,37]. This algorithm is indeed capable of providing polynomial-time approximation solutions for max cut problems. However, a notable limitation of semi-definite programming is its computational intensity, particularly for problems involving over 100,000 nodes [29], where local execution becomes impractical. In such scenarios, our method proves advantageous, offering a viable solution by enabling the processing of these large instances through cloud services.

In these experiments, our primary objective is to demonstrate that edge sampling can yield reasonable approximation ratios. Therefore, we confined our experimentation to smaller instances (with $|V| \leq 12912$) where obtaining optimal solutions is feasible. Nonetheless, given the consistent results across all tested instance sizes, we are confident that similar outcomes would be achievable with larger graphs.

5   Conclusion and Future Works

We introduce the application of graph sparsification as a pre-processing step for solving the maximum cut problem in cloud-based environments. Our experimental results demonstrate that this approach, when applied to classical and quantum-inspired solvers, consistently yields solutions with an approximation ratio of about 0.9, while simultaneously achieving a significant reduction in communication costs to cloud servers, ranging between 60% and 95%.

In our future research, we plan to expand our experiments to include quantum solvers. At present, quantum solvers are limited to addressing small-scale max cut instances. As a result, the communication overhead required to send the max cut problem to the solvers is not significantly high at this point. On the other hand, it is understood that a sparser graph results in shallower quantum circuits, thereby reducing the noise in quantum computations. The graph sparsification technique has already been used for solving max cut for the noisy data published under differential privacy [1]. Additionally, a recent work [23] highlights that an increased number of edges may result in higher complexity during circuit optimization processes. A dense input graph reduces the likelihood of achieving an efficient quantum circuit. In summary, we hypothesize that the graph sparsification could improve solution quality and simplify the process of optimizing quantum circuits.

Pre-processing techniques for combinatorial optimization problems utilizing machine learning algorithms have been proposed in previous studies [35,51]. However, our initial experiments suggest that directly applying these methods to the max cut problem may not yield the best results. We observed that a machine learning model trained on small graphs does not effectively transfer to larger graphs within this problem domain. As a result, our future work aims to develop a machine learning-based sparsification technique specifically tailored for the max cut problem.

References

1. Arora, R., Upadhyay, J.: On differentially private graph sparsification and applications. In: Advances in Neural Information Processing Systems, vol. 32 (2019)
2. Benczúr, A.A., Karger, D.R.: Approximating s-t minimum cuts in $O(n^2)$ time. In: STOC 1996, pp. 47–55 (1996)
3. Billionnet, A., Elloumi, S.: Using a mixed integer quadratic programming solver for the unconstrained quadratic 0–1 problem. Math. Program. **109**, 55–68 (2007)
4. Bonato, T., Jünger, M., Reinelt, G., Rinaldi, G.: Lifting and separation procedures for the cut polytope. Math. Program. **146**, 351–378 (2014)
5. Boros, E., Crama, Y., Hammer, P.L.: Upper-bounds for quadratic 0–1 maximization. Oper. Res. Lett. **9**(2), 73–79 (1990)
6. Burer, S., Monteiro, R.D., Zhang, Y.: Rank-two relaxation heuristics for max-cut and other binary quadratic programs. SIAM J. Optim. **12**(2), 503–521 (2002)
7. Charfreitag, J., Jünger, M., Mallach, S., Mutzel, P.: McSparse: exact solutions of sparse maximum cut and sparse unconstrained binary quadratic optimization problems. In: ALENEX 2022, pp. 54–66 (2022)

8. Codognet, P.: Constraint solving by quantum annealing. In: ICPP Workshops 2021, pp. 1–10 (2021)
9. Codognet, P.: Domain-wall/unary encoding in QUBO for permutation problems. In: QCE 2022, pp. 167–173 (2022)
10. Dahi, Z.A., Alba, E.: Metaheuristics on quantum computers: inspiration, simulation and real execution. Future Gener. Comput. Syst. **130**, 164–180 (2022)
11. Dunning, I., Gupta, S., Silberholz, J.: What works best when? A systematic evaluation of heuristics for max-cut and QUBO. INFORMS J. Comput. **30**(3), 608–624 (2018)
12. Elloumi, S., Faye, A., Soutif, E.: Decomposition and linearization for 0–1 quadratic programming. Ann. Oper. Res. **99**(1–4), 79–93 (2000)
13. Ferizovic, D.: A practical analysis of kernelization techniques for the maximum cut problem. Ph.D. thesis, Karlsruher Institut für Technologie (KIT) (2019)
14. Ferizovic, D., Hespe, D., Lamm, S., Mnich, M., Schulz, C., Strash, D.: Engineering kernelization for maximum cut. In: ALENEX 2020, pp. 27–41 (2020)
15. Fink, C.G., et al.: A centrality measure for quantifying spread on weighted, directed networks. Phys. A **626**, 129083 (2023)
16. Fixstars: About Amplify AE (2023). https://amplify.fixstars.com/ja/docs/amplify-ae/about.html
17. Gharibian, S., Le Gall, F.: Dequantizing the quantum singular value transformation: hardness and applications to quantum chemistry and the quantum PCP conjecture. In: STOC 2022, pp. 19–32 (2022)
18. Goemans, M.X., Williamson, D.P.: Improved approximation algorithms for maximum cut and satisfiability problems using semidefinite programming. J. ACM **42**(6), 1115–1145 (1995)
19. Grötschel, M., Nemhauser, G.L.: A polynomial algorithm for the max-cut problem on graphs without long odd cycles. Math. Program. **29**(1), 28–40 (1984)
20. Gurobi Optimization, LLC: Gurobi optimizer reference manual (2021)
21. Hammer, P.L., Hansen, P., Simeone, B.: Roof duality, complementation and persistency in quadratic 0–1 optimization. Math. Program. **28**, 121–155 (1984)
22. Herrero-Collantes, M., Garcia-Escartin, J.C.: Quantum random number generators. Rev. Mod. Phys. **89**(1), 015004 (2017)
23. Ito, T., Kakimura, N., Kamiyama, N., Kobayashi, Y., Okamoto, Y.: Algorithmic theory of qubit routing. In: Morin, P., Suri, S. (eds.) WADS 2023. LNCS, vol. 14079, pp. 533–546. Springer, Cham (2023). https://doi.org/10.1007/978-3-031-38906-1_35
24. JRT: Rudy: a rudimental graph generator by JRT (2023). https://web.stanford.edu/~yyye/yyye/Gset/
25. Karger, D.R.: Global min-cuts in RNC, and other ramifications of a simple min-cut algorithm. In: SODA 1993, pp. 21–30 (1993)
26. Karp, R.M.: Reducibility among combinatorial problems. In: Jünger, M., et al. (eds.) 50 Years of Integer Programming 1958-2008, pp. 219–241. Springer, Heidelberg (2010). https://doi.org/10.1007/978-3-540-68279-0_8
27. Khandekar, R., Rao, S., Vazirani, U.: Graph partitioning using single commodity flows. J. ACM **56**(4), 1–15 (2009)
28. Kikuchi, S., Togawa, N., Tanaka, S.: Dynamical process of a bit-width reduced Ising model with simulated annealing. IEEE Access **11**, 95493–95506 (2023)
29. Kim, S., Kojima, M.: Exact solutions of some nonconvex quadratic optimization problems via SDP and SOCP relaxations. Comput. Optim. Appl. **26**, 143–154 (2003)

30. King, R.: An improved approximation algorithm for quantum max-cut. Quantum **7**, 1180 (2022)
31. Lamm, S.: Scalable graph algorithms using practically efficient data reductions. Ph.D. thesis, Karlsruher Institut für Technologie (KIT) (2022)
32. Leskovec, J., Huttenlocher, D., Kleinberg, J.: Predicting positive and negative links in online social networks. In: WWW 2010, pp. 641–650 (2010)
33. Leskovec, J., Krevl, A.: SNAP Datasets: Stanford large network dataset collection (2014). http://snap.stanford.edu/data
34. Leskovec, J., Mcauley, J.: Learning to discover social circles in ego networks. In: Advances in Neural Information Processing Systems, vol. 25 (2012)
35. Li, M., Tu, S., Xu, L.: Generalizing graph network models for the traveling salesman problem with Lin-Kernighan-Helsgaun heuristics. In: NeurIPS 2023, pp. 528–539 (2023)
36. Liers, F., Pardella, G.: Partitioning planar graphs: a fast combinatorial approach for max-cut. Comput. Optim. Appl. **51**(1), 323–344 (2012)
37. Mahajan, S., Ramesh, H.: Derandomizing approximation algorithms based on semidefinite programming. SIAM J. Comput. **28**(5), 1641–1663 (1999)
38. Mallach, S., Junger, M., Charfreitag, J., Jordan, C.: (Prototype of a) maxcut and BQP instance library (2021). http://bqp.cs.uni-bonn.de/library/html/index.html
39. Mazumder, A., Sen, A., Sen, U.: Benchmarking metaheuristic-integrated quantum approximate optimisation algorithm against quantum annealing for quadratic unconstrained binary optimization problems. arXiv preprint arXiv:2309.16796 (2023)
40. McCormick, S.T., Rao, M.R., Rinaldi, G.: Easy and difficult objective functions for max cut. Math. Program. **94**, 459–466 (2003)
41. Mirka, R., Williamson, D.P.: An experimental evaluation of semidefinite programming and spectral algorithms for max cut. ACM J. Exp. Algorithmics **28**, 1–18 (2023)
42. Nemhauser, G.L., Trotter, L.E., Jr.: Vertex packings: structural properties and algorithms. Math. Program. **8**(1), 232–248 (1975)
43. Oku, D., Tawada, M., Tanaka, S., Togawa, N.: How to reduce the bit-width of an Ising model by adding auxiliary spins. IEEE Trans. Comput. **71**(1), 223–234 (2020)
44. Pardalos, P.M., Rodgers, G.P.: Computational aspects of a branch and bound algorithm for quadratic zero-one programming. Computing **45**(2), 131–144 (1990)
45. Punnen, A.P.: The Quadratic Unconstrained Binary Optimization Problem: Theory, Algorithms, and Applications. Springer, Cham (2022). https://doi.org/10.1007/978-3-031-04520-2
46. Rehfeldt, D., Koch, T., Shinano, Y.: Faster exact solution of sparse MaxCut and QUBO problems. Math. Program. Comput. **15**(3), 445–470 (2023)
47. Rosenberg, G., Vazifeh, M., Woods, B., Haber, E.: Building an iterative heuristic solver for a quantum annealer. Comput. Optim. Appl. **65**, 845–869 (2016)
48. Shih, W.K., Wu, S., Kuo, Y.S.: Unifying maximum cut and minimum cut of a planar graph. IEEE Trans. Comput. **39**(5), 694–697 (1990)
49. Shor, P.W.: Introduction to quantum algorithms. In: Proceedings of Symposia in Applied Mathematics, vol. 58, pp. 143–160 (2002)
50. Spielman, D.A., Srivastava, N.: Graph sparsification by effective resistances. In: STOC 2008, pp. 563–568 (2008)
51. Tayebi, D., Ray, S., Ajwani, D.: Learning to prune instances of k-median and related problems. In: ALENEX 2022, pp. 184–194 (2022)

Addressing Machine Unavailability in Job Shop Scheduling: A Quantum Computing Approach

Riad Aggoune[1] (iD) and Samuel Deleplanque[2(✉)] (iD)

[1] ITIS Department, Luxembourg Institute of Science and Technology,
Esch-sur-Alzette Luxembourg, Luxembourg
riad.aggoune@list.lu
[2] CNRS, Centrale Lille, JUNIA, Univ. Lille, Univ. Valenciennes, UMR 8520 IEMN,
41 boulevard Vauban, 59046 Lille Cedex, France
samuel.deleplanque@junia.com

Abstract. We consider solving the Job Shop Scheduling Problem (JSSP) with machine unavailability constraints using an analog quantum machine and running the quantum annealing metaheuristic. We propose a technique to handle these new constraints, whether the unavailability periods are known or variable, in order to integrate them into the same type of disjunctive model processed by the analog machine: Binary, Unconstrained, and Quadratic. We present results on small-scale instances corresponding to what these quantum machines can handle.

Keywords: JSSP · non-availability constraints · quantum computing · QUBO · quantum annealing

1 Introduction

Quantum optimization, leveraging quantum computers and algorithms to address complex optimization issues, stands as a highly promising area in quantum computing. As in the classical domain, two principal strategies are utilized to solve combinatorial problems in quantum optimization: exact methods like Grover's search algorithm [10] and meta-heuristics such as Quantum Annealing (QA) [11] and the Quantum Approximate Optimization Algorithm (QAOA) [8]. Exact and variational methods like QAOA can be processed on universal gate-based quantum computers, such as IBM machines. In contrast, QA is tailored for analog quantum computers, notably those produced by D-Wave.

Quantum Annealing, a key metaheuristic in quantum optimization, is particularly designed for combinatorial optimization problems, drawing from the principles of quantum mechanics and emulating the process of simulated annealing [12]. It utilizes quantum phenomena, such as superposition and quantum tunneling, to efficiently navigate through local minima and target the global minimum of a cost function. For heuristic approaches like QA, it is often necessary to transform the optimization problem into a format compatible with

M. Sevaux et al. (Eds.): MIC 2024, LNCS 14753, pp. 234–245, 2024.
https://doi.org/10.1007/978-3-031-62912-9_23

quantum computers. Quadratic Unconstrained Binary Optimization (QUBO) is generally the preferred format for mapping problems to quantum computers.

In this work, we study the resolution of the job shop scheduling problem with availability constraints by the quantum annealing metaheuristic. We first describe the problem in the following section and review the quantum-based solution methods recently proposed in the literature. Then, the QUBO formulation of the JSSP adapted from [1] is presented. Through minor modifications, we show how this QUBO can be adapted to integrate both fixed and flexible availability constraints. The paper concludes with numerical results obtained using D-Wave's quantum annealing machines and overall conclusions. This synthesis merges the concept of quantum annealing's efficacy with broader quantum optimization approaches and their application to specific problems like JSSP, highlighting the diverse methodologies and quantum computing platforms in use.

2 Problem Definition

The Job Shop Scheduling Problem with Availability Constraints (JSSP-AC) can be stated as follows: A set of n jobs $J = \{J_1, J_2, \ldots, J_n\}$ has to be processed on a set of m machines $M = \{M_1, \ldots, M_m\}$. Each job J_i consists of a linear sequence of n_i operations $(O_{i1}, O_{i2}, \ldots, O_{in_i})$. Each machine can process only one operation at a time and each operation O_{ij} with a processing time of p_{ij} time units needs exactly one machine. Each job visits the machines according to its own predefined routing. This problem generalizes the flow shop scheduling problem, in which all the jobs are processed following the same routing $(M_1, M_2, \ldots, M_m)$. There are k unavailability periods $\{h_{j1}, h_{j2}, \ldots, h_{jk}\}$ on each machine M_j. Two cases are considered in the paper: either the starting date S_{jk} of unavailability period h_{jk} of duration p'_{jk} is known in advance and fixed, or it is flexible and can vary within a time window. The objective is to determine the starting date of each operation O_{ij} so that the makespan noted C_{max} is minimized. The job shop scheduling problem with availability constraints is strongly NP-hard since the 2-machine flow shop scheduling problem is strongly NP-hard [4].

In what follows we first focus and the Job shop scheduling problem, then we generalize the approach to integrate the availability constraints.

The traditional solution approaches to solve JSSP include heuristics and meta-heuristics as well as exact methods, such as branch-and-bound and constraint programming [5]. The linear disjunctive model [15] for the JSSP can be expressed as follows. The starting times are represented by the integer variable vector, denoted by x. We use z to denote the binary variable vector, which satisfies the following conditions:

$$z_{ijk} = \begin{cases} 1 \text{ if the job } j \text{ precedes job } k \text{ on machine } i, \\ 0 \text{ otherwise.} \end{cases}$$

We note by $(\sigma_1^j, \ldots, \sigma_h^j, \ldots, \sigma_m^j)$ the processing order of job j through the machines. The minimization of the objective function (1) forces all the jobs to be finished as soon as possible.

$$\sum_{j \in J} x_{\sigma_m^j j} \tag{1}$$

Another objective extensively discussed and utilized within the literature is the concept of makespan. By adding constraint (2) to the model and replacing the objective function (1) with the minimization of the C_{max} variable, in order to ensure that the last job finishes as early as possible.

$$C_{max} \geq \sum_{j \in J} (x_{\sigma_m^j j} + p_{\sigma_m^j j}) \tag{2}$$

Constraints (3) forbid consecutive operations of one job to start before the previous one is finished.

$$x_{\sigma_h^j j} \geq x_{\sigma_{h-1}^j j} + p_{\sigma_{h-1}^j j} \qquad \forall j \in J, h = 2..m \tag{3}$$

Big $\mathcal{M}$ constraints (4) and (5) forbid to have more than one operation at a time on a given machine.

$$x_{ij} \geq x_{ik} + p_{ik} - \mathcal{M} z_{ijk} \qquad \forall j, k \in J, j < k, i \in M \tag{4}$$

$$x_{ik} \geq x_{ij} + p_{ij} - \mathcal{M}(1 - z_{ijk}) \qquad \forall j, k \in J, j < k, i \in M \tag{5}$$

3   Related Works

In the literature, the number of papers dedicated to quantum solutions for hard combinatorial optimization problems is growing fast. In particular, the job shop scheduling problem and its extensions is attracting more and more research works involving quantum computing. Those works can be classified according to the types of quantum computers and algorithms used to solve the problems: analog, universal computers, and simulators. In general, the solution approaches consist in first mapping the decision variables of the considered problems to the qubits of the quantum computer. Then, quantum algorithms are applied to make the qubits value evolve until solutions are found. Solving optimization problems with quantum computers is therefore strongly limited by the number of qubits available, among other hardware constraints.

Since the number of qubits is smaller in universal quantum computers, the studies of the JSSP involving those computers are scarce. The first one was developed by [2]. The authors have proposed four variational quantum heuristics for solving a JSSP with early and late delivery as well as production costs, adapted from a steel manufacturing process. They have compared the performance of the heuristics on two-machine flow shop instances using IBM gate-based computers with 5 to 23 qubits.

Recently, [14] have proposed a QAOA approach to the JSSP with a particular method for updating the parameters of the algorithm. The authors have also investigated the relationship between makespan and energy minimzation.

The number of research works involving quantum annealers and simulators is significantly higher. The first quantum computing approach for solving the JSSP was proposed by Venturelli *et al.* [19]. The authors proposed a time-indexed QUBO formulation and a quantum annealing solution for the makespan minimization. The method was implemented on a D-Wave quantum annealer, with 509 working qubits. The authors also proposed variable pruning techniques, through window shaving and immediate selections, to reduce the number of necessary qubits. The proposed QUBO model has been re-used in several studies, as listed below.

In [13] the authors have developed a hybrid quantum annealing heuristic to solve a particular instance of the job shop scheduling problem on the D-Wave 2000Q quantum annealing system that consists of 2041 qubits and a maximum of 6 connections between qubits. The proposed approach includes variable pruning techniques and a processing window heuristic. In [6], job shop instances with unitary operations have been tested on the D-Wave Advantage machine, built upon 5640 qubits and 15 possible connections between qubits. Extensive experiments with the reverse annealing procedure and comparisons with simulated annealing are also described. In [1], we have proposed a QUBO formulation for the minimisation of the total completion time in a job shop. The model was solved using the D-Wave hybrid solver and Advantage quantum annealing computer.

The flexible job shop, which is a generalization of the job shop problem with pools of parallel machines available for processing operations was considered in [7]. The authors proposed a QUBO derived from the one of [19] and an iterative procedure to solve relatively large size instances on a specialized hardware ([3]). Using the QUBO formulations proposed in [7], the authors in [18], also tackle the flexible job shop scheduling problem with the D-Wave solvers comparing various input models. Another QUBO formulation is proposed in [16] for assigning dispatching rules to the machines and scheduling the operations in a flexible job shop system. The problem is solved using the leap hybrid solver. In [17], the authors propose a QUBO formulation for the job shop scheduling with worker assignment considerations. Possible ways to approximate the makespan are discussed and instances solved with the Fujitsu Digital Annealer are described. In the same environment, the authors in [20] efficiently solve large instances of JSSP with a hybrid approach that combines constraints programming and QUBO models for one-machine problems.

The present paper also aims at extending the job shop scheduling model and solution approach, in particular the one proposed in [1], by considering additional constraints that are important in practice. To the best of our knowledge, it is the first study in the quantum optimization literature that integrates availability constraints on the machines of both fixed and flexible types.

4   QUBO Formulation

The QA metaheuristic, as executed on a D-Wave quantum machine, takes as input an unconstrained binary model, which can be quadratic. Either an Ising model ({$-1; +1$} variable values) or a QUBO model ({$0; 1$} variable values) can be provided. Since both models are isomorphic, and the machine is capable of converting QUBO into Ising, we focus on the classical binary variables in computing to more easily establish a connection with known MILP models.

We add some notations to those used in the linear formulation of the previous section. We use x to denote the binary variable vector, which, for each i, j and t, with $i = 1..n_i$, $j = 1..n$, $t = 1..T$, satisfies the following conditions:

$$x_{ij}^t = \begin{cases} 1 \text{ if the operation } i \text{ of the job } j \text{ starts in period } t, \\ 0 \text{ otherwise.} \end{cases}$$

The minimization of the Objective function $f(x)$ forces the last operations of all jobs to start globally as soon as possible (see expression (6)). Here, we adapt the objective function (1) from the integer formulation to a binary formulation that we develop for the QUBO:

$$f(x) = \sum_j \sum_t t.x_{n_i j}^t. \tag{6}$$

For optimizing the makespan, it is sufficient to add a virtual job consisting of a single operation that is executed instantly which will be connected to the last operations of the non-virtual jobs by precedence constraints ((10)). It then only remains to exclusively minimize the execution date of the virtual job as in function (7), where n_i of the virtual job $n + 1$ is equal to 1 since there is only one operation.

$$f(x) = \sum_t t.x_{(1)(n+1)}^t. \tag{7}$$

To force each operation to start exactly once through a relaxed constraints into the objective function, we apply a penalty $P1$ such that $P1(x) = \sum_i \sum_j P1(x, i, j)$ where each element is given by the expression (8).

$$P1(x, i, j) = (\sum_t x_{ij}^t - 1)^2, \qquad i = 1..n_i, j = 1..n. \tag{8}$$

We note $M_{ij}, i = 1..n_i, j = 1..n$, the required machine for the operation i of the job j. $P2(x)$ is the penalty that forbids to have more than one operation at a time on a given machine, such that $P2(x)$ is the sum of each element calculated by the quadratic expression (9).

$$P2(x, i, j, t, i', j', t') = x_{ij}^t x_{i'j'}^{t'},$$
$$(i, j, t) \cup (i', j', t') : i, i' = 1..n_i, j, j' = 1..n, (i, j) \neq (i', j'), \tag{9}$$
$$M_{ij} = M_{i'j'}, (t, t') \in T^2, 0 \leq t' - t < p_{ij}.$$

The last Penalty which is noted $P3(x)$ forbids consecutive operations to start before the previous one is finished. Each element of $P3(x)$ is calculated by the quadratic expression (10).

$$P3(x, i, j, t, t') = x_{ij}^t x_{i+1j}^{t'},$$

$$i = 1..(n_i - 1), j = 1..n, (t, t') \in T^2, t + p_{ij} > t'. \tag{10}$$

We can finally express the JSSP quadratically and without constraints through the QUBO formulation of the JSSP with its penalties balanced by 3 multipliers, λ_1, λ_2, and λ_3 (see expression 11) and its detailed form of equality (12).

$$f^{QUBO}(x) = f(x) + \lambda_1 P1(x) + \lambda_2 P2(x) + \lambda_3 P3(x). \tag{11}$$

$$f^{QUBO}(x) = \sum_j \sum_t t . x_{n_{ij}}^t + \lambda_1 \sum_j \sum_i (\sum_t x_{ij}^t - 1)^2$$

$$+\lambda_2 \sum_{(i,j,t) \cup (i',j',t') \in T1} x_{ij}^t x_{i'j'}^{t'} + \lambda_3 \sum_{(i,j,t,t') \in T2} x_{ij}^t x_{i+1j}^{t'}.$$

with: $\qquad$ (12)

$$T1 = (i, j, t) \cup (i', j', t') : i, i' = 1..n_i, j, j' = 1..n, (i, j) \neq (i', j'),$$

$$M_{ij} = M_{i'j'}, (t, t') \in T^2, 0 < t' - t < p_{ij}.$$

$$T2 = (i, j, t, t') : i = 1..(n_i - 1), j = 1..n, (t, t') \in T^2, t + p_{ij} > t'.$$

5   Non Fixed Resource Availability Constraints

Let's consider the problem of resource constraints due to unavailability, whether these are fixed or variable. The management of these resources proves to be intuitive when the problem is formulated as a QUBO. The UML activity diagram shown in Fig. 1 illustrates the methodology for developing the QUBO, with a particular emphasis on non-availability constraint management.

When a resource's unavailability is constant over time, it can be treated as a single operation already scheduled. Thus, it becomes possible to spread this constraint throughout all the other operations that cannot simultaneously use the resource. This consideration is expressed through elementary quadratic expressions of the form $x_{ij}^t x_{i'j'}^{t'}$, where i and j denote the fictive operation representing the resource's unavailability during a certain period p_{ij}. For any t included in this period, and for all operations characterized by i' and j' that use the same resource, we impose the constraint $x_{ij}^t x_{i'j'}^{t'} = 0$ with the related penalty.

When a resource's unavailability has to be scheduled, it should be considered as a unique operation of a project that can be scheduled at any time. If the objective is to minimize the makespan, the virtual operation related to the resource non-availability is integrated in the calculation of this makespan as a last operation of a job. Hence, the unavailability constraints, regardless of their nature, can be sequentially incorporated into the QUBO. We finally obtain a JSSP problem with additional jobs comprising single operations. The QUBO

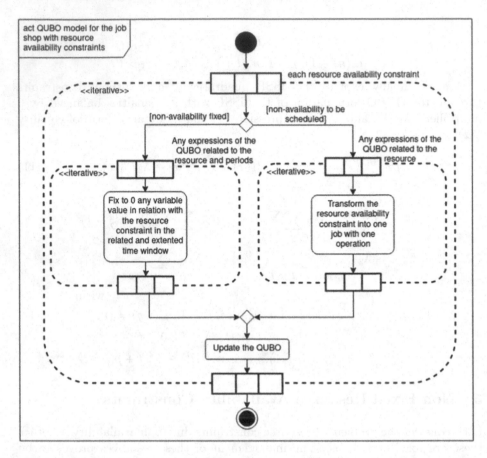

Fig. 1. The non-availability constraints are divided into two subsets: those whose unavailability window is already known, and those where this window is variable. In both cases, we consider each unavailability as a unique operation of a new Job. In the first case, it involves fixing the variables corresponding to these specific operations, and, in the second case, it involves considering the new operations as any other activities where optimization will lead them to finish globally as soon as possible.

model used is thus the same $f^{QUBO}(x)$ as the one given by expression (12) in the previous section.

In Fig. 2, we present an example of the Job Shop Scheduling Problem (JSSP) incorporating various types of unavailability constraints. The period labeled 'U1' denotes a time during which Machine 1 is unavailable, specifically in period 3. Conversely, the 'U2' period represents a variable unavailability duration. It is noteworthy that this variable unavailability period is strategically optimized prior to commencing the first operation of Job 3. As depicted, minimizing the makespan necessitates careful scheduling around the ends of these variable unavailability periods.

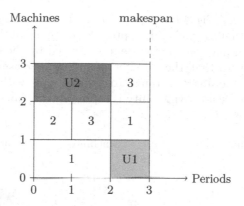

Fig. 2. A solution to JSSP with unavailability constraints, featuring three jobs and a total of six operations. Unavailability constraint U1 is associated with a fixed period of non-availability, whereas the period for U2 is variable.

6   Computational Experiments

In [1], we studied the impact that the number of periods in a JSSP instance has on its resolution. We noticed that iterating different experiments with a reduction in the number of periods until an infeasible solution is obtained showed a significant improvement in results, more than what could be expected from classical calculations, due to the narrowing of the solution space. In this study, we directly considered a relatively small number of periods. We achieved this result empirically and quickly, thanks to a small number of replications initially performed on a QUBO related to the same instance but with a larger number of periods.

We opted for the quantum quadratic unconstrained binary solver, which is non-hybrid, in contrast to the method of resolution discussed in [1], which relied on D-Wave's hybrid solution for solving constrained binary quadratic problems. The experiment was conducted on an initial instance of the JSSP with 3 jobs for a total of 7 operations. The size of the base instance was chosen to enable processing by the D-Wave Advantage machine, while also considering the ability to achieve optimality. The instance with a fixed period of unavailability is distinguished by the stopping of machine 2 during the second time period. The third case concerns a variable unavailability of machine 3 over a single time period. We aimed to minimize the makespan by adding a job with only a virtual operation at the beginning and another at the end of the experiment. This was done by adding precedence constraints, focusing the objective on minimizing the start date of the last virtual operation.

For these preliminary results, we achieved the optimal solution in all three cases, where each time 3 periods were necessary to meet all the constraints of the 3 problems. The figures respectively represent the optimal solutions for the JSSP case and for the case with variable unavailability. In the latter, the unavailability period was placed after all other operations, without affecting the makespan. Figure 3 and Fig. 4 correspond respectively to Figs. 5 and 4. We can observe the impact that the embedding process (mapping of the QUBO graph to the qubit graph) can have on the problem addressed by the quantum machine. This experiment shows that the number of qubits increases from 122 to 208 when moving from the JSSP instance to the instance with variable machine unavailability, where such unavailability is represented by an additional job with a single operation (Fig. 6).

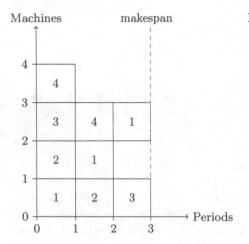

Fig. 3. JSSP without unavailability constraints solution obtained.

Fig. 4. JSSP with variable unavailability constraints solution obtained.

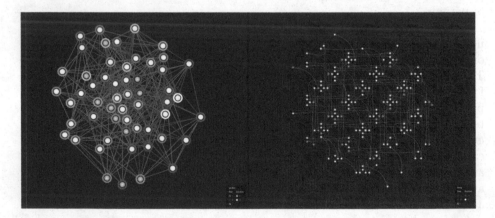

Fig. 5. Representation of the QUBO graph (left), corresponding to a scenario where all machines are continuously available, and its embedding (right) into the qubit graph of a the Advantage machine. This particular case involves 4 jobs and a total of 9 operations. The QUBO model has 48 variables, while the embedding on the QPU requires 122 qubits. (Visualization created using D-Wave Inspector).

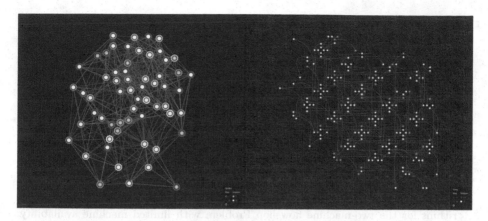

Fig. 6. Representation of the QUBO graph (left) illustrating a scenario in which a machine is unavailable during a specific period, and its corresponding embedding (right) into the qubit graph of the Advantage machine. This instance involves managing 4 jobs with a total of 9 operations. The QUBO model comprises 60 variables, and its embedding into the QPU utilizes 208 qubits. (Visualization created using D-Wave Inspector).

7   Discussion

In our study, we conducted an analysis of recent quantum computing strategies for the JSSP, particularly emphasizing QUBO models and their role in effectively incorporating practical constraints. We apply this method to a JSSP with machine unavailability constraints. These constraints can be either variable or fixed. Starting from a QUBO that models the JSSP and is well-suited for quadratic modeling, we treat these new constraints as unique operations of jobs that are added to the initial ones. Thus, the same QUBO formulation could be implemented to describe instances for the quantum machine.

Our research included the development and testing of this QUBO model using D-Wave's quantum annealing technology. We noted that current quantum hardware has limitations in managing the volume of variables produced by such models. Despite this, it's crucial to continue refining the modeling of real-world problems, as quantum approaches, while not currently outperforming classical methods, hold potential for future advancements, especially with the anticipated increase in available qubits. Our future research aims to explore methods to minimize variable count while still efficiently embedding necessary constraints. Additionally, the recent increase in the number of qubits, surpassing the 1000-qubit threshold in machines from IBM or Atom Computing, now allows for the consideration of solving small-scale instances like those addressed in this work. This is made possible through quantum algorithms such as QAOA implemented on these discrete (gate-based) machines.

References

1. Rao, P.U., Sodhi, B.: Scheduling with multiple dispatch rules: a quantum computing approach. In: Groen, D., de Mulatier, C., Paszynski, M., Krzhizhanovskaya, V.V., Dongarra, J.J., Sloot, P.M.A. (eds.) Computational Science – ICCS 2022. ICCS 2022. LNCS, vol. 13353, pp. 233–246. Springer, Cham (2022). https://doi.org/10.1007/978-3-031-08760-8_20
2. Amaro, D., Rosenkranz, M., Fitzpatrick, N., Hirano, K., Fiorentini, M.: A case study of variational quantum algorithms for a job shop scheduling problem. EPJ Quantum Technol. **9**, 100–114 (2022)
3. Aramon, M., et al.: Physics-inspired optimization for quadratic unconstrained problems using a digital annealer. Front. Phys. **7**, 48 (2019)
4. Błażewicz, J., Breit, J., Formanowicz, P., Kubiak, W., Schmidt, G.: Heuristic algorithms for the two-machine flowshop Problem with limited machine availability. Omega J. **29**, 599–608 (2001)
5. Da Col, G., Teppan, E.C.: Industrial-size job shop scheduling with constraint programming. Oper. Res. Perspect. **9** (2022)
6. Carugno, C., Ferrari Dacrema, M., Cremonesi, P.: Evaluating the job shop scheduling problem on a D-wave quantum annealer. Sci. Rep. **12**, 6539 (2022). https://doi.org/10.1038/s41598-022-10169-0
7. Denkena, B., Schinkel, F., Pirnay, J., Wilmsmeier, S.: Quantum algorithms for process parallel flexible job shop scheduling. CIRP J. Manuf. Sci. Technol. 12142 (2020)
8. Farhi, E., Goldstone, J., Gutmann, S.: A Quantum Approximate Optimization Algorithm, 2014. https://doi.org/10.48550/arxiv.1411.4028
9. Michael, G.R., Johnson, D.S.: Computers and Intractability: A Guide to the Theory of NP-Completeness. W. H. Freeman and Co., USA (1979)
10. Grover, L.K.: A fast quantum mechanical algorithm for database search. In: Proceedings of the Twenty-Eighth Annual ACM Symposium on Theory of Computing, pp. 212–219 (1996)
11. Tadashi, K., Nishimori, H.: Quantum annealing in the transverse ising model. Phys. Rev. E **58**(5), 5355 (1998)
12. Kirkpatrick, S., Gelatt, C.D., Jr., Vecchi, M.P.: Optimization by simulated annealing science vol. 220, p. 4598 (1983)
13. Kurowski, K., Węglarz, J., Subocz, M., Różycki, R., Waligóra, G.: Hybrid quantum annealing heuristic method for solving job shop scheduling problem. In: Krzhizhanovskaya, V.V., et al. (eds.) ICCS 2020. LNCS, vol. 12142, pp. 502–515. Springer, Cham (2020). https://doi.org/10.1007/978-3-030-50433-5_39
14. Kurowski, K., Pecyna T., Slysz R., Różycki, R., Waligóra, G., Węglarz, J.: Application of quantum approximate optimization algorithm to job shop scheduling problem. Eur. J. Oper. Res. **310**, 518–528 (2023). https://doi.org/10.1016/j.ejor.2023.03.013
15. Manne, A.S.: On the job-shop scheduling problem. Oper. Res. **8**(2), 219–223 (1960). https://doi.org/10.1287/opre.8.2.219
16. Rao, P.U., Sodhi, B.: Scheduling with multiple dispatch rules: a quantum computing approach. In: Groen, D., de Mulatier, C., Paszynski, M., Krzhizhanovskaya, V.V., Dongarra, J.J., Sloot, P.M.A. (eds.) Computational Science – ICCS 2022. ICCS 2022. LNCS, vol. 13353, pp. 233–246. Springer, Cham (2022). https://doi.org/10.1007/978-3-031-08760-8_20

17. Shimada, D., Shibuya, T., Shibasaki, T.: A decomposition method for makespan minimization in job-shop scheduling problem using ising machine. In: 2021 IEEE 8th International Conference on Industrial Engineering and Applications, pp. 307–314 (2021)
18. Schworm, P., Wu, X., Glatt, M., Aurich, J.C.: Solving flexible job shop scheduling problems in manufacturing with Quantum Annealing.. Prod. Eng. Res. Dev. **17**, 105–115 (2023). https://doi.org/10.1007/s11740-022-01145-8
19. Venturelli, D., Marchand, D.J., Rojo, G.: Quantum annealing implementation of job-shop scheduling, 2015. arXiv preprint: 1506.08479
20. Zhang, J., Lo Bianco, G., Beck, J.C.: Solving job-shop scheduling problems with QUBO-based specialized hardware. In: Proceedings of the International Conference on Automated Planning and Scheduling, vol. 32, no. 1, pp. 404–412 (2022)

Solving Edge-Weighted Maximum Clique Problem with DCA Warm-Start Quantum Approximate Optimization Algorithm

Huy Phuc Nguyen Ha[1], Viet Hung Nguyen[2], and Anh Son Ta[1(✉)]

[1] School of Applied Mathematics and Informatics, Hanoi University of Science
and Technology, Hanoi, Dai Co Viet, Vietnam
son.taanh@hust.edu.vn
[2] Université Clermont-Auvergne, CNRS, Mines de Saint-Etienne,
Clermont Auvergne INP, LIMOS, Clermont-Ferrand, France

Abstract. The Quantum Approximate Optimization Algorithm is a hybrid quantum-classic algorithm used for solving combinatorial optimization. However, this algorithm performs poorly when solving the constrained combinatorial optimization problem. To deal with this issue, we consider the warm-start Quantum Approximate Optimization Algorithm for solving constrained problems. This article presents a new method for improving the performance of the Quantum Approximate Optimization Algorithm, with the Difference of Convex Optimization. Our approach focuses on the warm-start version of the algorithm and uses the Difference of Convex optimization to find the warm-start parameters. To show our method's efficiency, we do several experiments on the edge-weighted maximum clique problem and see a good result.

Keywords: Maximum edge-weighted clique · QAOA · warm-start · DCA

1 Introduction

The Quantum Approximate Optimization Algorithm [8] is the hybrid classical-quantum algorithm for solving combinatorial optimization problems. This algorithm has many applications in solving real-life problems: portfolio optimization [14], schedule problem [17], wireless scheduling [18], artificial neural network [19], feature selection [1], and many other applications. The QAOA is computing the expectation from solutions generated from the quantum circuit and the optimization of this function is the NP-hard problem [10]. There are many recent improvements of the algorithm in many aspects: initial parameters method [2] and [21], mixer operator [9] and [22] for constraint preserving, finding the optimal solution of the expectation [6] and CVaR optimization [5]. These methods improve the performance of the QAOA in solving combinatorial optimization problems, especially the maximum cut problem. However, the improvement of

M. Sevaux et al. (Eds.): MIC 2024, LNCS 14753, pp. 246–261, 2024.
https://doi.org/10.1007/978-3-031-62912-9_24

the initial parameters [2] and the improvement of finding the optimal solution of the expectation [6] do not assure that the probability of the optimal solutions is greater than 0 because it considers all solutions with the same probability when these methods are used to solve the constrained problems. Furthermore, the CVaR optimization method in [5] is a bottleneck when choosing a suitable probability tail value. The QAOA limitation on solving constrained problems is also proved in [3] by showing that the QAOA can produce optimal solutions with negligible probability. The warm-start method can help us to improve the QAOA significantly because the relaxed version of the combinatorial optimization problems is close to the solution, and the probability of optimal value will be greater than 0. [9] introduces 2 warm-start methods: continuous value warm-start and rounded warm-start. The continuous value warm-start is shown for satisfying the adiabatic quantum computing theorem [8], which is shown for converging with an infinite number of layers. In contrast, the rounded warm-start method does not satisfy the adiabatic quantum computing theorem [8]. To find warm start parameters, [9] use the CPLEX solver for the portfolio optimization problem with a convex relaxed problem with box constraints for continuous value warm-start and use SDP for the maximum cut problem with a non-convex relaxed problem and use a regularized method for the solution from SDP for rounded warm-start. However, the box-constrained relaxed problem with indefinite quadratic programming is NP-hard [15]. In this article, we consider the relaxation problem with the feasible solution space approximated by a sphere. This problem can be solved quite efficiently by some available algorithms and the DCA shows its efficiency when solving this problem globally [12]. This article concentrates on improving the mixer and the initial state of the QAOA by using the Difference of Convex Algorithm (DCA), and our method is suitable for improving the performance of the warm start QAOA.

An effective method for nonconvex continuous optimization is DC programming and DCA, which Le Thi Hoai An and Pham Dinh Tao proposed in 1984 and significantly improved since 1994 (see [11–13] and references therein). They deal with a standard DC program of the following:

$$\alpha = \inf\{\{(x) := g(x) - h(x) : x \in \mathbb{R}^n\} \quad (P_{dc})$$

where g and h are lower semi-continuous proper convex functions on $\mathbb{R}^n$. The concept of DCA, based on the DC duality and the local optimality conditions, is straightforward: each iteration k of DCA approximates the concave part $-h$ by its affine majorization (that is, taking $y^k \in \partial h(x^k)$) and minimizes the resulting convex function:

$$\min\{g(x) - h(x^k) - \langle x - x^k, y^k \rangle : x \in \mathbb{R}^n\} \quad (P_k)$$

to obtain x^{k+1}. The convex DC components g and h, but not the DC function f itself, are used in the building of DCA. Additionally, a DC function f has an infinite number of DC decompositions $g - h$, which significantly influence the characteristics of DCA (such as the rate of convergence, robustness, efficiency, globality of computed solutions, etc.). In numerous applied sciences disciplines, the DCA has been effectively applied to non-convex real-world applications (e.g., [11–13]). It is one of the few effective non-smooth non-convex programming methods that enables the solution of large-scale DC programs. The constrained combinatorial optimization problem we consider in this article is the edge-weighted maximum clique [24]. This problem is to find a subset of vertices inducing a complete subgraph with the maximum total sum of edge weights. The maximum edge-weighted problem has many applications including computer vision and pattern recognition, bioinformatics, and retail industry. First, we transform the problem into the Ising model and use DCA optimization for the relaxed problem with sphere constraint to find the mixer and initial state with Egger's warm-start [9]. Next, we use the classical optimizer for optimizing QAOA circuit parameters. Finally, we determine the probability of optimal value to the maximum edge-weighted clique problem. To see the effectiveness of this method, we test with benchmark data on graph instances in [28] and compare with the Graph Neural Network [20] for continuous warm-start which is recently developed. The paper is organized as follows: Sect. 2 provides preliminary information about QAOA, the warm-start method for QAOA, and Trotterized Quantum Annealing. In Sect. 3, we introduce the DCA warm-start method with the sphere relaxed problem, and Sect. 4 reports numerical simulations using benchmark data to demonstrate the efficiency of our proposed method. Finally, we conclude our findings in Sect. 5.

2   Introduction to QAOA and Warm-Start Method

2.1   Introduction to QAOA

The Quantum Approximate Optimization Algorithm (QAOA) was first introduced in 2014 by Fahri [8]. QAOA is a hybrid classical-quantum algorithm for solving optimization problems with quadratic form $f(x) = \sum_{i,j} Q_{ij} x_i x_j$. It is designed to run on near-term quantum computers and works by encoding the problem as a cost function minimized via the use of a series of quantum gates and measurements. The algorithm alternates between classical optimization of parameters that control the quantum gates and quantum evolution under those gates. The algorithm's output is a quantum state that approximates the solution to the optimization problem. QAOA has been used for various optimization problems, including graph partitioning, MaxCut, portfolio optimization, and many other combinatorial optimization problems. The state function is defined as

$$|\psi(\beta, \gamma)\rangle = U(\beta_1)U(\gamma_1)...U(\beta_k)U(\gamma_k)|+\rangle^{\otimes n}$$

with $U(\beta) = e^{-i\mathbf{H}_B \beta}$ and $U(\gamma) = e^{-i\mathbf{H}_f \gamma}$. $U(\beta)$ and $U(\gamma)$ are parameterized quantum gates with H_f is problem Hamilton operator

$$H_f = \sum_{i,j=1}^{n} a_{i,j}\sigma_i^Z \sigma_j^Z + \sum_{i=1}^{n} b_i \sigma_i^Z + c$$

with σ_i^Z is the Pauli Z matrix and H_f corresponds to the objective function of combinatorial optimization problem, H_B is mixer Hamiltonian operator

$$H_B = \sum_{i=1}^{n} \sigma_i^X$$

with σ_i^X is Pauli X matrix. Generally, the mixer Hamiltonian is the operator used for preparing the initial state of the adiabatic quantum algorithm in solving combinatorial optimization and representing the transitions between quantum states when we consider the solution of the combinatorial optimization problem as quantum states. It helps explore the solution space efficiently. The vector $|+\rangle$ is defined by:

$$|+\rangle = \frac{1}{\sqrt{2}}(|0\rangle + |1\rangle)$$

with $|0\rangle = \begin{bmatrix} 1 \\ 0 \end{bmatrix}$ and $|1\rangle = \begin{bmatrix} 0 \\ 1 \end{bmatrix}$ are 2 qubits in a quantum computer. The initial state represents the initial probability of the solutions. The QAOA objective function is defined as follows:

$$\min_{\beta,\gamma\in[0,2\pi]} \langle\psi(\beta,\gamma)|H_f|\psi(\beta,\gamma)\rangle. \tag{1}$$

This means that to find the parameter for the QAOA circuit, we have to repeat the optimization algorithm many times [23]. The expectation can be represented as follows:

$$\langle\psi(\beta,\gamma)|H_f|\psi(\beta,\gamma)\rangle = \sum_{i=1}^{n} \lambda_i|\langle\psi(\beta,\gamma)|x_i\rangle|^2, \tag{2}$$

with x_i is the feasible solution represented by bitstring which is encoded in the quantum computer, λ_i is the value of objective function corresponding to x_i, $|\langle\psi(\beta,\gamma)|x_i\rangle|^2$ is the probability of x_i appears in the output of the algorithm. It is noticeable all of the solutions are encoded into columns of the identity matrix with size $2^n \times 2^n$ with n as the bitstring length corresponding to the feasible solutions of the combinatorial optimization problem and the values of the objective function is the eigenvalue of H_f. This algorithm makes use of classical computers to optimize the expectation function. It uses gradient and gradient-free optimizers such as gradient descent, COBYLA, SPSA, etc. with initial parameters (β_0,γ_0). Set $|\psi(\beta,\gamma)\rangle = |\psi\rangle$. We have the following relation between the expectation and the optimal solution probability when H_f eigenvalues are positive:

$$1 - \frac{\langle\psi|H_f|\psi\rangle - f(x_{opt})}{\Delta} \leq \sum_{f(x)=f(x_{opt})} |\langle\psi|x\rangle|^2 \leq 1 - \frac{\langle\psi|H_f|\psi\rangle - f(x_{opt})}{f_{max}}$$

With Δ is the spectral gap and f_{max} is the maximum value of $f(x)$ with $x \in \{0,1\}^n$.

Proof. We have the following formula:

$$
\langle \psi | H_f | \psi \rangle = \sum_{x \in \mathbf{B}^n} f(x) |\langle \psi | x \rangle|^2
$$

$$
= \sum_{f(x)=f(x_{opt})} f(x) |\langle \psi | x \rangle|^2 + \sum_{f(x) \neq f(x_{opt})} f(x) |\langle \psi | x \rangle|^2
$$

$$
\geq \sum_{f(x)=f(x_{opt})} f(x) |\langle \psi | x \rangle|^2 + \sum_{f(x) \neq f(x_{opt})} (f(x_{opt}) + \Delta) |\langle \psi | x \rangle|^2
$$

$$
\geq f(x_{opt}) + \Delta \left(1 - \sum_{f(x)=f(x_{opt})} |\langle \psi | x \rangle|^2 \right)
$$

We can indicate that:

$$
\sum_{f(x)=f(x_{opt})} |\langle \psi | x \rangle|^2 \geq 1 - \frac{\langle \psi | H_f | \psi \rangle - f(x_{opt})}{\Delta}
$$

Furthermore, we can see that:

$$
\langle \psi | H_f | \psi \rangle = \sum_{x \in \mathbf{B}^n} f(x) |\langle \psi | x \rangle|^2
$$

$$
= \sum_{f(x)=f(x_{opt})} f(x) |\langle \psi | x \rangle|^2 + \sum_{f(x) \neq f(x_{opt})} f(x) |\langle \psi | x \rangle|^2
$$

$$
\leq \sum_{f(x)=f(x_{opt})} f(x) |\langle \psi | x \rangle|^2 + \sum_{f(x) \neq f(x_{opt})} f_{max} |\langle \psi | x \rangle|^2
$$

$$
\leq f(x_{opt}) + f_{max} \left(1 - \sum_{f(x)=f(x_{opt})} |\langle \psi | x \rangle|^2 \right)
$$

We can see that:

$$
\sum_{f(x)=f(x_{opt})} |\langle \psi | x \rangle|^2 \leq 1 - \frac{\langle \psi | H_f | \psi \rangle - f(x_{opt})}{f_{max}}
$$

This bound works when the error between the expectation and the optimal value is smaller than the spectral gap. We also can transform all H_C with the arbitrary sign to a positive sign by adding $\sum_{i,j} |Q_{ij}|$. Furthermore, the upper bound can go to zero when the difference between the expectation and the optimal value is close to f_{max}.

The Difference Between QAOA and Quantum Annealing

In this section, we compare QAOA with Quantum Annealing [29] since although they are used to solve the combinatorial optimization and theoretical framework is adiabatic quantum computing, they have many differences when we implement. First, Quantum Annealing [29] and QAOA use the same theoretical framework as adiabatic quantum computing but the quantum computer of Quantum Annealing does not use quantum gates while the QAOA depends strongly on quantum gates. From the fact that QAOA uses the quantum gate, the state function it uses is the Trotter decomposition of Quantum Annealing [2] and QAOA uses the variational optimization to find a quantum state with classical optimization instead of depending on time as Quantum Annealing state function:

$$e^{-i[(1-\frac{t}{T})H_B+\frac{t}{T}H_f]}$$

with T is the time length of Quantum Annealing.

Trotterized Quantum Annealing

The Trotterized Quantum Annealing is introduced in [2] for initializing parameters for QAOA. This method uses the Trotter formula for the Quantum Annealing state function:

$$e^{-i[(1-\frac{j}{p})\Delta t H_B+\frac{i}{p}\Delta t H_f]} \approx e^{-i(1-\frac{i}{p})\Delta t H_B}e^{-i\frac{i}{p}\Delta t H_f} + O(\Delta t^2), \qquad (3)$$

and the initial parameters of the QAOA are $\gamma_j = \frac{j}{p}\Delta t$, $\beta_j = (1-\frac{i}{p})\Delta t$ for each layer with $\Delta t = \frac{T}{p}$ is the time step, T is the time that Quantum Annealing has the highest approximate ratio. This method shows its advantage in solving the maximum cut problem. The TQA produces suitable initial parameters when the Trotter error $O(\Delta t^2)$ is very small. This means that $T \ll p$. However, this can increase the complexity of calculating the parameters of the QAOA circuit. The TQA convergence depends on the initial state and the mixer of QAOA. If the initial state is close to the optimal solution superposition, the time of Trotterized Quantum Annealing will be shorter.

2.2 Introduction to the Warm-Start Method in Quantum Optimization

The quadratic unconstrained binary optimization has been studied in Combinatorial Optimization since the 1960s. The formulation of this problem can be seen as follow:

$$\min_{x\in\{0,1\}^n} x^T A x + bx$$

with x as the vector of decision variables. This problem belongs to the NP-hard class. If A is a positive semidefinite matrix, the relaxed version of QUBO is a convex optimization problem and can be solved easily by a classical optimizer. The relaxed version solution initializes the QAOA by replacing the initial states and operators instead of using the original QAOA mixer. The original QAOA mixer considers the transition of a solution by changing 1 bit in the solution's binary string. This mixer is similar to the regular graph adjacency matrix and uses spectral graph theory, we can indicate that the original QAOA considers all solutions with equal initial probabilities. This can lead to a high cost of computing and can lead to non-optimal solutions. The warm-start method [9] helps the QAOA find a favorable biased initial state with the corresponding mixer operator such that the QAOA converges to the optimal solution quickly. As a result, we replace the equal superposition state $|+\rangle^{\otimes n}$ by :

$$|\psi\rangle = \bigotimes_{i=0}^{n-1} R_Y(\theta_i)|0\rangle_n.$$

This state corresponds to the relaxed QUBO solution. $R_Y(\theta_i)$ is corresponding to the rotation of qubit i around the Y-axis of Bloch sphere with $\theta_i = 2\arcsin(\sqrt{c_i})$, $c_i \in [0, 1]$ is the solution of relaxed QUBO problem. We can see that the warm-start initial state represents the new initial probability of the solutions. Furthermore, the initial mixer is replaced by $\sum_{i=0}^{n-1} H_{WS,i}$ with $H_{M,i}$ is:

$$H_{WS,i} = \begin{bmatrix} 2c_i - 1 & -2\sqrt{c_i(1 - c_i)} \\ -2\sqrt{c_i(1 - c_i)} & 1 - 2c_i. \end{bmatrix}$$

This warm-start mixer Hamiltonian [9] helps to prepare the initial state. Furthermore, it helps the solutions transition to optimal solutions faster since the transition is decided based on the relaxed solution being close to the optimal solution. The warm-start mixer helps the QAOA optimization process more effectively from the biased initial state instead of equal superposition. If a component of the relaxed solution vector is equal to 0 or 1, we can initiate it with $|0\rangle$ or $|1\rangle$. This can lead to nonoverlap of continuous solution on discrete version. In this case, we use [9]'s regularization method with $\epsilon \in [0, 0.5]$.

3   Warm Start Method for QAOA with Non-convex Relaxed Quadratic Binary Optimization Problem

3.1   General Warm-Start Method with DCA

Considering the following quadratic binary optimization:

$$\min f(x) = x^T Q x + px + c \qquad (4)$$
$$\text{s.t } x \in \{0, 1\}^n \qquad (5)$$

with Q is an indefinite or not semi-positive definite matrix. In general, solving a box-constrained problem with an indefinite quadratic function is NP-hard and it has many local optimal solutions [15] that can worsen the warm-start QAOA. Instead of considering the box-constrained problem, we consider the relaxed problem with sphere constraint. The sphere constraint contains the box constraints and the non-convex optimization problem with sphere constraint can be solved easily with DCA. We use the warm-start method for QAOA with the relaxed version of the problem with sphere constraint. First, we transform the problem into the Ising model instance by using $x = \frac{z+e}{2}$ with $e = \begin{bmatrix} 1 & 1 & \ldots & 1 \end{bmatrix}$. We have the following formulation:

$$\min_{z \in \{-1,1\}^n} z^T A z + b z + c$$

with $A = \frac{1}{4}Q$, $b = \frac{1}{2}Qe + \frac{1}{2}p$. The relaxed problem we consider is the quadratic minimization over a sphere. The relaxed version of this problem can be written as follows:

$$\min z^T A z + b z + c$$
$$||z||^2 = n.$$

This problem is a non-convex optimization problem. Let E is the constraint $||z||^2 = n$. We can decompose the objective function as follow:

$$\frac{1}{2} z^T A z + b^T z + c + \chi_E(z) = \frac{1}{2} z^T (A + \rho \mathbb{I} - \rho \mathbb{I}) z + \chi_E(z) + b^T z + c$$
$$= \frac{1}{2} \rho ||z||^2 + b^T z + c + \chi_E(z) - \frac{1}{2} z^T (\rho \mathbb{I} - A) z.$$

From this decomposition, we can apply the DC algorithm by solving the following problem in every iteration:

$$z_{k+1} = \min \left\{ \frac{1}{2} \rho ||z||^2 + b^T z + c + \chi_E(z) - \frac{1}{2} z^T ((\rho \mathbb{I} - A) z_k) \right\}$$

with ρ is the positive number such that $(\rho \mathbb{I} - A)$ is positive definite. This problem is equivalent to

$$\min_{z \in E} \left\| z - \frac{(\rho \mathbb{I} - A) z_k - b}{\rho} \right\|^2.$$

This problem show that z_{k+1} is the projection of $\frac{(\rho \mathbb{I} - A) z_k - b}{\rho}$ on E:

$$z_{k+1} = P_E \left(\frac{(\rho \mathbb{I} - A) z_k - b}{\rho} \right).$$

In this algorithm, we can execute in two ways with $z_0 \in \mathbb{R}^n$:

- if $||(\rho \mathbb{I} - A) z_k - b|| \leq \rho \sqrt{n}$ take $z_{k+1} = \frac{(\rho \mathbb{I} - A) z_k - b}{\rho}$
- Otherwise $z_{k+1} = \sqrt{n} \frac{(\rho \mathbb{I} - A) z_k - b}{||(\rho \mathbb{I} - A) z_k - b||}$.

The algorithm with stop if $||z_{k+1} - z_k|| \leq \epsilon$. To find the global solution to the relaxed problem with sphere constraint, we use the restart method introduced in [12]. This method finds the suitable initial point for the quadratic problem with sphere constraint. After solving the relaxed problem with sphere constraint, we transform the problem's solution to $x = \frac{z \pm e}{2}$. The relaxed problem with sphere constraint result may have some component that is out of range $[-1, 1]$ and we use the regularization method by using a function to map z_i^* value from outside range $[-1, 1]$ to inside $[-1, 1]$. We use the Gaussian error function $erf(x) = \frac{2}{\sqrt{\pi}} \int_0^x e^{-t^2} dt$ before transform the solutions back to $[0, 1]$:

$$x^* = \begin{cases} \frac{erf(z_i^*)+1}{2} & \text{if } z_i^* < -1 \\ x_i^* & \text{if } 0 \leq x_i^* \leq 1 \\ \frac{erf(z_i^*)+1}{2} & \text{if } z_i^* > 1 \end{cases}$$

This method can make z_i^* from outside $[-1, 1]$ to a value close to -1 or 1 which can make the warm-start state function overlap with the optimal solution superposition. After this step, we transform the component of x_i^* to the quantum circuit by using the equation $\alpha_i = 2 \arcsin(\sqrt{x_i^*})$ and we have the following initial state and the mixer operator:

$$|\psi\rangle = \bigotimes_{i=0}^{n-1} R_Y(\alpha_i)|0\rangle_n.$$

$$H_{WS} = \sum_{i=0}^{n-1} H_{WS,i}$$

$$\text{with } H_{WS,i} = \begin{bmatrix} 2x_i^* - 1 & -2\sqrt{x_i^*(1-x_i^*)} \\ -2\sqrt{x_i^*(1-x_i^*)} & 1 - 2x_i^*. \end{bmatrix}$$

This regularization method helps the initial state overlap with the optimal solution superposition since all the warm-start parameters are in the range $(0, 1)$. Furthermore, we see that the warm-start QAOA also has another role is rounding the solution from sphere constrained relaxed problem since the mechanic of the QAOA takes advantage of superposition to find the optimal solution with the regularized relaxed problem's solution embedded into warm-start QAOA parameters and the quantum circuit measure the probability of the optimal solution.

3.2   Quadratic Formulation of Edge-Weighted Max Clique Problem

The edge-weighted max clique problem is finding the clique of a graph with the maximum weight. The formulation of this problem can be written as follows:

$$\max \frac{1}{2} x^T A_G x$$

$$x_i + x_j \leq 1 \text{ with } (i, j) \notin E(G)$$

$$x_i \in \{0, 1\}, \, i \in V(G).$$

In this section, we consider the penalized quadratic version of this problem:

$$\max \frac{1}{2} x^T (A_G - M A_{\bar{G}}) x$$
$$x \in \{0, 1\}^n,$$

with A_G is the adjacency matrix of G, M is a very large number and $A_{\bar{G}}$ is the complement graph $\bar{G}$ adjacency matrix. In the relaxed version of the weighted quadratic max clique with sphere constraint, we use the DCA method [12]. We transform x into z by $x = \frac{z+e}{2}$. The relaxed problem with sphere constraint problem is:

$$\max \frac{1}{8} (z + e)^T (A_G - M A_{\bar{G}})(z + e)$$
$$||z||^2 = n$$

with $e = \begin{bmatrix} 1 & 1 & 1 \dots 1 \end{bmatrix}$, n is the number of nodes in a graph. After finding the solution from the DCA for the relaxed problem with sphere constraint, we have the solution z^*. After this step, we apply the process in Sect. 3.1 to solve the maximum edge-weighted clique problem.

4   Numerical Simulation

Comparative Algorithm. We compare our method with the Graph Neural Network warm-start method for the QAOA [20] showing its recent efficiency in solving the maximum cut problem. This method uses the Graph Neural Network with an unsupervised learning process to predict the probability of the nodes in a graph. The loss function of the Graph Neural Network has the following form:

$$f_{loss} = \frac{1}{2} x^T (-A_G + M A_{\bar{G}}) x$$
$$x \in [0, 1]^n$$

with x as the probability vector of nodes in the graph, and M as an arbitrarily large number. The probability from the prediction is used to warm-start the QAOA with the continuous warm-start method. Furthermore, we compare our method with the original QAOA with original mixer.

Data. In this section, we utilize graphs from the data provided in Fuchs et al.'s work [28] to demonstrate the algorithm's advantages concerning edge-weighted maximum clique. The graphs utilized encompass weighted graphs featuring 14 nodes, encompassing a variety of Erdos-Renyi graphs, Barabási-Albert graphs, and Watts-Strogatz graphs. Our experiment is conducted on the Qiskit-aer simulator.

Metric. To measure the method's efficiency, we use 2 main metrics as follows:

- Optimal solution probability: Measure the probability of warm-start QAOA to obtain the optimal solution. This metric has the following formula:

$$\sum_{f(x)=f(x_{opt})} |\langle \psi(\beta, \gamma | x \rangle|^2$$

  The probability of optimal solution also represents the fidelity of the quantum state to the optimal solution superposition.
- Run time of the algorithm: Measure the warm start QAOA runtime. This metric can show how long it takes to solve the problem.

Result. Initially, we focus on the probability of optimal value for QAOA with random parameters with a single layer, as outlined in Table 1 for the maximum edge weight clique. We can see that 2 warm-start methods and the original QAOA can find the optimal value for every instance. We consider the probability and time of finding the optimal solution. A noteworthy observation from these tables is that DCA warm-start QAOA exhibits higher probabilies of optimal value in the edge-weighted maximum clique compared to the GNN warm-start QAOA and original QAOA when utilizing the same initial parameters. Additionally, the runtime for warm-start QAOA is consistently lower than that of the GNN warm-start QAOA and original QAOA across all graphs. This difference can be attributed to warm-start QAOA's ability to generate a variational state close to the state with the lowest energy of the problem Hamiltonian, in contrast to the GNN warm-start QAOA and original QAOA, beginning with an initial state far from the minimum energy state. These results are consistent across various instances. Furthermore, some instances employing GNN warm-start QAOA display a probability of optimal value close to zero due to the performance of the Graph Neural Network on the edge-weighted maximum clique. The Graph Neural Network predicts the nodes in the graph with low accuracy, reducing the probability of attaining optimal value. Subsequently, we compare the outcomes of Trotterized Quantum Annealing for DCA warm-start QAOA, GNN warm-start QAOA, and original QAOA, as depicted in Table 2. Our approach consistently demonstrates a higher probability of optimal value than GNN warm-start QAOA with Trotterized Quantum Annealing initialization and original QAOA. The Graph Neural Network's poor prediction on every node can lead to the acceptance of non-feasible solutions leading to the decrease of optimal solution probability. Additionally, our method exhibits lower computation times in every case presented in Table 2 compared to the GNN warm-start QAOA with TQA initialization because the GNN inference process repeats thousands of times and takes a long time to complete. Furthermore, it also produces the initial state for the warm-start QAOA is far from the optimal solution superposition which makes the warm-start QAOA take more time to find the optimal solution. Furthermore, the original QAOA considers all solutions with equal initial probability and the transition between feasible and non-feasible solutions which makes the algorithm use more time to find the optimal value with low probability. We show the probability distribution of our method in Fig. 1 and Fig. 2. In 2 figures, we group all non-feasible solutions into 1 label. We can see that our method brings high probability for optimal solution. Especially in Fig. 2, the probability of non-feasible solution is less than the feasible solutions.

Table 1. Comparision between DCA warm-start and standard QAOA with the maximum edge weight clique

Graph	Opt value	DCA warm-start		GNN warm-start [20]		Original QAOA	
		Opt sol prob	Run time	Opt sol prob	Run time	Opt sol prob	Run time
w.ba.n14.k4.4	3.7088	0.0543	50	0.0002	135	2.3e−05	240
w.ba.n14.k4.3	5.9049	0.0520	17	7.15e−05	106	0.0002	257
w.ba.n14.k4.2	2.8242	0.0162	17	0.0002	98	5.2e−05	180
w.ba.n14.k4.0	4.3356	0.0770	28	0.0002	104	1.1e−05	90
w.ba.n14.k2.1	2.0473	0.0902	46	0.0006	114	0.0006	55
w.ba.n14.k2.2	2.8433	0.0201	52	0.0008	110	0.0002	62
w.ba.n14.k2.3	1.9396	0.0471	49	0.0007	122	0.0015	52
w.er.n14.k2.0	2.2057	0.0210	12	0.0009	121	6.1e−5	90
w.er.n14.k2.1	2.0843	0.0550	20	0.0007	115	6.1e−5	77
w.er.n14.k2.3	2.7976	0.4000	51	0.0008	118	6.1e−5	77
w.er.n14.k2.4	0.8850	0.0060	27	0.0019	117	6.1e−5	77
w.er.n14.k4.3	2.9713	0.0500	8	0.0002	105	6.1e−5	77
w.er.n14.k4.4	2.4563	0.0210	24	0.0006	109	6.1e−5	240
w.ws.n14.k2.0	1.4297	0.2310	18	0.0007	100	6.1e−5	77
w.ws.n14.k2.1	0.9782	0.0060	9	0.0020	114	6.1e−5	77
w.ws.n14.k2.2	0.7848	0.0270	10	0.0023	115	6.1e−5	77
w.ws.n14.k2.3	0.9184	0.0110	9	0.0021	107	6.1e−5	71
w.ws.n14.k4.0	2.4070	0.0060	31	0.0007	115	6.1e−5	71
w.ws.n14.k4.1	2.0749	0.0064	66	0.0008	120	0.0010	56
w.ws.n14.k4.4	2.3099	0.0100	70	0.0007	118	6.1e−5	600

Table 2. Comparision between DCA warm-start and standard QAOA with the maximum edge weight clique with Trotterized Quantum Annealing initial parameters

Graph	Opt value	DCA warm-start		GNN warm-start [20]		Original QAOA	
		Opt sol prob	Run time	Opt sol prob	Run time	Opt sol prob	Run time
w.ba.n14.k4.4	3.7088	0.0042	12	0.0021	96	0	1020
w.ba.n14.k4.3	5.9049	0.0720	39	0.0011	87	0	1443
w.ba.n14.k4.2	2.8242	0.0039	9	0.0015	83	0	1380
w.ba.n14.k4.0	4.3356	0.0767	23	0.0026	86	1.44e−05	1454
w.ba.n14.k2.1	2.0473	0.2011	40	0.0046	85	0.0002	65
w.ba.n14.k2.2	2.8433	0.0941	45	0.0066	100	2.1530e−06	62
w.ba.n14.k2.3	1.9396	0.1223	43	0.0078	100	0.0007	66
w.er.n14.k2.0	2.2057	0.0044	7	0.0044	89	0.0038	1200
w.er.n14.k2.1	2.0843	0.0290	14	0.0050	79	0	1435
w.er.n14.k2.3	2.7976	0.1609	118	0.0055	86	0	1391
w.er.n14.k2.4	0.8850	0.0892	60	0.0072	86	0.0007	1380
w.er.n14.k4.3	2.9713	0.0840	24	0.0016	86	0.0003	1378
w.er.n14.k4.4	2.4563	0.0323	31	0.0053	86	0.0001	1366
w.ws.n14.k2.0	1.4297	0.1826	6	0.0038	76	0.0113	1375
w.ws.n14.k2.1	0.9782	0.0330	32	0.0115	78	0	77
w.ws.n14.k2.2	0.7848	0.0788	57	0.0109	83	0.0010	1434
w.ws.n14.k2.3	0.9184	0.0318	27	0.0106	91	0.0001	710
w.ws.n14.k4.0	2.4070	0.0276	9	0.0047	85	2.2867e−06	1411
w.ws.n14.k4.1	2.0749	0.0057	59	0.0047	118	0.0007	65
w.ws.n14.k4.4	2.3099	0.0651	22	0.0046	85	0.0003	1417

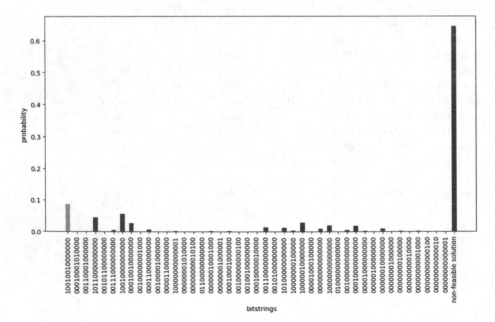

Fig. 1. DCA warm-start QAOA with random initial point of instance w.ba.n14.k2.1

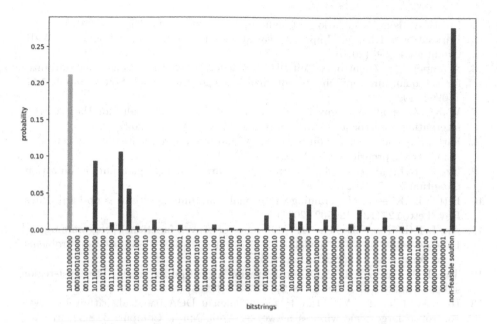

Fig. 2. DCA warm-start QAOA with TQA of instance w.ba.n14.k2.1

5   Conclusion and Feature Work

In this article, we have proposed a new method for warm-starting the QAOA by using the DCA for the relaxed problem with sphere constraint. This method provides the initial state and mixer operator for the QAOA by solving the relaxed problem with sphere constraint and shows a good result when compared with the GNN warm-start method and original QAOA. To see the DCA warm-start method advantages, we do the numerical simulation carefully with graphs from the previous article. The DCA warm-start method shows its effectiveness in time and the optimal solution probability. In the future, we will test this algorithm on large-scale problems and other real-life problems.

Acknowledgement. This research is funded by Hanoi University of Science and Technology (HUST) under project number T2023-TĐ-001.

References

1. Gloria, T., Dacrema, M.F., Cremonesi, P.: Feature selection for classification with QAOA. In: 2022 IEEE International Conference on Quantum Computing and Engineering (QCE). IEEE (2022)
2. Sack, S.H., Serbyn, M.: Quantum annealing initialization of the quantum approximate optimization algorithm. Quantum **5**, 491 (2021)
3. Anshu, A., Metger, T.: Concentration bounds for quantum states and limitations on the QAOA from polynomial approximations. Quantum **7**, 999 (2023)

4. Lucas, A.: Ising formulations of many np problems. Front. Phys. **5** (2014)
5. Barkoutsos, P.K., et al.: Improving variational quantum optimization using CVaR. Quantum **4**, 256 (2020)
6. Acampora, G., Chiatto, A., Vitiello, A.: Genetic algorithms as classical optimizer for the quantum approximate optimization algorithm. Appl. Soft Comput. **142**, 110296 (2023)
7. Wang, Z., et al.: X y mixers: analytical and numerical results for the quantum alternating operator ansatz. Phys. Rev A **101**(1), 012320 (2020)
8. Farhi, E., Goldstone, J., Gutmann, S.: A quantum approximate optimization algorithm. arXiv preprint arXiv:1411.4028 (2014)
9. Egger, D.J., Marecek, J., Woerner, S.: Warm-starting quantum optimization. Quantum **5**, 479 (2021)
10. Bittel, L., Kliesch, M.: Training variational quantum algorithms is np-hard. Phys. Rev. Lett. **127**(12), 120502 (2021)
11. An, L.T.H., Tao, P.D.: The DC (difference of convex functions) programming and DCA revisited with DC models of real world non convex optimization problems. Ann. Oper. Res. **133**(2), 23–46 (2005)
12. Tao, P.D., An, L.T.H.: A DC optimization algorithm for solving the trust-region subproblem. SIAM J. Optim. **8**(2), 476–505 (1998)
13. Hoai An, L.T., Ta, A.S., Tao, P.D.: An efficient DCA based algorithm for power control in large scale wireless networks. Appl. Math. Comput. **318**(1), 215–226 (2018)
14. Brandhofer, S., et al.: Benchmarking the performance of portfolio optimization with QAOA. Quantum Inf. Process. **22**(1), 25 (2022)
15. Pardalos, P.M.: Global optimization algorithms for linearly constrained indefinite quadratic problems. Comput. Math. Appl. **21**(6–7), 87–97 (1991)
16. An, L.T.H., Tao, P.D.: A branch and bound method via DC optimization algorithms and ellipsoidal technique for box constrained nonconvex quadratic problems. J. Global Optim. **13**(2), 171–206 (1998)
17. Kurowski, K., Pecyna, T., Slysz, M., Rozycki, R., Waligora, G., Wglarz, J.: Application of quantum approximate optimization algorithm to job shop scheduling problem. Eur. J. Oper. Res. **310**(2), 518–528 (2023)
18. Choi, J., Oh, S., Kim, J.: Quantum approximation for wireless scheduling. Appl. Sci. **10**(20), 7116 (2020)
19. Torta, P., Mbeng, G.B., Baldassi, C., Zecchina, R., Santoro, G.E.: Quantum approximate optimization algorithm applied to the binary perceptron. Phys. Rev. B **107**(9), 094202 (2023)
20. Jain, N., Coyle, B., Kashefi, E., Kumar, N.: Graph neural network initialisation of quantum approximate optimisation. Quantum **6**, 861 (2022)
21. Shaydulin, R., Safro, I., Larson, J.: Multistart methods for quantum approximate optimization. In: 2019 IEEE High Performance Extreme Computing Conference (HPEC), pp. 1–8. IEEE (2019)
22. Hadfield, S., Wang, Z., O'gorman, B., Rieffel, E.G., Venturelli, D., Biswas, R.: From the quantum approximate optimization algorithm to a quantum alternating operator ansatz. Algorithms **12**(2), 34 (2019)
23. Grange, C., Poss, M., Bourreau, E.: An introduction to variational quantum algorithms for combinatorial optimization problems. 4OR, **21**(3), 363–403 (2023)
24. Hosseinian, S., Fontes, D.B.M.M., Butenko, S.: A nonconvex quadratic optimization approach to the maximum edge weight clique problem. J. Global Optim. **72**, 219–240 (2018)

25. Buchheim, C., et al.: An exact algorithm for nonconvex quadratic integer minimization using ellipsoidal relaxations. SIAM J. Optim. **23**(3), 1867–1889 (2013)
26. Pham Dinh, T., Nguyen Canh, N., Le Thi, H.A.: An efficient combined DCA and B&B using DC/SDP relaxation for globally solving binary quadratic programs. J. Global Optim. **48**, 595–632 (2010)
27. Ta, A.S., Khadraoui, D., Tao, P.D., et al.: Solving partitioning-hub location- routing problem using DCA. J. Ind. Manag. Optim. **8**(1), 87–102 (2011)
28. Fuchs, F.G., Kolden, H.Ø., Aase, N.H., Sartor, G.: Efficient encoding of the weighted max k-cut on a quantum computer using QAOA. SN Comput. Sci. **2**(2), 89 (2021)
29. Kadowaki, T., Nishimori, H.: Quantum annealing in the transverse Ising model. Phys. Rev. E **58**(5), 5355 (1998)

Comparing Integer Encodings in QUBO for Quantum and Digital Annealing: The Travelling Salesman Problem

Philippe Codognet[✉][ID]

JFLI - CNRS, Sorbonne University, University of Tokyo, Tokyo, Japan
codognet@is.s.u-tokyo.ac.jp

1 Introduction

In the domain of combinatorial optimization problems and decision science, the use of quantum computers to solve concrete problems has started to raise interest, in particular with the use of quantum and quantum-inspired annealing systems. Quantum Annealing (QA) has been proposed by [7] and [4] and incorporated as the core computational mechanism in so-called Ising Machines [13] such as D-Wave quantum computers and quantum-inspired/digital systems such as Fujitsu Digital Annealer. In QA, problems are encoded in quantum Hamiltonians (energy functions) and quantum dynamics is used to find solutions (ground states of minimal energy). From a metaheuristic viewpoint, QA can be seen as similar to simulated annealing but with a different manner to escape local minima. Indeed, if in simulated annealing escaping from local minima is done by accepting with a certain probability non-decreasing moves, this is performed in QA by the phenomenon of *quantum tunnelling* which makes it possible to traverse energy potential barriers in the energy landscape as long as they are not too large, i.e. high and narrow peaks do not cause a problem.

Interestingly, the formulation of problems as Ising models in the QA paradigm is related to the Quadratic Unconstrained Binary Optimization (QUBO), a modelling paradigm rooted in pseudo-Boolean optimization which became a general modelling language for combinatorial problems in the last two decades.

Although classical graph-based combinatorial problems are naturally and rather univocally modelled in Ising/QUBO [10], more complex constrained optimization problems are sometimes difficult to formulate. Indeed, two things have to be taken care of: the encoding of integers by Boolean variables and the representation of constraints in QUBO. It is therefore important to compare the performance of different QUBO models of the same problem, in order to investigate the path towards the most efficient QUBO formulation.

The Travelling Salesman Problem (TSP) is a classical constraint optimization problem that can be modelled in QUBO and solved by quantum annealing or quantum-inspired annealing. Several models of the TSP are possible in QUBO, and we will consider here those based on the so-called permutation formulation. This can be done by first choosing an encoding for integer variables and then by

M. Sevaux et al. (Eds.): MIC 2024, LNCS 14753, pp. 262–267, 2024.
https://doi.org/10.1007/978-3-031-62912-9_25

encoding the permutation constraint in the objective function as a *penalty*, as constraints have to be represented in QUBO by (quadratic) penalty expressions, value of which will be minimal when the constraint is satisfied. Two main schemes exist for encoding integers in QUBO: the classical one-hot encoding and the recently proposed domain-wall encoding [2]. Each encoding will lead to very different penalty terms in the objective function, and will produce different QUBO matrices. We experimented with several TSP instances from TSPLIB, and we will present and compare in this paper the results for one-hot and domain-wall models that have been implemented for two quantum-inspired annealing QUBO solvers based on clusters of GPUs: the Fixstars Amplify Annealer Engine and the ABS QUBO solver.

2   Quantum Annealing and QUBO

In the Quantum Annealing paradigm, combinatorial optimization problems can be described by Ising models and Ising Hamiltonians, the ground states of which correspond to the minimal solutions of the original problem, see for instance [10]. Ising models are indeed equivalent to formulations in Quadratic Unconstrained Binary Optimization (QUBO), a paradigm which is conceptually very simple but has shown to be quite powerful at modelling various types of combinatorial problems, see for instance [5,9]. Therefore QUBO has become in the last years the standard input language for all quantum and quantum-inspired annealing hardware and software solvers.

Consider n Boolean variables $x_1, ..., x_n$, a QUBO problem consists in minimizing an *objective function* defined by a quadratic expression: $\sum_{i \leq j} q_{ij} x_i x_j$

It is therefore usual to represent a QUBO problem by a vector x of n binary decision variables and a square $n \times n$ matrix Q with coefficients q_{ij}, as the problem can be written: minimize $y = x^t Q x$, where x^t is the transpose of x.

Observe that, as x_i are Boolean variables, $x_i^2 = x_i$, thus this quadratic formulation also includes a linear part: the diagonal of the Q matrix.

3   The Travelling Salesman Problem in QUBO

The Travelling Salesman Problem (TSP) [6] is one of the most well-known combinatorial problem and it was one of the first problems to be proven NP-hard. Let us consider a graph $< G, E >$ with a set of n nodes G and a set of undirected labelled edges E, the TSP consists in finding an hamiltonian cycle of minimal cost in G. For the sake of simplicity, we consider that G is defined by a $n \times n$ distance matrix D, with possibly arbitrary large coefficients.

An Ising or QUBO model for the TSP is easily derived from the classical integer linear programming formulation, cf. [10], consisting of an objective function representing the cost of the tour and a set of constraints ensuring that the tour indeed forms a cycle, that is, that each node is visited exactly once. It is equivalent to require that the nodes in the tour form a *permutation* of $\{1, \ldots, n\}$, if nodes are represented by integers in $\{1, \ldots, n\}$. Representing such a constraint in QUBO will depend on the way integers are encoded by Boolean variables, for which two schemes are mainly used in the QA community: the classical *one-hot* encoding and the more recent *unary/domain-wall* encoding.

In the *one-hot* encoding formulation of the TSP, we consider n^2 Boolean variables x_{it} which have value 1 if node i is visited at time t and 0 otherwise. The objective function (tour cost) to minimize is then expressed by:

$$\sum_{(i,j)\in G\times G} D_{ij} \sum_{t=1}^{n-1} x_{i,t}\, x_{j,t+1} \quad + \quad \sum_{k=1}^{n} D_{k,1}\, x_{k,n}$$

The permutation constraint will be enforced by encoding in QUBO the set of constraints $\sum_{i=1}^{n} x_{ij} = 1$ and $\sum_{j=1}^{n} x_{ij} = 1$. To encode such constraints in QUBO, we remark that $\sum_{k=1}^{n} x_k = 1 \iff (\sum_{k=1}^{n} x_k - 1)^2 = 0$, which gives the (quadratic) *penalty* expression: $-\sum_{k=1}^{n} x_k + 2\sum_{k<k'} x_k x_{k'}$

This expression is minimal when the constraint is satisfied.

Adding together the penalties for the $2 \times n$ pseudo-Boolean constraints gives a quadratic penalty term corresponding to the permutation constraint in QUBO:

$$\sum_{i=1}^{n}\sum_{j<j'} x_{ij}x_{ij'} + \sum_{j=1}^{n}\sum_{i<i'} x_{ij}x_{i'j} - \sum_{i=1}^{n}\sum_{j=1}^{n} x_{ij}$$

Another integer encoding, the *domain-wall* encoding has been proposed in [2] in an Ising setting, and uses $n-1$ Boolean variables to encode an integer with domain $\{0,\dots,n-1\}$. Converted to a Boolean setting, domain-wall reduced to the well-known *unary* encoding on a fixed number of bits: a number n is encoded by n bits set to 1, followed by zeros. This is also called *thermometer* encoding, and gives a unique unary encoding for each integer.

In the *unary/domain-wall* encoding formulation of the TSP, we consider each node of the graph to be represented by an integer x_i with domain $\{0,\dots,n-1\}$ and we consider $n \times (n-1)$ Boolean variables x_{ij} where $x_{i,0},\dots,x_{i,n-2}$ is the unary/domain-wall encoding of x_i. To be a valid unary/domain-wall encoding, we need to enforce the following constraint [3]:

$\forall i \in \{0,\dots,n-1\}, \forall j \in \{0,\dots,n-3\}\ \ x_{i,j} \geq x_{i,j+1}$.

For instance, 11100 is a valid unary/domain-wall encoding and represents the integer value 3, while 11011 and 00011 are not valid unary/domain-wall encodings. As the Boolean constraint $x \geq y$ can be represented in QUBO by the (quadratic) penalty $y - xy$, this corresponds to the following penalty, for each original integer variable x_i: $\sum_{i=0}^{n-3}(x_{i,j+1} - x_{i,j}\, x_{i,j+1})$

To represent the objective function of the TSP, we also need to define the fact that a node x_i is visited by the tour at time t, which can be done in unary/domain-wall encoding by considering the Boolean formula: $x_{i,t} - x_{i,t+1}$.

With the convention that $x_{i,-1} = 1$, the objective function (tour cost) to minimize is then expressed by:

$$\sum_{(i,j)\in G\times G} D_{ij} \sum_{t=0}^{n-2}(x_{i,t-1} - x_{i,t})(x_{j,t} - x_{j,t+1}) \quad + \quad \sum_{k=1}^{n} D_{k,1}\, x_{k,n-1}$$

The quadratic penalty corresponding to the permutation constraint on n integer variables x_i with domains $\{0, \ldots, n-1\}$ and unary/domain-wall encoded by $x_{i,0} \ldots x_{i,n-2}$ has been defined in [3] by remarking that the following property holds: $\forall j \in \{0, n-2\}$, $\sum_{i=1}^{n} x_{ij} = (n-1) - j$

We thus have $n-1$ pseudo-Boolean linear equations (one for each index j), for which the quadratic penalty can be defined as a generalization of the transformation for the one-hot constraint: $(2(j-n)+3) \sum_{i=0}^{n-1} x_{ij} + 2 \sum_{i<i'} x_{ij} x_{i'j}$

The quadratic penalty corresponding to the permutation constraint is then just the sum of all the penalties corresponding to these $n-1$ equations:

$$\sum_{j=0}^{n-2} \left((2(j-n)+3) \sum_{i=0}^{n-1} x_{ij} + 2 \sum_{i<i'} x_{ij} x_{i'j} \right)$$

4   Experimental Results

The QUBO model for the TSP with one-hot encoding is well-known, but the QUBO model with unary/domain-wall encoding is new. We will compare their characteristics and performances in this section.

When integrating penalties in the objective function to represent constraints in QUBO, one has to consider multiplicative coefficients in order to make the penalty part compatible with the original objective part to minimize. Indeed, penalty coefficients have to be large enough to ensure that no feasible solution is lost [8,15]. For the TSP model with one-hot encoding, a standard method consists in taking the maximal value in the distance matrix as penalty coefficient. We decided to use this method for the one-hot QUBO model because, although other methods have been proposed, they usually result in larger coefficients with no clear improvement on the performance [1].

For the unary/domain-wall encoding, choosing correct penalty coefficients is a bit more complicated, as we have two different term parts in the penalty: $\sum_{i=0}^{n} 3(x_{i,j+1} - x_{i,j} x_{i,j+1})$ and $\sum_{j=0}^{n-2} \left((2(j-n)+3) \sum_{i=0}^{n-1} x_{ij} + 2 \sum_{i<i'} x_{ij} x_{i'j} \right)$. The term coefficients of the first part are -1 and +1, while for the second part they range up to $2n$. Therefore we need to multiply the penalty terms coming from the first part by an additional $2n$ factor. Then, as for one-hot, all penalty terms should be multiplied by the maximum value in the distance matrix.

The charateristics of the QUBO upper-diagonal matrices obtained with both models are presented in the table below, for several instances taken from the TSPLIB library[1]. We can observe than the unary/domain-wall matrices are smaller and more sparse (even proportionally), but have larger coefficients.

These problems are too large for being solved directly by the quantum processing unit (QPU) of D-Wave Advantage system [12], which have up to 5600 qubits but do not implement a complete connection graph between qubits (only a 15-node connectivity). Using the Hybrid Solver of the D-Wave quantum computer, which is mixing classical decomposition of larger problems and execution

---

[1] http://comopt.ifi.uni-heidelberg.de/software/TSPLIB95/.

of smaller problems on the QPU, makes it possible to find the optimal solution for very small TSP instances (up to size 13, in a few seconds of computation time) but it cannot find the the optimal solution for the smallest instances that we consider here (gr17 and ulysses16), even after 60 s of computation time which is the overall time-limit that we take for all solvers.

instance	one-hot encoding				unary encoding			
	vars	max coef	min coef	number of coefs. <> 0	vars	max coef	min coef	number of coefs <> 0
ulysses16	256	5579	−5578	7936 = 24.12%	240	85521	−89248	5624 = 19.45%
ulysses22	484	5579	−5578	20812 = 17.73%	462	119139	−122716	14817 = 13.85%
gr17	289	1490	−1490	9537 = 22.76%	272	24335	25330	6409 = 17.26%
gr21	441	1730	−1730	18081 = 18.56%	420	35745	−36330	12495 = 14.13%
gr24	576	778	−778	27072 = 16.29%	552	18141	−18672	18756 = 12.29%
gr48	2304	2166	−2166	218880 = 8.24%	2256	102389	−103968	152520 = 5.99%
fri26	676	560	−560	34476 = 14.07%	650	14388	−14560	22607 = 10.69%
att48	2304	5324	−5324	218880 = 8.24%	2256	253752	−255552	160872 = 6.32%
hk48	2304	5468	−5468	218880 = 8.24%	2256	257502	−262464	160680 = 6.31%
eil51	2601	172	−172	262701 = 7.76%	2550	8660	−8772	278512 = 7.62%
berlin52	2704	3432	−3432	278512 = 7.62%	2652	177610	−178464	203086 = 5.77%

We thus decided to use so-called "quantum-inspired" or "digital" annealing systems which are GPU-based QUBO solvers that somehow simulate the quantum annealing process: Fixstars Amplify Annealing Engine (AE), which is a commercial product with no real description of its internal computation model [11] and ABS QUBO Solver, whose solving algorithm is described in [14,16]. The table below presents, for each solver, the annealing time (in seconds) to reach the optimal solution (Time-To-Solution) or the best non-optimal solution found within the timeout of 60 s (when numbers are postfixed with a '*').

instance	optimal solution	Amplify AE (TTS/Best sol)		ABS Solver (TTS/Best sol)	
		one-hot encoding	unary encoding	one-hot encoding	unary encoding
ulysses16	6859	0.315	2.013	0.070	0.495
ulysses22	7013	4.350	7114*	0.146	7036*
gr17	2085	0.262	6.807	0.052	0.651
gr21	2707	0.208	23.34	0.053	0.997
gr24	1272	0.255	1294*	0.083	1278*
gr48	5046	16.25	7296*	7.446	6078*
fri26	937	3.089	0.851	0.363	0.124
att48	10628	10705*	15507*	10661*	12745*
hk48	11461	11479*	16513*	25.89	13959*
eil51	426	430*	611*	17.19	527*
berlin52	7542	7694*	10898*	7679*	10826*

5   Conclusion

We have presented two models of the TSP in QUBO: one with one-hot encoding (already well-known) and one with unary/domain-wall encoding (new), together with their performance on two quantum-inspired annealing solvers. Experiments on various instances from TSPLIB show that the one-hot encoding is clearly more performant than the unary/domain-wall encoding, with both solvers.

References

1. Ayodele, M.: Penalty weights in QUBO formulations: permutation problems. In: Pérez Cáceres, L., Verel, S. (eds.) EvoCOP 2022. LNCS, vol. 13222, pp. 159–174. Springer, Cham (2022). https://doi.org/10.1007/978-3-031-04148-8_11
2. Chancellor, N.: Domain wall encoding of discrete variables for quantum annealing and QAOA. Quantum Sci. Technol. **4**, 045004 (2019)
3. Codognet, P.: Domain-wall / unary encoding in QUBO for permutation problems. In: IEEE Quantum Computing and Engineering (QCE), pp. 167–173 (2022)
4. Farhi, E., Goldstone, J., Gutmann, S., Lapan, J., Lundgren, A., Preda, D.: A quantum adiabatic evolution algorithm applied to random instances of an np-complete problem. Science **292**(5516), 472–475 (2001)
5. Glover, F.W., Kochenberger, G.A., Du, Y.: Quantum bridge analytics I: a tutorial on formulating and using QUBO models. Ann. OR **314**, 141–183 (2022)
6. Jünger, M., Reinelt, G., Rinaldi, G.: Chapter 4 the traveling salesman problem. In: Network Models, Handbooks in Operations Research and Management Science, vol. 7, pp. 225–330. Elsevier (1995)
7. Kadowaki, T., Nishimori, H.: Quantum annealing in the transverse Ising model. Phys. Rev. E **58**, 5355–5363 (1998)
8. Kochenberger, G., Glover, F., Alidaee, B., Rego, C.: A unified modeling and solution framework for combinatorial optimization problems. OR Spectrum **26**, 237–250 (2004)
9. Kochenberger, G.A., et al.: The unconstrained binary quadratic programming problem: a survey. J. Comb. Optim. **28**(1), 58–81 (2014)
10. Lucas, A.: Ising formulations of many NP problems. Front. Phys. **2** (2014)
11. Matsuda, Y.: Research and development of common software platform for ising machines. In: 2020 IEICE General Conference (2020). (in Japanese)
12. McGeoch, C., Farré, P.: The Advantage system: Performance update (2021). Technical report, D-Wave Inc., 01-10-2021
13. Mohseni, N., McMahon, P., Byrnes, T.: Ising machines as hardware solvers of combinatorial optimization problems. Nat. Rev. Phys. (2022). published online 04/05/2022
14. Nakano, K., et al.: Diverse adaptive bulk search: a framework for solving QUBO problems on multiple GPUs. In: 2023 IEEE International Parallel and Distributed Processing Symposium Workshops (IPDPSW). IEEE (2023)
15. Verma, A., Lewis, M.: Penalty and partitioning techniques to improve performance of QUBO solvers. Discret. Optim. **44**, 100594 (2022)
16. Yasudo, R., et al.: GPU-accelerated scalable solver with bit permutated cyclic-min algorithm for quadratic unconstrained binary optimization. J. Parallel Distrib. Comput. **167**, 109–122 (2022)

Solving Quadratic Knapsack Problem with Biased Quantum State Optimization Algorithm

Huy Phuc Nguyen Ha[1], Viet Hung Nguyen[2], and Anh Son Ta[1(✉)]

[1] School of Applied Mathematics and Informatics, Hanoi University of Science and Technology, Hanoi, Dai Co Viet, Vietnam
son.taanh@hust.edu.vn
[2] Université Clermont-Auvergne, CNRS, Mines de Saint-Etienne, Clermont Auvergne INP, LIMOS, Clermont-Ferrand, France

Abstract. The Quantum Approximate Optimization Algorithm is the hybrid classic-quantum algorithm that is used for solving the combinatorial optimization problem. However, the algorithm performs poorly in the constrained combinatorial optimization problem because it considers all solutions with the same initial probability. In this article, we propose a new quantum state that improves the QAOA performance and does not require slack variables for inequality constraints. We also introduce some properties of our new quantum state in solving a constrained combinatorial optimization problem. To see our method's efficiency, we use our method to solve the quadratic knapsack problem and compare it with the quantum state method of Hao et al. [5] which is one of the most recent and effective methods for solving the constrained combinatorial optimization.

Keywords: Quadratic knapsack problem · QAOA · biased quantum state

1 Introduction

The Quantum Approximate Optimization Algorithm (QAOA) [9] is the hybrid quantum-classical algorithm used for solving the combinatorial optimization problem. QAOA addresses the optimization challenge by creating a parameter-based quantum state through a series of quantum circuit layers. These layers alternate phase and mixing operations, adjusting parameters to maximize a specific measure of solution effectiveness. QAOA shows potential in various fields like optimization, finance, and machine learning, and efforts have been made to adapt it for quantum chemistry applications. However, QAOA has poor performance in constrained optimization problems since it considers all solutions including infeasible solutions with the same initial probability. From this disadvantage, many methods have been used for improving the performance of the QAOA in the constrained binary optimization problem including Quantum Alternating Operator Ansatz [15], QAOA with Grover Mixer [16]. The Quantum Alternating Operator Ansatz is the QAOA's improvement with the mixer operator's change. The new mixer operator in Quantum Alternating Operator Ansatz is used to restrict the initial state of the constrained problem into feasible solution space. Quantum Alternating Operator Ansatz is used for many problems including maximum independent set [1], minimum exact cover problem [2]. The Grover mixer is the method using the feasible solution representation matrix for the initial operator. This method restricts the initial state of the constrained problem into a feasible solution space with the same probability for every feasible solution. These methods depend heavily on the structure of the problem's constraints to find a suitable mixer operator. To overcome this dependence Hao et al. [5] introduce an in-constrained state function. This method does not require finding a suitable mixer for the problem using the original quantum state of the QAOA and calculating the solutions in a feasible set. The main advantage of this state function is that it does not depend on the structure of the problem's constraints. The disadvantage of the three methods is that converging to the optimal value with a low-depth circuit is hard to obtain since it considers all the feasible solutions with the same initial probability as the original QAOA state function. Furthermore, finding the optimal value of expectation of the QAOA is an NP-hard problem. One of the approaches to solve this problem is to find an initial state of the QAOA to reduce the non-feasible solutions and improve the approximate ratio of the QAOA with a low-depth circuit with a biased quantum state [4]. This method has been applied with good results for the maximum cut problem with positive weight on edges [4] and job shop scheduling problems [12]. However, [12] considers the problem with equality constraints which is easy to transform to QUBO, and [4] considers the maximum cut problem that has no constraint. However, many constrained combinatorial optimization problems have inequality constraints. In general, the inequality constraints can be transformed into the equality constraints by using the slack variables which can increase the complexity of the problem. The penalty method also distracts the QAOA from finding the optimal value because of the penalty term. In this article, we introduce a new quantum state for the QAOA that is suitable for inequality constraints and does not require additional variables to transform the problem into QUBO. We apply our state function to the quadratic knapsack problem. The quadratic knapsack problem can be solved by many methods, including heuristic and exact methods [18,19] on classical computers. To the best of our knowledge, our article is the first article to apply QAOA in solving the quadratic knapsack problem. This article focuses on solving the quadratic knapsack problem with the QAOA with a new quantum state that helps the algorithm converge to the optimal value with a low-depth quantum circuit. The paper is organized as follows: Sect. 2 provides preliminary information about QAOA, the quadratic knapsack problem. In Sect. 3, we introduce the biased quantum state for constrained problems and show the convergence of the biased quantum state. Section 4 reports numerical simulations using randomly generated data methods in [13] and compares with [5] state function to demonstrate the efficiency of our proposed method. We choose the [5] state function since it does not depend on the problem structure, unlike Quantum Alternating Operator Ansatz and QAOA with Grover mixer that depend on instances. Finally, we conclude our findings in Sect. 5.

2   Preliminary

2.1   Introduction to QAOA

Quantum Approximate Optimization Algorithm (QAOA) was first introduced in 2014 by Fahri [9]. QAOA is a hybrid classical-quantum algorithm for solving optimization problems. It is designed to run on near-term quantum computers and works by encoding the problem as a cost function minimized via the use of a series of quantum gates and measurements. The algorithm alternates between classical optimization of parameters that control the quantum gates and quantum evolution under those gates. The algorithm's output is a quantum state that approximates the solution to the optimization problem. QAOA has been used for various optimization problems, including graph partitioning, MaxCut, portfolio optimization, and many other combinatorial optimization problems. The state function is defined as

$$|\psi(\beta,\gamma)\rangle = U(\beta_1)U(\gamma_1)...U(\beta_k)U(\gamma_k)|+\rangle^{\otimes n}$$

with $U(\beta) = e^{-i\mathbf{H}_B\beta}$ and $U(\gamma) = e^{-i\mathbf{H}_f\gamma}$. $U(\beta)$ and $U(\gamma)$ are parameterized quantum gates with H_f is problem Hamilton operator

$$H_f = \sum_{i,j=1}^{n} a_{i,j}\sigma_i^Z\sigma_j^Z + \sum_{i=1}^{n} b_i\sigma_i^Z + c$$

with σ_i^Z is the Pauli Z matrix and H_f corresponds to the objective function of combinatorial optimization problem, H_B is mixer Hamiltonian operator

$$H_B = \sum_{i=1}^{n} \sigma_i^X$$

with σ_i^X is Pauli X matrix. The vector $|+\rangle$ is defined by:

$$|+\rangle = \frac{1}{\sqrt{2}}(|0\rangle + |1\rangle)$$

with $|0\rangle = \begin{bmatrix}1\\0\end{bmatrix}$ and $|1\rangle = \begin{bmatrix}0\\1\end{bmatrix}$ are 2 qubits in a quantum computer. The QAOA objective function is defined as follows:

$$\min_{\beta,\gamma\in[0,2\pi]} \langle\psi(\beta,\gamma)|H_f|\psi(\beta,\gamma)\rangle. \tag{1}$$

The expectation can be represented as follows:

$$\langle\psi(\beta,\gamma)\,|H_f|\,\psi(\beta,\gamma)\rangle = \sum_{i=1}^{n} \lambda_i |\langle\psi(\beta,\gamma)|x_i\rangle|^2, \qquad (2)$$

with x_i is the feasible solution represented by bitstring which is encoded in the quantum computer, λ_i is the value of objective function corresponding to x_i, $|\langle\psi(\beta,\gamma)|x_i\rangle|^2$ is the probability of x_i. It is noticeable all of the solutions are encoded into columns of the identity matrix with size $2^n \times 2^n$ with n as the bitstring length corresponding to the feasible solutions of the combinatorial optimization problem and the values of the objective function is the eigenvalue of H_f. This algorithm makes use of classical computers to optimize the expectation function. It uses gradient and gradient-free optimizers such as gradient descent, COBYLA, SPSA, etc. with initial parameters (β_0, γ_0). The disadvantage of the QAOA can be seen in the initial state which considers all the solutions with the same initial probability. This can lead to the low performance of this algorithm when it is used to solve a constrained problem.

2.2 Introduction to Quadratic Knapsack Problem and Its Reformulations for QAOA

In this article, we introduce the quadratic knapsack problem (QKP): n items are given where item j has a positive integer weight w_j. In addition, we are given a cost matrix $Q = \{Q_{ij}\}$ that Q_{ii} is the profit when we select item i and $Q_{ij} + Q_{ji}$ when we select item i and j. The QKP calls for selecting an item subset whose overall weight does not exceed a given knapsack capacity W, so as to maximize the overall profit:

$$\max x^T Q x$$
$$\text{s.t}\, w^T x \leq W$$
$$x \in \{0,1\}^n$$

with w is the vector of item size, Q is the cost matrix with Q_{ij} is the cost of taking item i and j. The problem we consider is the quadratic knapsack problem which has many applications: finance, logistics, and telecommunications [17]. The quadratic knapsack problem can be transformed into QUBO problem by 6 ways [14]:

– Method 1:

$$-x^T Q x + \lambda \left(W - \sum_{i=1}^{N} w_i x_i - \sum_{k=1}^{M} 2^{k-1} y_k \right)^2 \qquad (3)$$

with y_k is additional variable, $M = \lceil log_2(W+1) \rceil$ is the number of additional variables. This method encodes the capacity with binary variables.

– Method 2:

$$-x^T Qx + \lambda \left(\left(W + 1 - 2^{M-1} \right) y_M + \sum_{k=1}^{M-1} 2^{k-1} y_k - \sum_{i=1}^{N} w_i x_i \right)^2 \qquad (4)$$

Unlike Method 1, we adopt a different strategy by employing slack variables to represent the remaining capacity instead of encoding the total weight of the items. To facilitate this, an offset of $2^{M-1} - 1$ is incorporated, along with a set of binary auxiliary variables. In this scenario, $M = \lceil log_2(W+1) \rceil$ auxiliary binary variables are once again necessary.

– Method 3:

$$-x^T Qx + \lambda \left(W - \sum_{i=1}^{N} w_i x_i - \sum_{k=1}^{M} (k-1) y_k \right)^2 \qquad (5)$$

This QUBO formulation resembles Type 1, but it incorporates a one-hot encoding instead of a binary one. In this context, $M = \max_{i=1}^{n} w_i$ auxiliary variables are imperative. It's noteworthy that any solution exhibiting a discrepancy between the total weight and capacity greater than M is suboptimal, as one can include additional items without violating the capacity constraint. It is essential to highlight that this QUBO formulation is effective only when all Q_{ij} values are non-negative. If any Q_{ij} is non-positive, adjusting M to $W + 1$ becomes necessary. Additionally, it is pertinent to acknowledge that strict enforcement of the one-hot encoding is not mandatory in this case.

– Method 4:

$$-x^T Qx + \lambda \left(\sum_{k=1}^{M} (W - k + 1) y_k - \sum_{i=1}^{N} w_i x_i \right)^2 \qquad (6)$$

This expression closely resembles Method 2 but with a deviation in the encoding method. Instead of utilizing a binary encoding, a one-hot encoding is employed. In this context, we introduce $M = \max_{i=1}^{n} w_i$ auxiliary variables, employing a similar technique. It is essential to emphasize that for this approach to be effective, all Q_{ij} must be non-negative. If this condition is not met, $M = W + 1$ auxiliary variables become necessary. Notably, the enforcement of the one-hot encoding remains flexible in this scenario.

– Method 5:

$$-x^T Qx + \lambda \left(W - W_{\text{offset}} - \sum_{i=1}^{N} w_i x_i \right)^2 \qquad (7)$$

with W_{offset} is the difference between the capacity and the total weight of the solution. This formulation does not require additional binary variables, but it has the W_{offset} which is hard to determine.

– Method 6:

$$-x^T Q x + \lambda_1 \left(W - \sum_{i=1}^{N} w_i x_i - \sum_{k=1}^{M} (k-1) y_k \right)^2 + \lambda_2 \left(\sum_{k=1}^{M} y_k - 1 \right)^2 \quad (8)$$

This category represents an expansion of Method 3 by incorporating an additional penalty term with a weight of λ_2 to ensure the implementation of one-hot encoding as per Method 3. The initial penalty term carries a weight of λ_1. Similar to the previous type, if all Q_{ij} values are non-negative, $M = \max_{i=1}^{n} w_i$ auxiliary variables are required; otherwise, $M = W + 1$ auxiliary variables are necessary.

We can see that 6 methods require more than n qubits to transform the problem into QUBO. This increases the complexity of the quantum optimization algorithm when transforming the problem into a QUBO instance. Furthermore, they also distract the algorithm from finding optimal solutions because of the penalty terms. To solve the problem without additional slack variables, we use the Hao et al. [5] in-constraint quantum state function. The in-constrained quantum state function is suitable for constrained problems. This state function is introduced based on the fact that employing the expectation value of the objective function (alternatively, the energy of the Hamiltonian encoding the objective) as the optimization objective proves effective for unconstrained problems in variational quantum algorithms. However, when addressing constrained problems, incorporating constraints into the Hamiltonian via penalty terms presents a challenge. In such cases, optimizing the energy may fail to accurately represent the original problem's objective. The state function has the following form:

$$|\psi\rangle = \frac{\sum_{x_i \in F} \langle \psi(\beta, \gamma) | x_i \rangle}{||\sum_{x_i \in F} \langle \psi(\beta, \gamma) | x_i \rangle||}, \quad (9)$$

with F as the feasible space of the problem, x_i is the binary bitstring representing the problem's solution. The state function involves the straightforward process of eliminating the amplitudes of infeasible bases and subsequently normalizing the remaining state. We can compute the expectation value using post-processed samples when executing this process on a quantum device. This can be achieved either by iterating over the Pauli terms in the Hamiltonian or by preparing a state that closely approximates the samples. Furthermore, the advantages of this state function are not requiring the penalty function and requiring the number of qubits equal to the dimension of the problem. In the next section, we introduce the biased quantum state function for solving the quadratic knapsack problem.

3   Introduction to Biased Quantum State for Constrained Quadratic Binary Optimization

In this section, we introduce the instruction of the quantum state based on QAOA for constrained quadratic binary optimization. The good quantum state has to boost the optimal solution probability and decrease the non-feasible solution. In general, we consider the cost operator H_C has the following form:

$$H_C = \sum_{x_i \in \{0,1\}^n} f(x_i)|x_i\rangle\langle x_i| \tag{10}$$

with eigenstate $|x_i\rangle$, x_i is the binary string, and eigenvalue $f(x_i)$ with $f(x_1) \geq f(x_2) \geq \ldots \geq f(x_k)$ with k is the number of feasible solutions. We want to use a function such that it increases the overlap of the state function with the superposition of the optimal value. First, we consider the penalty function of the constrained problem as follows:

$$g(x) = f(x) - \chi_F(x) \tag{11}$$

with $\chi_F(x)$ is the indicator function:

$$\chi_F(x) = \begin{cases} 0, & \text{if } x \in F \\ \infty, & \text{otherwise} \end{cases}. \tag{12}$$

If $x \notin F$, $g(x)$ goes to $-\infty$. The aim of our method and penalty function for QAOA have the same role is to define which solution is not in the feasible space. Next, we construct the diagonal matrix H that represents the value of the objective function with the indicator as follows:

$$H_{ii} = g(x_i)$$
$$H_{ij} = 0 \text{ if } i \neq j$$

This matrix is similar to the H_C matrix with feasible solutions. The difference between H and H_C is the non-feasible solutions are set as minus infinity. Without losing generality, we consider the following order of the H eigenvalues $f(x_{opt}) = g(x_1) \geq g(x_2) \geq \ldots \geq g(x_{2^n})$. Let $h(H)$ is the operation on the state function introduce the following state:

$$h(H)\psi(\beta, \gamma) = \sum_{i=1}^{2^n} h(g(x_i))\langle\psi(\beta, \gamma)|x_i\rangle|x_i\rangle \tag{13}$$

whose normalized state function has the following form:

$$\frac{h(H)\psi(\beta, \gamma)}{\|h(H)\psi(\beta, \gamma)\|} = \frac{\sum_{i=1}^{2^n} h(g(x_i))\langle\psi(\beta, \gamma)|x_i\rangle|x_i\rangle}{\sum_{i=1}^{2^n} h(g(x_i))|\langle\psi(\beta, \gamma)|x_i\rangle|^2} \tag{14}$$

In this article, we introduce the new initial state for the QAOA by using its cost operator with the function $h(H) = e^H$. We have the following state:

$$|\psi_{new}(\beta, \gamma)\rangle = \sum_{x_i \in \mathbb{B}^n} \frac{a_i e^{g(x_i)/2}|x_i\rangle}{\sqrt{\sum_{x_i \in \mathbb{B}^n} |a_i|^2 e^{g(x_i)}|x_i\rangle}}. \tag{15}$$

with a_i is equal to $\langle x_i | \psi(\beta, \gamma) \rangle$, and $\mathbb{B} = \{0, 1\}$. This function is suitable for every problem, including the constrained problem since it considers the bias of the solutions depending on their value. The higher the value, the higher the probability. If $x \notin F$, $g(x)$ goes to $-\infty$ which can indicate that $e^{g(x)}$ goes to 0. This leads to the elimination of the non-feasible solutions in the initial probability since the original quantum state contains all solutions based on the quantum entanglement and the Hadamard gate considers every qubit with the same probability. We can see that this state function focuses on the solutions that are near the optima rather than considering all solutions with the same probability. This can help the QAOA to converge the optimal solutions. To see our method convergence, we consider the following propositions:

Proposition 1. *Suppose $|\psi_1\rangle$ and $|\psi_2\rangle$ are 2 vectors such that $|||\psi_1\rangle||$, $|||\psi_2\rangle||$ $\leq a$ and $|||\psi_1\rangle - |\psi_2\rangle|| \leq b$ and A is the bounded linear operator with $||A||$ is the maximal eigenvalue in absolute value of A, then we have the following inequality:*

$$|\langle\psi_1|A|\psi_1\rangle - \langle\psi_2|A|\psi_2\rangle| \leq 2||A||ab. \tag{16}$$

Proof. Applying the triangle inequality, we have:

$$|\langle\psi_1|A|\psi_1\rangle - \langle\psi_2|A|\psi_2\rangle| \leq |\langle\psi_1|A|\psi_1\rangle - \langle\psi_1|A|\psi_2\rangle| + |\langle\psi_2|A|\psi_1\rangle - \langle\psi_2|A|\psi_2\rangle|$$
$$\leq ||\langle\psi_1|A(|\psi_1\rangle - |\psi_2\rangle)||| + ||\langle\psi_2|A(|\psi_1\rangle - |\psi_2\rangle)|||$$
$$\leq 2||A||ab$$

From proposition 1, we have to find the Euclidean distance of the QAOA state to the superposition of the optimal solutions since it represents the convergence speed of the state function to the superposition of the optimal solutions. The algorithm can converge after finite steps if the Euclidean distance is small. We can see that the $|\psi_{new}(\beta, \gamma)\rangle$ eliminates all the non-feasible solutions which leads to the reduction of the problem Hamiltonian from 2^n values to k values such that satisfies the constraints of the problem. From that fact, the Hamiltonian we consider the values correspond to the feasible solutions. Without losing generality, we consider the following order of the H's eigenvalues $f(x_{opt}) = g(x_1) \geq g(x_2) \geq \ldots \geq g(x_{2^n})$. We have to estimate the distance of the $|\psi_{new}(\beta, \gamma)\rangle$ and the superposition of the optimal solution. We set the superposition of the optimal solution by vector $|x_{opt}\rangle$, the distance we consider is:

$$|||\psi_{new}(\beta, \gamma)\rangle - |x_{opt}\rangle|| \leq \epsilon, \tag{17}$$

we have to estimate the ϵ to indicate the convergence of the state function $|\psi_{new}(\beta, \gamma)\rangle$ after being optimized. Set the coefficient of the optimal solution superposition in the state $e^H |\psi(\beta, \gamma)\rangle$ as α with $\alpha = e^{g(x_{opt})} |\langle\psi(\beta, \gamma)|x_{opt}\rangle|^2$. We notice that

$$|\alpha - ||e^H \psi(\beta, \gamma)||| = |||\alpha|x_{opt}\rangle|| - ||e^H \psi(\beta, \gamma)||| \leq e^{g(x_2)}. \tag{18}$$

As a result, we have the following estimation:

$$\left|\left|\left||x_{opt}\rangle - \frac{e^H|\psi(\beta,\gamma)\rangle}{||e^H|\psi(\beta,\gamma)\rangle||}\right|\right|\right| \leq \left|\left|\left||x_{opt}\rangle - \frac{e^H|\psi(\beta,\gamma)\rangle}{\alpha}\right|\right|\right| + \left|\left|\left|\frac{e^H|\psi(\beta,\gamma)\rangle}{\alpha} - \frac{e^H|\psi(\beta,\gamma)\rangle}{||e^H|\psi(\beta,\gamma)\rangle||}\right|\right|\right|$$

$$\leq \frac{2e^{g(x_2)}}{\alpha} = \frac{2e^{(g(x_2)-g(x_1))}}{|\langle\psi(\beta,\gamma)|x_{opt}\rangle|^2}.$$

With $g(x_2) - g(x_1) \ll -1$, we can have the following inequality:

$$\left|\left|\left||x_{opt}\rangle - \frac{e^H|\psi(\beta,\gamma)\rangle}{||e^H|\psi(\beta,\gamma)\rangle||}\right|\right|\right| \leq \epsilon \qquad (19)$$

Furthermore, considering the expectation difference between the optimal value superposition and the quantum state, we have:

$$|\langle x_{opt}|H_C|x_{opt}\rangle - \langle\psi_{new}(\beta,\gamma)|H_C|\psi_{new}(\beta,\gamma)\rangle| \leq 2||H_C||||\psi_{new}(\beta,\gamma)\rangle - |x_{opt}\rangle||$$

$$= 2||H_C||\frac{2e^{(g(x_2)-g(x_1))}}{|\langle\psi(\beta,\gamma)|x_{opt}\rangle|^2}$$

We can assume that all the feasible solution to the problem has a positive value because we can transform the arbitrary sign eigenvalue of the cost operator into positive by adding a value that makes all the values positive. We have the following inequality:

$$|\langle x_{opt}|H_C|x_{opt}\rangle - \langle\psi_{new}(\beta,\gamma)|H_C|\psi_{new}(\beta,\gamma)\rangle| \leq \frac{4f(x_1)e^{(g(x_2)-g(x_1))}}{|\langle\psi(\beta,\gamma)|x_{opt}\rangle|^2} \qquad (20)$$

$g(x_2) - g(x_1) = f(x_2) - f(x_1)$ if $x_2 \in F$ which is also known as spectral gap of H_C. We can assume that we can find a parameter such that the probability $|\langle\psi(\beta,\gamma)|x_{opt}\rangle|^2$ is greater than 0 significantly and the difference can go to 0 and $g(x_1) - g(x_2) \gg 1$. In other words, the algorithm converges to the optimal value. To conclude, our quantum state function for QAOA has 2 advantages from the previous articles [4] and Hao et al. in-constraint quantum state function:

- The state of the optimal solution has the highest bias in the quantum state, which can lead to the convergent of the QAOA [4]. In this section, we have proved the convergent with spectral gap condition.
- The number of qubits required for this function equals the number of cost function variables without using the penalty method similar to Hao et al. in-constraint quantum state function. We compare the number of qubits using in our methods with 6 QUBO formulations when solving the quadratic knapsack problem from Sect. 3: Our biased state function requires n qubits, Method 1 requires $n + \lceil log_2(W + 1) \rceil$, Method 2 requires $n + \lceil log_2(W + 1) \rceil$, Method 3 requires $n + W + 1$ in worst cases, Method 4 requires $n + W + 1$ in worst cases, Method 5 requires n qubits but finding W_{offset} is hard and Method 6 requires $n + W + 1$ in worst cases. With small-scale quantum computers, the QAOA can not be implemented for 6 reformulations.

4   Numerical Simulation

Comparative method. In this section, we compare our method with Hao et al. in-constraint quantum state function:

$$|\psi\rangle = \frac{\sum_{x_i \in F} \langle \psi(\beta, \gamma) | x_i \rangle}{||\sum_{x_i \in F} \langle \psi(\beta, \gamma) | x_i \rangle||}. \tag{21}$$

This state function is suitable for solving constrained combinatorial optimization problems. It considers only the solutions in the feasible space. We use this state function to compare with our state function because both of them use the same number of qubits and this state function does not depend on problem structure.

Data. In this section, our data is generated by method from [13]. We consider the elements in Q to be in $[-5, 5]$ and $[-10, 5]$ and the weight vector w has the component in $[1, 50]$, capacity W is chosen equal to $\frac{\sum_{i=1}^{n} w_i}{2}$. The dimension of the problem we consider is 20. We divide them into 2 types: Type 1 and Type 2. Type 1 with the elements in Q to be in $[-5, 5]$ and Type 2 with the elements in Q to be in $[-10, 5]$. Every instance is named as $QKP - i$ with i as the order of the instance.

Metric. In this article, we consider 2 metrics to compare the performance of 2 methods. optimal solution probability and approximate ratio. The optimal solution probability represents the probability of optimal solution when using 2 methods and it can be understood as the fidelity of the quantum state with optimal solution superposition

$$\text{Opt solution prob} = |\langle x_{opt} | \psi_{new}(\beta, \gamma) \rangle|^2. \tag{22}$$

The approximate ratio represents the ratio between the expectation and the optimal value when using 2 methods.

$$\text{approx ratio} = \frac{\langle \psi_{new}(\beta, \gamma) | H_C | \psi_{new}(\beta, \gamma) \rangle}{f(x_{opt})}. \tag{23}$$

Backend and Optimizer for Quantum Circuit. The backend we use is qiskit-aer simulator and the classical optimizer we use is differential evolution. This method is suitable for optimizing continuous functions. The initial parameters for the quantum circuit are found by using the method based on Trotterized Quantum Annealing [3].

Result. In Table 1 and Table 2, we compare our method with Hao et al. in-constraint quantum state function on instances. We can see that our quantum state has a better approximate ratio and fidelity than the Hao et al. in-constraint quantum state function. Furthermore, some instances of our method have the probability of the optimal solution being less than 1 because the spectral gap of the problem operator is not larger than 1 significantly. Most instances use the Hao et al. in-constraint quantum state function have a low approximate ratio and the optimal solution probability equal to 0. This can be indicated by the fact that even though the Hao et al. in-constraint quantum state function has feasible solutions, the value is too far from the optimal value which distracts the in-constraint state function from finding the optimal solution. Furthermore, the performance of the in-constraint state function [5] depends heavily on the QAOA state function. Our state function can overcome this issue because our state function focuses on finding optimal solutions.

Table 1. Comparision between 2 quantum states for quadratic knapsack problem with Type 1 instance

Graph	Biased quantum state		In constrained quantum state [5]	
	Opt solution prob	Approx ratio	Opt solution prob	Approx ratio
QKP-1	0.8092	0.9894	0	−0.0186
QKP-2	0.9983	0.9997	0	−0.3197
QKP-3	0.8805	0.9972	0	0.2770
QKP-4	1	1	0.0013	−0.3058
QKP-5	1	1	0	0.1631
QKP-6	1	1	0	−0.3758
QKP-7	1	1	0	0.0896
QKP-8	1	1	0	0.1184
QKP-9	1	1	0.0048	0.0928
QKP-10	1	1	0	0.1966
QKP-11	0.6015	0.9829	0	0.0937
QKP-12	0.9998	0.9999	0	−0.3609
QKP-13	1	1	0.0009	−0.2551
QKP-14	1	1	0	0.0003
QKP-15	1	1	0	−0.6909

Table 2. Comparision between 2 quantum states for quadratic knapsack problem with Type 2 instance

Graph	Biased quantum state		In constrained quantum state [5]	
	Opt solution prob	Approx ratio	Opt solution prob	Approx ratio
QKP-16	1	1	0	−9.2341
QKP-17	0.9374	0.9900	0	−7.6625
QKP-18	0.8491	0.9745	0	−13.6539
QKP-19	1	1	0	−22.4807
QKP-20	1	1	0.0019	−0.7900
QKP-21	0.9993	0.9995	0	−16.5703
QKP-22	1	1	0	−3.4833
QKP-23	1	1	0.0019	−6.1926
QKP-24	0.5559	0.9596	0	−6.9568
QKP-25	1	1	0	−12.3328
QKP-26	1	1	0	−13.1611
QKP-27	0.9965	0.9983	0	−0.7308
QKP-28	1	1	0	−12.0275
QKP-29	1	1	0	−17.1578
QKP-30	1	1	0	−13.6000

5   Conclusion and Feature Work

In this article, we propose a new quantum state for constrained problems that do not require slack variables and increase the approximate ratio of the quantum optimization algorithm. We also point out when the state function is close to the optimal solution superposition. Our state is close to the optimal solution superposition when the spectral gap is significantly large and the probability of optimal solution with QAOA state function is larger than 0 significantly. We also see that the two main advantages of our state function in solving constrained problems are reducing the number of qubits and creating a high bias for the optimal solution state. To show our state function efficiency, we use randomly generated data with the method in [13]. Furthermore, we compare our state function with the in-constraint state function [5] and show that our state function performs better in both approximate ratio and optimal solution probability.

Acknowledgement. This research is funded by Hanoi University of Science and Technology (HUST) under project number T2023-TĐ-001.

References

1. Saleem, Z.H.: Max-independent set and the quantum alternating operator ansatz. Int. J. Quantum Inf. **18**(04), 2050011 (2020)
2. Wang, S.-S., Liu, H.-L., Song, Y.-Q., Gao, F., Qin, S.-J., Wen, Q.-Y.: Quantum alternating operator ansatz for solving the minimum exact cover problem. Phys. A **626**, 129089 (2023)
3. Sack, S.H., Serbyn, M.: Quantum annealing initialization of the quantum approximate optimization algorithm. Quantum **5**, 491 (2021). 11
4. Amaro, D., Modica, C., Rosenkranz, M., Fiorentini, M., Benedetti, M., Lubasch, M.: Filtering variational quantum algorithms for combinatorial optimization. Quantum Sci. Technol. **7**(1), 015021 (2022)
5. Hao, T., Shaydulin, R., Pistoia, M., Larson, J.: Exploiting in-constraint energy in constrained variational quantum optimization. In: 2022 IEEE/ACM Third International Workshop on Quantum Computing Software (QCS), pp. 100–106. IEEE (2022)
6. Barkoutsos, P.K.l., et al.: Improving variational quantum optimization using CVaR. Quantum **4**, 256 (2020)
7. Acampora, G., Chiatto, A., Vitiello, A.: Genetic algorithms as classical optimizer for the quantum approximate optimization algorithm. Appl. Soft Comput. **142**, 110296 (2023)
8. Wang, Z., et al.: X y mixers: analytical and numerical results for the quantum alternating operator ansatz. Phys. Rev. A **101**(1), 012320 (2020)
9. Farhi, E., Goldstone, J., Gutmann, S.: A quantum approximate optimization algorithm. arXiv preprint arXiv:1411.4028 (2014)
10. Egger, D. J., Marecek, J., Woerner, S.: Warm-starting quantum optimization. Quantum **5**, 479 (2021)
11. Bittel, L., Kliesch, M.: Training variational quantum algorithms is NP-hard. Phys. Rev. Lett. **127**(12), 120502 (2021)
12. Amaro, D., Rosenkranz, M., Fitzpatrick, N., Hirano, K., Fiorentini, M.: A case study of variational quantum algorithms for a job shop scheduling problem. EPJ Quantum Technol. **9**(1), 5 (2022)
13. Moler, C., Van Loan, C.: Nineteen dubious ways to compute the exponential of a matrix, twenty-five years later. SIAM Rev. **45**(1), 3–49 (2003)
14. Punnen, A.P., Pandey, P., Friesen, M.: Representations of quadratic combinatorial optimization problems: a case study using quadratic set covering and quadratic knapsack problems. Comput. Oper. Res. **112**, 104769 (2019). 9
15. Hadfield, S., et al.: From the quantum approximate optimization algorithm to a quantum alternating operator ansatz. Algorithms **12**(2), 34 (2019)
16. Bartschi, A., Eidenbenz, S.: Grover mixers for QAOA: shifting complexity from mixer design to state preparation. In: 2020 IEEE International Conference on Quantum Computing and Engineering (QCE), pp. 72–82. IEEE (2020)
17. Fomeni, F.D., Kaparis, K., Letchford, A.N.: A cut-and-branch algorithm for the quadratic knapsack problem. Discret. Optim. **44**, 100579 (2022)
18. Kellerer, H., et al.: Multidimensional Knapsack Problems. In: Knapsack Problems. Springer, Berlin, Heidelberg (2004). https://doi.org/10.1007/978-3-540-24777-7_9
19. Pisinger, D.: The quadratic knapsack problem-a survey. Discret. Appl. Math. **155**(5), 623–648 (2007)

Quantum Optimization Approach for Feature Selection in Machine Learning

Gérard Fleury[2], Bogdan Vulpescu[1], and Philippe Lacomme[2(✉)]

[1] Laboratoire de Physique de Clermont, Campus Universitaire des Cézeaux, 4 Avenue Blaise Pascal, 63178 Aubière, France
bogdan.vulpescu@clermont.in2p3.fr
[2] LIMOS - UMR CNRS 6158, Université Clermont Auvergne, 1 rue de la Chebarde, 63177 Aubière, France
{gerard.fleury,phiilppe.lacomme}@isima.fr

Abstract. This is intended to be a technical companion presenting some achievements recently published about the usage of quantum algorithms for the selection of relevant features in a given data set. Based on the paradigm of machine learning, such methods use the concept of mutual information between pairs of observables and between observables and the inferred class, in the special case of a simple classification task. Those probabilistic quantities have been discussed a number of times in several works on information theory. Starting from the paper (Mücke et al., 2023), we provide some further inside about the technical details of their work, with an additional test done on a gate processor using the same binary quadratic approximation model.

Keywords: QUBO · feature selection · machine learning · QAOA · quantum annealing

1 Feature Selection

1.1 State of the Art

Dimensionality reduction is a data strategy management that consists in identifying the minimum features to reduce as far as possible the input data dimensions with the objective to favor resolution of numerous machine learning models. Feature selection consists in keeping only a subset of the available features and it is especially relevant for data coming from sources producing redundant or unnecessary features.

Feature selection has received attention for years and a specific trend of researches is concerned by the reduction to the minimal number of features referred to as dimensionality reduction by (Van Der Maaten et al., 2009). Recently, (Mücke et al., 2023) showed how to express the problem of feature selection in the form of a quadratic unconstrained binary optimization problem (QUBO). This was then implemented on a quantum annealing system (D-Wave), showing promising results in comparison with other ways to select the features.

M. Sevaux et al. (Eds.): MIC 2024, LNCS 14753, pp. 281–288, 2024.
https://doi.org/10.1007/978-3-031-62912-9_27

1.2 QUBO Feature Selection

In a simple binary classification problem (supervised), the data set is described by the X_j variables, or attributes (the features), with $j = 1, ..., N$ ($N = 10$ in this example) and the target class variable Y. The data set is composed of n observations (samples, $n = 10000$ in this example) denoted by $\left(x^i, y^i\right)$ with $i = 1, .., n$ where $x^i = \left(x_1^i, \ldots, x_N^i\right)$ represents one single observation which is classified in the y^i class. The goal is to find a subset S of features $\left(x^k, y^k\right)$ with $k \in S$ and $S \subset [N]$ (with $[N] = \{1, 2, \ldots, N\}$), such that the inference capability of a classification tool which learns from this data set will show at least the same performance, although an improvement should be in general expected.

Reducing the number of features used for the training means less hardware computing resources for less complex deep layer networks, for instance, but it can also improve the performance of the generalization by extracting only essential information from a smaller number of uncorrelated features.

In (Mücke et al., 2023) it is explained that the binary decision variables of the corresponding QUBO model are chosen to be the indicators for those selected features, 0 if the feature is not selected and 1 if it is. A matrix Q is constructed from the mutual information, both the relevance of the individual features for the classification decision (called I_i, , corresponding to the diagonal elements of the matrix) and between pairs of different features (called R_{ij}, , for the non-diagonal elements of the matrix). A tuning parameter between those two contributions can finally determine how many features we want to keep in the data set.

Let us note that R_{ij} and I_i are conditional probabilities that can be computed in different way. In this paper R_{ij} and I_i have been computed using the same (Mücke et al., 2023) proposal, which is basically a Kullback-Leibler divergence. In a previous study (Nguyen, 2014) the Shannon entropy was used, the two approaches being mathematically equivalent and leading to the same results.

In the QUBO model, the features to be selected will be indicated by the decision variable $x = (x_1, \ldots, x_N)$ with $x_i \in \{0; 1\}, N = 10$, such that $x_i = 1$ if $i \in S$ (the feature is selected) and $x_i = 0$ if $i \notin S$ (the feature is not selected).

The objective function is a weighted sum between the relevance of the features, expressed by the quantities I_i and the feature pairs redundancies, expressed by R_{ij} (see the referenced paper), with some weighting parameter α, which will determine how many features will be selected at the end. The goal is to find a vector of decision values $x = (x_1, \ldots, x_n)$ which minimizes this objective function:

$$\min_x Q(x, \alpha) = \min_x [-\alpha . \sum_{i=1}^{N} I_i . x_i + (1 - \alpha) . \sum_{i=1}^{N} \sum_{j=1}^{N} R_{ij} . x_i . x_j]$$

which is obviously a quadratic unconstrained binary optimization problem (QUBO).

A QUBO problem can be mapped on a system of n qubits whose states "0" and "1" correspond to the two states of the computational basis of the Z Pauli operator. The objective (or cost) function will be given by the expectation value of an operator built with products of Z operators which describe the coupling between pairs of qubits (the redundancy between pairs of features) and single qubit interaction with some external field (expressing the relevance of the individual features). This complex operator is the

Hamiltonian of the system of qubits and it is used to build the evolution operator in order to make the initial random quantum states evolve towards the states which correspond to the minimum of the expectation value of the Hamiltonian, or the energy of the system of interacting qubits.

2 Hamiltonian Modelization and Resolution

2.1 The Hamiltonian

The binary variables of the QUBO model take values $\{0, 1\}$, while measuring qubits in the computational basis corresponding to the Pauli Z operator projects a state of a single qubit in one of the Z eigenstates $|0\rangle$ or $|1\rangle$, corresponding to the eigenvalues $+1$ and -1, respectively. Therefore it is necessary to do the transformation $x \rightarrow \frac{1}{2}Id - \frac{1}{2}Z$ which assigns to a binary variable a one-qubit operator having the eigenvalues $\{0, 1\}$. In this way, we can calculate the total Hamiltonian as follows.

For the linear part of the problem Hamiltonian we have $H_{P,L} = -\alpha. \sum\limits_{i=1}^{n} l_i.x_i$ leading to:

$$H_{P,L} = -\alpha. \sum_{i=1}^{N} l_i.\frac{1}{2}(Id - Z_i) = -\frac{\alpha}{2}. \sum_{i=1}^{N} l_i.Id + \frac{\alpha}{2} \sum_{i=1}^{N} l_i.Z_i$$

and after considering only the non-constant terms:

$$H_{P,L} = +\frac{\alpha}{2}. \sum_{i=1}^{N} l_i.Z_i$$

For the quadratic part of the Hamiltonian we have

$$H_{P,Q} = (1 - \alpha). \sum_{i=1}^{N} \sum_{j=1}^{N} R_{ij}.x_i.x_j$$

leading to:

$$H_{P,Q} = (1 - \alpha). \sum_{i=1}^{N} \sum_{j=1}^{N} R_{ij}.\frac{1}{2}(Id - Z_i)\frac{1}{2}(Id - Z_j)$$

$$H_{P,Q} = \frac{(1-\alpha)}{4}. \sum_{i=1}^{N} \sum_{j=1}^{N} R_{ij}.(Z_iZ_j - Z_i - Z_j + Id)$$

$$H_{P,Q} = \frac{(1-\alpha)}{4}. \sum_{i=1}^{N} \sum_{j=1}^{N} R_{ij}.(Z_iZ_j - Z_i - Z_j) + \frac{(1-\alpha)}{4}. \sum_{i=1}^{N} \sum_{j=1}^{N} R_{ij}.Id$$

and considering only the non-constant terms:

$$H_{P,Q} = \frac{(1 - \alpha)}{4}. \sum_{i=1}^{N} \sum_{j=1}^{N} R_{ij}.(Z_iZ_j - Z_i - Z_j)$$

The complete problem Hamiltonian will therefore be:

$$H_P = H_{P,L} + H_{P,Q} = \frac{\alpha}{2}\cdot\sum_{i=1}^{N} I_i.Z_i + \frac{(1-\alpha)}{4}\cdot\sum_{i=1}^{N}\sum_{j=1}^{N} R_{ij}.(Z_iZ_j - Z_i - Z_j)$$

which can be further arranged in order to separate single and quadratic terms in the Z_i operators:

$$H_P = \sum_i \left(\frac{\alpha}{2}I_i - \frac{1-\alpha}{4}R_i^{row} - \frac{1-\alpha}{4}R_i^{col}\right)Z_i + \frac{1-\alpha}{4}\sum_{i,j} R_{ij}Z_iZ_j$$

where we have introduced notations for the sums:

$$R_i^{row} = \sum_j R_{ij}, \quad R_i^{col} = \sum_j R_{ji}.$$

2.2 Finding the Optimal Solution

The approach in (Mücke et al., 2023) is to use the quantum adiabatic evolution method which can be implemented on the D-Wave quantum processors. The idea is to start from a ground state of a rather simple Hamiltonian H_D and to "slowly" evolve towards the final Hamiltonian describing our problem, H_P. According to the adiabatic theorem of quantum mechanics, under this conditions there is a good chance that at the end of this time evolution the system will be still in the ground state of the final Hamiltonian, which now represents the solution we were looking for.

The adiabatic quantum optimization has been proven to be equivalent to standard computation (Aharonov, 2007) and if the evolution is "infinitely slow" it can be shown that the optimal solution is always found. In practice, "infinitely slow" depends on the energy spectrum of the problem Hamiltonian and is related to the time constant of the quantum transition to the next lower energy level, assuming there is no crossing of the energy levels during the adiabatic evolution. The main difference from the classical simulated annealing is that escaping from a local minimum is "naturally" obtained by the quantum tunneling of the system of qubits through an energy barrier towards a neighboring lower minimum.

In 2014 (Farhi et al., 2014) introduced the QAOA method (Quantum Approximate Optimization Algorithm), which alternates the application of H_P (the problem Hamiltonian) and H_D (the diffusion or mixing Hamiltonian) as evolution operators $U_P = e^{-i\gamma H_P}$ and $U_D = e^{-i\beta H_D}$ several times, with parameter values expressed by the vectors $\vec{\gamma}$ and $\vec{\beta}$, respectively:

$$\left|\psi\left(\vec{\gamma},\vec{\beta}\right)\right\rangle = \left[e^{-i.\vec{\beta}_k.H_D}.e^{-i.\vec{\gamma}_k.H_P}.e^{-i.\vec{\beta}_{k-1}.H_D}.e^{-i.\vec{\gamma}_{k-1}.H_P}\ldots.e^{-i.\vec{\beta}_1.H_D}.e^{-i.\vec{\gamma}_1.H_P}\right]|s\rangle$$

In one version of the H_D operator (called the mixing operator), the initial state $|s$ encodes an equal superposition of all possible solutions, i.e. $|s\rangle = \frac{1}{\sqrt{2^n}}\cdot\sum_{j=0}^{2^N-1} |j\rangle$, where n is the number of qubits and $|j\rangle = |x_{j,1}x_{j,2}\ldots x_{j,N}\rangle$ with $x_{j,k} \in \{0,1\}, k = 1\ldots N$ (where N is the number of features).

This algorithm is called hybrid, in the sense that it requires a classical optimization method after each application of the evolution operators (gradient descent, genetic algorithm or any other method) to act on the pair of vectors $\left(\vec{\gamma}, \vec{\beta}\right)$ and find the values which minimize the expectation value of the problem Hamiltonian, obtained from quantum measurements, $\left\langle \psi\left(\vec{\gamma}, \vec{\beta}\right)|H_P|\psi\left(\vec{\gamma}, \vec{\beta}\right)\right\rangle$.

3 Numerical Experiments

3.1 Preliminaries

The original data and results from (Mücke et al., 2023) are available at the following web address:

https://github.com/Castle-Machine-Learning/feature-selection-data

A binary classified "synthesized" data set with 10 features sampled 10000 times is extracted and is available as a CSV file here:

https://www.isima.fr/~lacomme/feature/

The conditional probabilities R_{ij} and I_i are combined in a single matrix stored in a plain text file MI.txt (MI from Mutual Information), which is available at the same address and can be used for future researches on the topic. The code that extracts from the data the conditional probabilities used to calculate the mutual information matrix is also provided as a Python script.

3.2 Numerical Experiments

Two solving methods have been used for this optimization problem:

- Experiments on the D-Wave quantum annealer (like in the original work) using the **hybrid_binary_quadratic_model_version2** solver.
- Experiments on the IBM Quantum platform with a QAOA circuit, programmed with Qiskit.

Previous to the real quantum sampler from D-Wave, we have checked the solution with the simulated annealer provided by the "dwave-neal" implementation and with an exact solver available through the "dimod" shared API for the D-Wave samplers. As our problem was formulated as a BQM model (binary quadratic model), we have queried the D-Wave system for the available real samplers for our common user account with a 60 s/1 month trial credit. The only available sampler was the "hybrid_binary_quadratic_model_version2".

The latter numerical experiments have been carried out considering an objective function that is composed of two terms. Let us note $\overline{g}_s$ the estimator $\left\langle \psi\left(\vec{\gamma}, \vec{\beta}\right)|H_P|\psi\left(\vec{\gamma}, \vec{\beta}\right)\right\rangle_s$ using s shots and $\overline{d}_s$ the estimator of the average $\left\langle \psi\left(\vec{\gamma}, \vec{\beta}\right)|H_P|\psi\left(\vec{\gamma}, \vec{\beta}\right)\right\rangle_{j=1,d}$ of the distribution lower than d, where d is the lower decile of the total distribution of the final state energy obtained in the measurements. In the minimization procedure the function passed to be minimized is the sum:

$$\left\langle \psi\left(\vec{\gamma}, \vec{\beta}\right)|H_P|\psi\left(\vec{\gamma}, \vec{\beta}\right)\right\rangle_s + \left\langle \psi\left(\vec{\gamma}, \vec{\beta}\right)|H_P|\psi\left(\vec{\gamma}, \vec{\beta}\right)\right\rangle_{j=1,d}.$$

The classical minimization part of the algorithm used a genetic algorithm implemented with the Python Generic Algorithm from the **PyGAD** library, parametrized for 50 generations, 25 solutions per population and a tournament based procedure for parent-selection. All the experiments have been carried out with a QAOA circuit depth $p = 2$ and with a balance between the feature relevance and redundancy $\alpha = 0.7$ and used a Qiskit noiseless quantum simulator.

The D-Wave processor uses a continuous evolution of the state vector, starting with one well defined state and finishing, hopefully, in the desired state which minimizes the energy (or cost) function. QAOA starts with an equal superposition of all possible states and finishes in another superposition, having the optimal state appearing with higher probabilities among other feasible states (Table 1).

Table 1. Comparison between the best solutions found with both D-Wave and QAOA

	Selected Features
D-Wave solution	$\|0000110101, S = \{5, 6, 8, 10\}$
QAOA solution (Qiskit)	$\|0000110101, S = \{5, 6, 8, 10\}$

The total number of features is 10 meaning that we have a total of 1024 possible selections in the problem (from no feature at all to all 10 features): the optimal feature selection is coming out as 0000110101, with a cost of -0.527796 and has a probability of 0.09% to be found just by a random sampling. At the end of the optimization, QAOA produces a probability distribution of the feasible selections where about 0.37% of the distribution is concentrated on 0000110101.

The convergence of QAOA can be analyzed considering the $\left(\vec{\gamma}, \vec{\beta}\right)$ evolution during the optimization. Table 2 pushes us into considering we have a strong convergence during the optimization.

A large part of the distribution has been shifted towards lower values of the energy and larger probabilities are now concentrated to the best solutions: Fig. 1 and Fig. 2 allow to estimate the difference between frequencies using the initial $\left(\vec{\gamma}, \vec{\beta}\right)$ parameters (Fig. 1) and the frequencies using the final parameters $\left(\vec{\gamma}, \vec{\beta}\right)$ at the end of the genetic algorithm (Fig. 2).

After optimization, the experiments show that 10% of the distribution is now concentrated in the energy range $[-0.527796; -0.516076]$ and the decile is at about 2.22% of the optimal solution having the value -0.527796. The quartile has the value -0.498029 and is about 5.63% of the optimal solution. Before optimization, the decile has the value -0.492178 and is about 6.7% of the optimal solution and the quartile is -0.42971, meaning that we have a gap about of 18.58% within the optimal solution.

The weighting parameter α allows to tune the relative importance of I_i and R_{ij}, larger values of α leading to solutions with a larger number of retained features, as shown in Table 3.

Table 2. Example of convergence curve for QAOA ($p = 2$ and $\alpha = 0.7$)

Iterations	β_1	β_2	γ_1	γ_2
1	4.45341296	6.18663462	5.60029776	3.34755873
2	0.97346632	5.43377067	0.07123081	3.43671716
3	3.07716687	4.5265709	0.92788789	3.7159407
...				
	0.53259607	3.93097723	7.81304791	5.54586062
	5.40755342	3.93097723	7.86559953	5.54586062
	6.60296413	3.93097723	8.06180695	5.51719293
	7.51464180	3.93097723	8.06668202	5.5959680
20	7.41270086	3.93543961	8.06180695	3.97896011

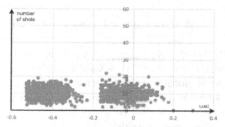

Fig. 1. Initial distribution of shots for random $\left(\vec{\gamma}, \vec{\beta}\right)$ parameters (10000 shots for sampling of the final state), with $p = 2$ and $\alpha = 0.7$

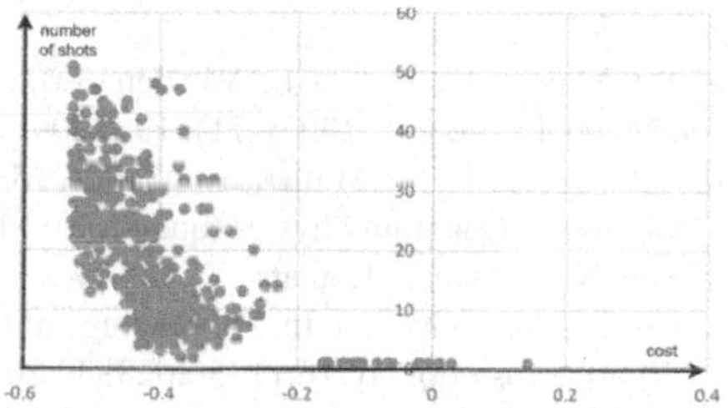

Fig. 2. Final distribution of shots after QAOA optimization (10000 shots for the sampling), with $p = 2$ and $\alpha = 0.7$

Table 3. Evolution of the solution found with α

α	Selected Features	Solution cost
0.3	0000010001, $S = \{6, 10\}$	−0.1780446
0.7	0000110101, $S = \{5, 6, 8, 10\}$	−0.5277972
0.9	1010110101, $S = \{1, 3, 5, 6, 8, 10\}$	−0.7463387

It is possible to conclude that QAOA succeeds in concentrating a large part of the distribution on high quality solution, increasing significantly the probability to find either the optimal one or a solution close to the optimal one.

4   Concluding Remarks

In this article we solve the QUBO problem which finds an optimal selection of features from a data set used for supervised machine learning, both on a D-Wave annealer as in the original paper (Mücke et al., 2023) and on a quantum gate processor from IBM. The results prove that it is possible to solve the feature selection problem using a QAOA based approach, which offers the possibility to solve any Hamiltonian which is not limited to only quadratic terms: quantum gates processor offer a larger field of problems to be solved.

Disclosure of Interests.   Authors have no conflict of interest and do not received financial support of both D-Wave and IBM.

References

Van Der Maaten, L., Postma, E., Van den Herik, J., et al.: Dimensionality reduction: a comparative. J. Mach. Learn. Res. **10**(66–71), 13 (2009)

Mücke, S., Heese, R., Müller, S., Wolter, M., Piatkowski, N.: Feature selection on quantum computers. Quantum Physics (quant-ph); Machine Learning (2023). arXiv:2203.13261

Nguyen, X.V., Chan, J., Romano, S., Bailey, J.: Effective global approaches for mutual information based feature selection. In: Proceedings of the 20th AKM SIGKDD International Conference (2014). https://doi.org/10.1145/2623330.2623611

https://github.com/dwave-examples/mutual-information-feature-selection

Aharonov, D., van Dam, W., Kempe, J., Landau, Z., Lloyd, S., Regev, O.: Adiabatic quantum computation is equivalent to standard quantum computation. SIAM J. Comput. **37**(1), 166–194 (2007). arXiv:quant-ph/0405098

Albash, T., Lidar, D.A.: Adiabatic quantum computing. Quantum Physics (quant-ph) (2018). arXiv:1611.04471 [quant-ph]

Hadfield, S., Wang, Z., O'Gorman, B., Rieffel, E.G., Venturelli, D., Biswas, R.: From the quantum approximate optimization algorithm to quantum alternating operator ansatz. Quantum Physics (quant-ph) (2019). arXiv:1709.03489

Farhi, E., Goldstone, J., Gutmann, S.: A quantum approximate optimization algorithm (2014). arXiv:1411.4028

International Conference on Variable Neighborhood Search (ICVNS)

Advanced Algorithms for the Reclaimer Scheduling Problem with Sequence-Dependent Setup Times and Availability Constraints

Oualid Benbrik[1]($\boxtimes$)(iD), Rachid Benmansour[1,2](iD), Abdelhak Elidrissi[3](iD),
and Angelo Sifaleras[4](iD)

[1] SI2M Laboratory INSEA, Rabat, Morocco
{obenbrik,r.benmansour}@insea.ac.ma

[2] LAMIH CNRS UMR 8201, INSA Hauts-de-France, Polytechnic University
of Hauts-de-France (UVHC), Campus Mont Houy, 59313 Valenciennes Cedex 9,
France

[3] Rabat Business School, International University of Rabat, Parc Technopolis,
Rabat, Morocco
abdelhak.elidrissi@uir.ac.ma

[4] Department of Applied Informatics, School of Information Sciences,
University of Macedonia, 156 Egnatias Str., 54636 Thessaloniki, Greece
sifalera@uom.gr

Abstract. Scheduling of reclaimers activities in dry bulk terminals significantly impact terminal throughput, a crucial performance indicator for such facilities. This study addresses the Reclaimer Scheduling Problem (RSP) while considering periodic preventive maintenance activities for reclaimers. These machines are integral for reclaiming dry bulk materials stored in stockyards, facilitating their loading onto vessels via shiploaders. The primary aim of the objective function entails the minimization of the overall completion time, commonly referred to as the makespan. Since this problem is $\mathcal{NP}$-hard, we propose a novel greedy constructive heuristic. The solutions obtained from this heuristic serve as the starting point for an efficient General Variable Neighborhood Search (GVNS) algorithm to handle medium-scale instances resembling real stockyard configurations. Computational experiments are conducted by comparing the proposed methods across various problem instances. The results demonstrate that the developed GVNS, coupled with the constructive heuristic for initial solution finding, efficiently improves scheduling efficacy. Thus, it emerges as a new state-of-the-art algorithm for this problem.

Keywords: Reclaimer Scheduling · Bulk Ports · Sequence-Dependent
Setup Times · Availability Constraints · Machine Eligibility
Restrictions · Variable Neighborhood Search · Heuristic

© The Author(s), under exclusive license to Springer Nature Switzerland AG 2024
M. Sevaux et al. (Eds.): MIC 2024, LNCS 14753, pp. 291–308, 2024.
https://doi.org/10.1007/978-3-031-62912-9_28

1   Introduction and Literature Review

Bulk terminals play a pivotal role in global trade by facilitating the efficient handling and storage of large quantities of commodities, such as coal, minerals, grains, raw materials, and so on. These terminals serve as crucial nodes in the logistics chain, ensuring the seamless flow of goods between various modes of transportation. The importance of bulk terminals cannot be overstated, given their pivotal role in maritime transport, which handles approximately 80% of the world's trade volume, as reported by the United Nations Conference on Trade and Development (UNCTAD 2022) [15]. Despite their indispensable contribution to global trade, bulk terminals have not received proportionate attention in the research literature when compared to container terminals. While container terminals have been extensively studied, the operational challenges specific to bulk terminals have been relatively understudied. However, recent research is placing a growing emphasis on understanding and addressing the distinctive challenges faced by bulk terminals.

The overall configuration of dry bulk terminals involves a designated berth area where vessels anchor for the loading or unloading of materials, utilizing shiploaders or cranes. Complementing this, the terminal features a yard where bulk cargoes are managed, either through addition as stockpiles using stacker machines or complete reclamation using reclaimer machines, facilitating subsequent delivery to ships at the berths. The research at hand is prompted by a keen interest in the operational intricacies of bulk ports, with a specific emphasis on the Newcastle Coal Infrastructure Group (NCIG) terminal, a notable coal export terminal in Australia [7]. The NCIG stockyard incorporates diverse stockpads, each tailored with specific positions for unloaded coal. Rail tracks are strategically positioned between parallel stockpads, accommodating stacker-reclaimer (SR) machinery for effective material handling.

The effective scheduling of reclaimers constitutes crucial aspects of resource management in dry bulk terminals, directly influencing terminal throughput—a key performance indicator for these facilities. Despite its paramount significance, research on this subject is relatively underdeveloped, with a limited number of papers addressing the RSP. To the best of our knowledge, Hu and Yao [10] were the pioneers in addressing the SR scheduling problem at an iron ore terminal. They concentrated on minimizing the makespan for a given set of handling operations using a genetic algorithm (GA). Similarly, Angelelli et al. [1] conducted a study on bulk material reclamation in stockyards. They presented and analyzed multiple variants of an abstract scheduling problem for the reclaiming operations and demonstrated the $\mathcal{NP}$-hardness of these variants. Kalinowski et al. [11] extended the work presented by Angelelli et al. [1], relaxing the assumption that all stockpiles must be stacked at the beginning of the planning period. They further investigated the dynamic version of the problem, although they did not consider the setup times (i.e., traveling time) of reclaimers. Recently, Ünsal [16] delved into the RSP within a realistic world setting. He posited that the problem is a variant of the parallel machine scheduling problem and presented two versions—one with stacking operations and one without. The author developed an arc-time indexed Mixed Integer Programming (MIP) formulation to solve the problem.

The loading and unloading process at the yard-side presents risks to critical equipment like the SR, necessitating periodic preventive maintenance to prevent breakdowns and accidents [2]. This maintenance, including inspections, lubrication, and safety testing, is essential for terminal reliability but leads to downtime affecting stockpile handling. Benbrik et al. [3] pioneered the integration of preventive maintenance into reclaimer scheduling, developing mathematical formulations for the RSP. They explored two cases: one with two stockpads and one reclaimer, resulting in two novel formulations, and another with three stockpads and two reclaimers, leading to a unique model. Their formulations, solved using CPLEX, successfully handled small instances but struggled with medium ones. Therefore, this paper extends the scope of the second case of the configuration addressed in [3], with the primary objective of solving this problem with a real configuration of the stockyard involving multiple stockpads.

The main contributions of this paper are as follows:

- Investigation of a real configuration of a coal export terminal involving the minimization of the makespan.
- Development of a novel and innovative greedy constructive heuristic. Additionally, the design of an efficient GVNS metaheuristic for solving medium-sized instances of the problem within a reasonable computational time.
- Provision of empirical results from numerical experiments for reasonable computing times, considering both the literature and industrial practices.

The remaining sections of this paper are organized as follows. Section 2 presents the problem addressed in this study. Section 3 introduces a version of the MIP model previously developed for solving the problem. In Sect. 4, we describe the proposed greedy constructive heuristic procedures. Section 5 presents the GVNS approach employed in this research. Numerical experiments are conducted in Sect. 6. Finally, Sect. 7 concludes the paper by summarizing the findings and discussing future perspectives.

2   Problem Description

This paper addresses the scheduling problem related to the reclamation of stockpiles using a set of identical reclaimer machines, denoted as $\mathcal{M} = \{M_1, M_2 \ldots, M_m\}$. The operational layout consists of parallel stockpads on the yard-side of a dry bulk export terminal, represented by $\mathcal{P} = \{\mathcal{P}_1, \mathcal{P}_2, \ldots, \mathcal{P}_{m+1}\}$. Each reclaimer machine is mounted on a rail track between two adjacent stockpads. Let $\mathcal{P}_z = \{n_{(z-1)} + 1, \ldots, n_z\}$ represent the set of stockpiles in stockpad $\mathcal{P}_z$, with $n_0 = 0$ and n_z denoting the number of stockpiles in $\mathcal{P}_z$. The set $\mathcal{N} = \{J_1, J_2, \ldots, J_n\}$ encompasses all stockpiles across all stockpads, where $n = \sum_{z=1}^{m+1} n_z$. Each stockpile i, where $i \in [\![1, n]\!]$, possesses a length denoted by L_i. The time required to reclaim a stockpile, p_i, is determined as the ratio of its length to the reclamation speed s (i.e., $p_i = L_i/s$). Introducing sequence-dependent setup times, denoted as $t_{i,j}$, accounts for the travel time between two consecutive stockpiles. The setup time is the duration between completing the reclamation of $J_i \in \mathcal{N}$ and commencing the reclamation of the subsequent stockpile $J_j \in \mathcal{N}$. We assume $t_{0,j} = 0$, signifying no setup before processing the first reclaiming job. Additionally, the triangle property holds for setup times, ensuring $t_{i,l} + t_{l,j} \geq t_{i,j}$ for any three distinct jobs J_i, J_j, and J_l. Furthermore, strict adherence to the eligibility restrictions of the machines is enforced; each machine possesses the capability to pivot its boom, facilitating the processing of adjacent stockpiles along the rail track, while reclamation of stockpiles from other stockpads is not allowed. On a reclaimer machine M_k, stockpile reclamation occurs during the interval between consecutive preventive periodic maintenance activities, with the length denoted as T_k. Each maintenance activity has a duration of σ. Reclaiming tasks are prohibited during maintenance activities, and no breakdowns occur after maintenance. The problem is denoted as Reclaimer Scheduling Problem with Preventive Periodic Maintenance Activities (RSP-PPMA), with the objective of finding a feasible schedule to minimize the makespan. In this problem, we refer to each reclaimer as a machine, and each operation of reclaiming stockpile as a job.

Importantly, this paper expands upon the previous work conducted by Benbrik et al. [3]. The main notations used to describe the problem are listed in Table 1.

Figure 1 displays a graphical representation of a feasible solution to the addressed RSP-PPMA. In the figure, green rectangles represent T_k, indicating the duration between two consecutive maintenance activities on machine $M_k \in \mathcal{M}$. The maintenance activities are denoted by PM in yellow rectangles, with σ representing the duration of a maintenance activity. Jobs are scheduled within batches denoted as $\{B_1, B_2, \ldots, B_b, \ldots, B_n\}$, each having a duration of $T_k + \sigma \ \forall k \in [\![1, m]\!]$.

Table 1. Notations.

Sets and Indices	
m	Number of machines
n_z	Number of jobs in stockpad z
n	Number of jobs ($n = \sum_{z=1}^{m+1} n_z$)
$\mathcal{M}$	Set of machines ($M_1, M_2, \ldots, M_k, \ldots, M_m$)
$\mathcal{P}$	Set of stockpads ($\mathcal{P}_1, \mathcal{P}_2, \ldots, \mathcal{P}_z, \ldots, \mathcal{P}_{m+1}$)
$\mathcal{P}_z$	Set of jobs in stockpad z ($\mathcal{P}_z = \{n_{(z-1)} + 1, \ldots, n_z\}$ with $n_0 = 0$)
$\mathcal{N}$	Set of jobs in all stockpads ($\mathcal{N} = \{J_1, J_2, \ldots, J_n\}$)
Parameters	
A	Large positive integer
p_i	Processing time of job $J_i \in \mathcal{N}$
$t_{i,j}$	Travel time of machine between stockpile J_i and stockpile J_j, i.e., setup time
T	Time interval between two consecutive maintenance activities
σ	Duration of a maintenance activity
Decision variables	
$C_{i,k}$	The completion time of job $J_i \in \mathcal{N}$ on machine $M_k \in \mathcal{M}$
C_{max}	Maximum completion time (makespan)

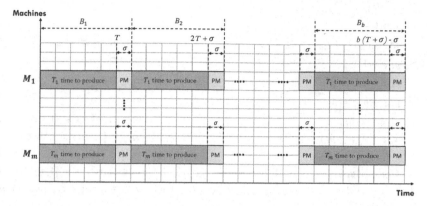

Fig. 1. Graphical representation of a feasible solution. (Color figure online)

3 Mathematical Formulation

This section introduces a complex version of the MIP formulation previously proposed by Benbrik et al. [3], tailored for addressing the RSP-PPMA. The model is specialized for scenarios involving three parallel stockpads and two reclaimer machines. The scheduling strategy involves grouping jobs into batches denoted as $\mathcal{B} = \{B_1, B_2, \ldots, B_b, \ldots, B_n\}$ for the two reclaimer machines, with the overarching objective of minimizing the makespan. Each batch on machine $M_k \in \mathcal{M}$ has a capacity constraint denoted as T_k, and the time allocated for each maintenance activity is represented by σ. Throughout this formulation, the assumption $T_k = T, \forall k \subset [\![1, m]\!]$ is adopted, where the notation $[\![X, Y]\!]$ is used to indicate the interval of all integers between X and Y included. This problem can be seen, as a variant of the parallel machine scheduling problem [16].

The binary decision variables in this formulation are denoted as follows:

$$x_{i,j} = \begin{cases} 1 & \text{if job } J_j \text{ follows job } J_i \text{ in the sequence} \\ 0 & \text{otherwise} \end{cases}$$

$$y_{i,k} = \begin{cases} 1 & \text{if job } J_i \text{ is processed on machine } M_k \in \mathcal{M} \\ 0 & \text{otherwise} \end{cases}$$

$$\alpha_{i,b}^k = \begin{cases} 1 & \text{if job } J_i \text{ is in batch } b \in \mathcal{B} \text{ on machine } M_k \in \mathcal{M} \\ 0 & \text{otherwise} \end{cases}$$

$$(MIP) \min C_{max} \tag{1}$$

$$\text{s.t. } C_{max} \geq C_{i,k} \qquad\qquad \forall i \in [\![1,n]\!], \forall k \in [\![1,m]\!] \tag{2}$$

$$y_{i,1} = 1 \qquad\qquad \forall i \in \mathcal{P}_1 \tag{3}$$

$$y_{i,2} = 1 \qquad\qquad \forall i \in \mathcal{P}_3 \tag{4}$$

$$y_{i,1} + y_{i,2} = 1 \qquad\qquad \forall i \in \mathcal{P}_2 \tag{5}$$

$$C_{i,k} \geq p_i y_{i,k} \qquad\qquad \forall i \in [\![1,n]\!], \forall k \in [\![1,m]\!] \tag{6}$$

$$x_{i,j} + x_{j,i} \geq y_{i,k} + y_{j,k} - 1 \qquad\qquad \forall i,j \in [\![1,n]\!], i \neq j, \forall k \in [\![1,m]\!] \tag{7}$$

$$C_{i,k} \leq A y_{i,k} \qquad\qquad \forall i \in [\![1,n]\!], \forall k \in [\![1,m]\!] \tag{8}$$

$$C_{j,k} + A(3 - x_{i,j} - y_{i,k} - y_{j,k}) \geq C_{i,k} + t_{i,j} + p_j \qquad \forall i,j \in [\![1,n]\!], i < j, \forall k \in [\![1,m]\!] \tag{9}$$

$$C_{j,k} + t_{j,i} + p_i \leq C_{i,k} + A(2 - y_{i,k} - y_{j,k} + x_{i,j}) \qquad \forall i,j \in [\![1,n]\!], i < j, \forall k \in [\![1,m]\!] \tag{10}$$

$$\sum_{b=1}^{n} \alpha_{i,b}^{k} = y_{i,k} \qquad\qquad \forall i \in [\![1,n]\!], \forall k \in [\![1,m]\!] \tag{11}$$

$$C_{i,k} \geq (b-1)\alpha_{i,b}^{k}(T_k + \sigma) + t_{n+1,i}\alpha_{i,b}^{k} + p_i y_{i,k} \qquad \begin{array}{l} \forall i \in [\![1,n]\!], \forall b \in [\![2,n]\!], \\ \forall k \in [\![1,m]\!] \end{array} \tag{12}$$

$$C_{i,k} \leq b\alpha_{i,b}^{k} T_k + (b-1)\sigma - t_{i,n+1} + A(1 - \alpha_{i,b}^{k}) \qquad \forall i,b \in [\![1,n]\!], \forall k \in [\![1,m]\!] \tag{13}$$

$$C_{i,k} \geq 0 \qquad\qquad \forall i \in [\![1,n]\!], \forall k \in [\![1,m]\!] \tag{14}$$

$$x_{i,j} \in \{0,1\} \qquad\qquad \forall i,j \in [\![1,n]\!] \tag{15}$$

$$y_{i,k} \in \{0,1\} \qquad\qquad \forall i \in [\![1,n]\!], \forall k \in [\![1,m]\!] \tag{16}$$

$$\alpha_{i,b}^{k} \in \{0,1\} \qquad\qquad \forall i,b \in [\![1,n]\!], \forall k \in [\![1,m]\!] \tag{17}$$

The objective function (1) aims to minimize the makespan of each reclaimer machine. Constraint set (2) ensures that the makespan of an optimal schedule is not less than the completion time of all jobs that have been executed on each machine. The sets of constraints (3)–(5) impose eligibility restrictions on the reclaimer machines. Constraints (3) and (4) specify that all jobs in the stockpad $\mathcal{P}_1$ and the last stockpad $\mathcal{P}_3$ are processed on machines M_1 and M_2, respectively. Constraint set (5) guarantees that each job J_i in stockpad $\mathcal{P}_2$ is assigned to exactly one reclaimer machine. Constraint set (6) calculates the completion time of job J_i on each machine. Constraint set (7) ensures that no two jobs J_i and J_j can overlap in time. Constraints (8)–(10) specify that a job can be processed only if the machines are available. Constraints (9) and (10) indicate that no two jobs J_i and J_j scheduled on the same reclaimer machine (i.e., $y_{i,k} = y_{j,k} = 1$) can overlap in time. Constraint set (11) ensures that each job J_i is assigned to exactly one batch on a corresponding machine M_k. Constraints (12) and (13) guarantee that every job J_i processed by machine M_k must be executed within batch B_b of this machine. Constraint (12) ensures that in batch B_b of machine M_k, with $b \in [\![2,n]\!]$, the scheduling of any job J_i in this batch is performed after the end of preventive maintenance and the setup time $t_{(n+1),i}$ spent after maintenance. Additionally, constraint (13) requires that each job J_i processed by machine M_k should be finished before the starting time of preventive maintenance activity and its related setup time $t_{i,(n+1)}$. Constraint set (14) defines the completion time of job J_i on machine M_k as a positive continuous variable. Constraints (15), (16), and (17) define the variables $x_{i,j}$, $y_{i,k}$, and $\alpha_{i,b}^k$ as binaries.

4   Greedy Constructive Heuristic Procedure

In this section, we present a novel constructive heuristic algorithm tailored for solving the RSP-PPMA (Algorithm 1). Constructive methods, widely used for generating feasible solutions from scratch in optimization problems, serve as the foundation for our innovative approach. The algorithm begins with an initialization phase, assigning jobs from the first and last stockpads, $\mathcal{P}_1$ and $\mathcal{P}_{m+1}$, to machines M_1 and M_m, respectively. The core of our constructive heuristic procedure (Phase 2) is an iterative process that traverses each intermediate stockpad $\mathcal{P}_z$, where $z \in [\![2, m]\!]$. Within this phase, a job J_i is selected from a specific position (pos_i) within stockpad $\mathcal{P}_z$. The selected job is simultaneously allocated to both the current machine M_z and the preceding machine M_{z-1}, adhering to eligibility constraints. Allocation decisions depend on the comparison of the last completion times for jobs assigned to M_z and M_{z-1}. Based on this evaluation, the job either remains at M_z or is transferred to M_{z-1}, striving for a locally optimal job-machine allocation. This iteration continues until all positions within the stockpad have been considered. Moving to Phase 3, the algorithm generates prioritized job sequences π_k for each machine M_k, where $k \in [\![1, m]\!]$. This prioritization is achieved by sorting jobs using the Longest Processing Time (LPT) rule. The resulting prioritized sequences π_k are then concatenated, forming an ordered list that guides the creation of the final schedule π as a feasible solution to the RSP.

The preventive maintenance assignment scheduler algorithm (PMAS) (Algorithm 2) is devised to incorporate Preventive Maintenance (PM) considerations into the initial job sequence π generated by the constructive heuristic procedure. In Phase 4 of the Algorithm 1, PMAS is applied to seamlessly integrate PM and calculate the makespan with the final sequence π_{PM}. The input includes the initial sequence π and the batch duration representing the time between two consecutive preventive maintenance activities (i.e., T). The algorithm iterates through each machine M_k, applying the Algorithm 1 to generate individual sequences π_k. Subsequently, the completion times for each job in π_k are calculated, and jobs exceeding the batch duration are moved to the next batch. The algorithm maintains track of completion times, and the last completion time of each machine is reported (i.e., makespan). The final sequence π_{PM} is generated by incorporating the PM strategy, ensuring that jobs are appropriately scheduled within batches. This is achieved by updating π_{PM} through the union operation with the sequences $\pi_{k,\text{PM}}$ for each machine M_k. The output includes π_{PM} and the makespan C_{max}, providing a feasible solution to the RSP that seamlessly integrates both job sequencing and PM considerations.

Algorithm 1. Constructive heuristic algorithm (Constructive)

    Input: $z, n_z, \mathcal{P}_z, m$
1: **Phase 1: Initialization**
2: **if** $z = 1$ **then**
3:     Assign($M_1, \mathcal{P}_1$)
4: **end if**
5: **if** $z = m + 1$ **then**
6:     Assign($M_m, \mathcal{P}_{(m+1)}$)
7: **end if**
8: **for** $k = 2$ **to** $m - 1$ **do**
9:     $\pi_k \leftarrow \emptyset$
10: **end for**
11: **Phase 2: Constructive Heuristic Procedure**
12: $z \leftarrow 2$, $pos \leftarrow 1$
13: **while** $(pos \leq n_z)$ **do**   ▷ $z \in [\![2, m]\!]$
14:     **for each** $z \in [\![2, m]\!]$ **do**
15:         Choose the job J_i situated at the position pos within the stockpad $\mathcal{P}_z$.
16:         Assign(M_{z-1}, J_i)
17:         Assign(M_z, J_i)
18:         **if** $\left(C_{max}(M_z) < C_{max}(M_{z-1}) \right)$ **then**
19:             Assign(M_z, J_i)
20:             Remove(M_{z-1}, J_i)   ▷ Local search procedure
21:         **else**
22:             Assign(M_{z-1}, J_i)
23:             Remove(M_z, J_i)
24:         **end if**
25:     **end for**
26:     $pos \leftarrow pos + 1$
27: **end while**
28: **Phase 3: Schedule jobs using the** *LPT* **rule**
29: $\pi \leftarrow \emptyset$
30: **for each** $k \in [\![1, m]\!]$ **do**
31:     $\pi_k \leftarrow$ Sort(M_k, LPT)
32:     $\pi \leftarrow \pi \cup \pi_k$
33: **end for**
34: **Phase 4: Integration of preventive maintenance and makespan calculation using PMAS**
35: $(\pi_{\text{PM}}, C_{max}) \leftarrow$ PMAS(π, T)
    Output: π_{PM}, C_{max}

Algorithm 2. Preventive maintenance assignment scheduler algorithm (PMAS)

    Input: π, T ▷ Initial sequence and batch duration
1: $\pi_{\text{PM}} \leftarrow \emptyset$ ▷ Initialize final sequence
2: **for** $k = 1$ **to** m **do**
3:     $\pi_k \leftarrow$ Constructive($\mathcal{N}, \mathcal{P}, \mathcal{M}$)   ▷ Individual sequences for each M_k
4:     $C_{max} \leftarrow 0$
5:     $batch_k \leftarrow 1$
6:     **for each** $J_i \in \pi_k$ **do**
7:         Calculate the completion time $C_{i,k}$ of job J_i
8:         **if** $C_{i,k} > batch_k \times T_k$ **then**
9:             Move the current job J_i to the next batch
10:             Update the completion time $C_{i,k}$
11:             $batch_k \leftarrow batch_k + 1$
12:         **end if**
13:     **end for**
14:     Report the last completion time $C_{i,k}$ of machine M_k
15:     **if** $C_{i,k} > C_{max}$ **then**
16:         Update C_{max} with the last completion time $C_{i,k}$
17:     **end if**
18:     $\pi_{\text{PM}} \leftarrow \pi_{\text{PM}} \cup \pi_{k,\text{PM}}$ ▷ Sequence with PM
19: **end for**
    Output:
20: Final sequence π_{PM} and Makespan C_{max}

---

5   General Variable Neighborhood Search

Advanced optimization techniques, such as Simulated Annealing (SA), Tabu Search (TS), and Variable Neighborhood Search (VNS), constitute sophisticated metaheuristic approaches. These methods are strategically crafted to navigate beyond local optima within the search space while steering other heuristics. The VNS is a metaheuristic method initially introduced by Mladenović and Hansen [12]. It incorporates a local search procedure and dynamically adjusts neighborhood structures throughout the solution process. Over time, VNS has transcended its status as just a metaheuristic, transforming into a general frame-work for heuristic development. Numerous variants have emerged from the original schema [9], rendering this methodology a robust and potent tool in the optimization context. The VNS method has been widely employed in a diverse set of optimization problems (see Sifaleras and Konstantaras [14]; Benmansour and Sifaleras [5]; Benmansour et al. [4]; Benmansour et al. [6]; Elidrissi et al. [8]).

In this paper, we utilize the primary framework of the methodology known as GVNS. However, as mentioned earlier, various well-known variants exist along-side the general design. Some of the classical and widely adopted ones include Basic VNS (BVNS), Reduced VNS (RVNS), and Variable Neighborhood Descent (VND). For a recent survey on VNS, we direct the reader to the work by Hansen et al. [9]. The GVNS proposed in this paper for the RSP-PPMA generally incorporates advanced local search procedures, such as pipe VND, sequential VND, cyclic VND, and others, in the algorithm's improvement phase (cf. Hansen et al., [9]).

5.1   Neighborhood Structures

The efficiency of the GVNS metaheuristic relies on defining suitable neighborhood structures. In this paper, we categorize permutation-based neighborhoods into two types: intra-machine and inter-machine. Intra-machine neighborhoods concentrate on single-reclaimer machine impacts, comprising the Intra-machine exchange (N_1), Intra-machine insert (N_2), and Intra-machine reverse (N_3) neighborhoods. The Intra-machine exchange involves swapping positions of two jobs $\pi_{k,\mathrm{PM}}^{J_i}$ and $\pi_{k,\mathrm{PM}}^{J_j}$ on a single machine $M_k \in \mathcal{M}$. Intra-machine insert moves a job within the same machine, which involves taking a job from its current position and inserting it into another position within the same machine $M_k \in \mathcal{M}$. Intra-machine reverse chooses two jobs $\pi_{k,\mathrm{PM}}^{J_i}$ and $\pi_{k,\mathrm{PM}}^{J_j}$ and reverses the order of the jobs between them on a machine. Inter-machine neighborhoods, affecting two machines, consist of the Inter-machine exchange (N_4) and Inter-machine insert (N_5). The Inter-machine exchange chooses jobs $\pi_{k-1,\mathrm{PM}}^{J_i}$ and $\pi_{k+1,\mathrm{PM}}^{J_j}$, where $J_i, J_j \in P_k \cup P_{k+1}$, and exchanges them, while Inter-machine insert removes a job from one machine and inserts it into another, adhering to eligibility restrictions.

5.2   Variable Neighborhood Descent

We propose incorporating the neighborhood structures detailed in Sect. 5.1 collectively within the context of a VND heuristic to locally refine a given solution π_{PM}. The general framework of the VND is delineated in Algorithm 3. It initiates with an initial solution π_{PM} derived from a constructive heuristic, extensively discussed in Sect. 4. Subsequently, the algorithm persistently endeavors to construct an improved solution from the current state π_{PM} by exploring its neighborhood $N_l(\pi_{PM})$. Thus, the effectiveness of the VND variants proposed in this study relies on the sequence of five neighborhood structures $(N_1, N_2, N_3, N_4, N_5)$ and the chosen search strategy (*first* or *best* improvement), known as the neighborhood change step. The approach chosen for transitioning between neighborhoods is the pipe strategy. The steps of this strategy are outlined in Algorithm 4. Following preliminary tests, we opted for the best improvement in the search strategy, and in our proposed GVNS, the selected neighborhood order is $N_4(\pi_{PM}), N_5(\pi_{PM}), N_3(\pi_{PM}), N_1(\pi_{PM}), N_2(\pi_{PM})$.

Algorithm 3. Variable Neighborhood Descent VND

Data: π_{PM}, l_{max}
Result: π_{PM}
1: **while** There is no improvement **do**
2:    $l \leftarrow 1$
3:    **while** $l \leq l_{max}$ **do**
4:       $\pi'_{PM} \leftarrow$ Local Search(π_{PM}, N_l)
5:       ChangeNeighborhood-Pipe(π_{PM}, π'_{PM}, l)
6:    **end while**
7: **end while**
8: **return** π_{PM}

Algorithm 4. ChangeNeighborhood-Pipe (π_{PM}, π'_{PM}, l)

1: **if** $\left(C_{max}(\pi'_{PM}) < C_{max}(\pi_{PM})\right)$ **then**
2:    $\pi_{PM} \leftarrow \pi'_{PM}$
3: **else**
4:    $l \leftarrow l + 1$
5: **end if**
6: **return** π_{PM}

5.3   Shake Strategy

The shaking phase, pivotal for escaping local optima during convergence, is integral to the algorithm's effectiveness (Mladenović and Hansen, [12]). This phase systematically generates k random jumps from the current solution π_{PM}. Based on our experiments, we adopted a diversification approach involving the random selection from a set of predefined neighborhood structures, namely N_3, N_1, and

N_2. Subsequently, we apply the selected structure k times, where $1 \leq k \leq k_{max}$. It's worth noting that introducing additional neighborhood structures in the shaking method has been observed to detrimentally impact result quality. The procedures of the shaking phase are outlined in Algorithm 5.

Algorithm 5. Shaking

    Data: π_{PM}, k
1:   $p \leftarrow$ randomInteger$(1, 3)$
2: **for** $j = 1$ to k **do**
3:     **if** $(p = 1)$ **then**
4:        Generate a random $\pi'_{PM} \in N_3(\pi_{PM})$
5:     **end if**
6:     **if** $(p = 2)$ **then**
7:        Generate a random $\pi'_{PM} \in N_1(\pi_{PM})$
8:     **end if**
9:     **if** $(p = 3)$ **then**
10:       Generate a random $\pi'_{PM} \in N_2(\pi_{PM})$
11:    **end if**
12: **end for**
13:    **return** π'_{PM}

5.4   GVNS for the RSP-PPMA

In this section, we present the overall pseudocode of GVNS as it is implemented to solve the RSP-PPMA, which is presented in Algorithm 6. This scheme has three input parameters: the initial solution (π_{PM}), the maximum perturbation level (k_{max}), and the maximum computing time (T_{max}). The parameters (T_{max}) and (k_{max}), determined after preliminary experimentation, will be provided in Sect. 6.2. The diversifcation and intensifcation ability of GVNS relies on the shaking phase and VND, respectively. Shaking step of GVNS consists of three neighbohood structures N_3, N_1, and N_2. In the VND step, the five proposed neighborhood structures are used. The stopping criterion is a CPU time limit T_{max}. It is worth noting that the construction of the initial solution lies outside the GVNS framework. Typically, this initial solution can be generated randomly, following the common practice in the VNS community. However, a more sophisticated constructive procedure, as supported by literature (see Sánchez-Oro et al. [13]), can significantly enhance the quality of the best solution. A well-designed starting point is often more promising than a simple random solution. In Sect. 4, we detailed the constructive procedures proposed for the RSP-PPMA, which furnish the initial solutions for the GVNS.

Algorithm 6. General variable neighborhood search GVNS

Data: $\pi_{\text{PM}}, k_{max}, T_{max}$
Result: π_{PM}
1: while $CPU \leq T_{max}$ do
2:     $k \leftarrow 1$
3:     while $k \leq k_{max}$ do
4:         $\pi'_{\text{PM}} \leftarrow \text{Shaking}(\pi_{\text{PM}}, k)$
5:         $\pi''_{\text{PM}} \leftarrow \text{VND}(\pi'_{\text{PM}})$
6:         if $\left(C_{max}(\pi''_{\text{PM}}) < C_{max}(\pi_{\text{PM}})\right)$ then
7:             $\pi_{\text{PM}} \leftarrow \pi''_{\text{PM}}$
8:             $k \leftarrow 1$
9:         else
10:             $k \leftarrow k + 1$
11:         end if
12:     end while
13: end while
14: return π_{PM}

---

6   Computational Results

To evaluate and showcase the effectiveness of the proposed algorithm on problem instances of different sizes, extensive computational experiments were carried out. The MIP model for the RSP-PPMA, as employed in a prior study by Benbrik et al. [3], was implemented using the CPLEX 22.1 MIP solver with default configurations. Simultaneously, all other algorithms were coded in C++.

During the experiments, a personal computer with an Intel(R) Core(TM) i7-7700HQ CPU operating at 2.8 GHz and 8 GB of RAM was used. The MIP formulations are analyzed based on the following metrics: the objective value (*Opt*) of the test instances solved to optimality within 1800 s, the time required for solving these optimally solved instances (*CPU*) in seconds (s), the objective function value of the instances unsolved within 1800 s (instances with feasible solutions), denoted as *Best Integer*, and the optimality gap for the test instances which could not be solved within 1800 s, denoted as *Gap(%)*. Importantly, optimal solutions were only achievable for small instances with $n = 15$ jobs and $m = 2$ stockpads due to the $\mathcal{NP}$-hard nature of the RSP-PPMA. As a result, the constructive heuristics and GVNS versions were adapted for medium instances. It is crucial to emphasize that, to mitigate the influence of stochastic variations, 10 runtime executions of the GVNS were performed for each problem instance. Consequently, the Best (Best.), Maximum (Max.), and Average (Avg.) objective function values were determined from these 10 runs and reported. Additionally, the average computation times (CPU) were calculated based on the 10 runs, with each run's computation time corresponding to the moment when the best solution encountered during that specific run was identified.

6.1   Benchmark Instances

The characteristics of the test instances derived from the scheduling environment of the NCIG terminal in Australia. These instances possess the following characteristics:

- The processing times of reclaiming stockpiles p_i, the setup times $t_{i,j}$, the time interval T, and the duration of a maintenance activity σ are generated following the approach proposed by Benbrik et al. [3]. Specifically, $p_i \sim U(60, 140)$, $t_{i,j} = \beta \times \min(p_i, p_j)$, where $\beta \sim U(0.05, 0.15)$, $\sigma \sim U(20, 90)$, and $T = \max\left(\max_{i \in N} p_i, 4 \times \sum_{i=1}^{n} p_i/n\right)$.
- The number of jobs in the stockpads is categorized into two sets of test problem instances. For small problem instances, the number of jobs n is chosen from $\{10, 15\}$, while the number of machines m is fixed at 2 (i.e., $\mathcal{P} = \mathcal{P}_3$). In the case of medium problem instances, the number of jobs n varies from 30 to 100. Specifically, n takes values from $\{30, 40, 50, 60\}$ when m equals 3 (i.e., $\mathcal{P} = \mathcal{P}_4$). It is essential to note that, for each combination of values (n, m), a total of 10 distinct problem instances were generated for both the small and medium-scale cases.
- In total, 60 unique problem instances were generated for every combination of values $(n, T, \sigma, \mathcal{P})$. These instances were evenly distributed across the two problem categories, with 20 instances designated for the small problem set, and 40 instances assigned to the medium problem set.

6.2   Tuning Parameters

A series of experiments were conducted to identify optimal parameter values for the GVNS algorithm. The algorithm relies on two key tuning parameters: k_{max} denotes the maximum perturbation level, and T_{max} represents the maximum time allotted to the GVNS. After preliminary experimentation, a thoughtful selection was made for the parameter configuration. Specifically, k_{max} is set to 20, chosen for its ability to strike a balance between solution quality and computational time (CPU). For small-sized instances $\left(n \in \{10, 15\}, T, \sigma, \mathcal{P} = \mathcal{P}_3\right)$, T_{max} is set to the computation time required to find an optimal solution using the CPLEX solver. For medium-sized instances, T_{max} is determined by the formula $T_{max} = (n \times m)/5$, indicating a polynomial increase in time with the growth of jobs and machines.

6.3   An Analysis of the Effectiveness of the Proposed Constructive Heuristic for Small Problems

In this section, we conduct an analysis of the performance of the developed greedy constructive heuristics for small-scale instances. For this heuristic, we calculate the percentage deviation for every problem instance from its optimum using the following formula:

$$Dev = 100 \times \left(\frac{C_{max}^{H} - opt}{opt}\right)$$

Here, C_{max}^{H} represents the makespan achieved by the greedy constructive heuristic, and opt denotes the optimum value obtained through the MIP formulation proposed in Benbrik et al. [3].

As anticipated, the computational experiment results presented in Table 2 demonstrate that the *Dev* values of the proposed greedy constructive heuristic, aimed at minimizing the makespan for small problem instances, generally fall within the range of 1.78 % to 13.29 % for the combination of values $(n = 10, T, \sigma, \mathcal{P} = \mathcal{P}_3)$, and 7 % to 22.72 % for the combination of values $(n = 15, T, \sigma, \mathcal{P} = \mathcal{P}_3)$.

In addition to evaluating the efficacy of heuristic based on their performance in addressing medium problem instances, the obtained outcomes for small-scale problems are deemed reasonably satisfactory and promising. This heuristic, furthermore, can be regarded as a robust initial solution for the GVNS metaheuristics.

Table 2. Evaluation of the heuristic algorithm for the RSP-PPMA in small scale instances for $\mathcal{P} = \mathcal{P}_3$.

Problem Instance				MIP				Constructive Heuristic		
$\mathcal{P}$	n	T	σ	Objective value		*Gap* (%)	*CPU* (sec)	C_{max}^{H}	*CPU* (sec)	*Dev* (%)
				Opt	*Best integer*					
$\mathcal{P}_3$	10	370.00	22	**550.70**	–	0.0	1.36	583.70	0.001	5.99
	10	378.00	54	**627.90**	–	0.0	2.87	639.90	0.001	1.91
	10	366.61	31	**557.45**	–	0.0	7.07	602.45	0.001	8.07
	10	378.61	47	**612.40**	–	0.0	1.40	639.60	0.001	4.44
	10	295.20	60	**525.10**	–	0.0	0.69	594.90	0.001	13.29
	10	355.40	87	**672.90**	–	0.0	0.83	684.90	0.001	1.78
	10	323.80	65	**580.50**	–	0.0	2.02	627.50	0.001	8.10
	10	339.80	81	**606.80**	–	0.0	1.01	664.00	0.001	9.43
	10	424.40	22	**650.10**	–	0.0	1.41	684.30	0.001	5.26
	10	404.00	60	**582.45**	–	0.0	1.27	621.90	0.001	6.77
	15	321.00	87	**813.00**	–	0.0	3.51	952.85	0.001	17.20
	15	395.86	31	**837.11**	–	0.0	87.20	1003.21	0.001	19.84
	15	397.43	22	**920.51**	–	0.0	145.25	984.91	0.001	7.00
	15	349.71	60	**893.03**	–	0.0	824.38	1025.83	0.001	14.87
	15	416.43	31	**874.23**	–	0.0	54.61	1052.41	0.001	20.38
	15	394.00	58	**880.20**	–	0.0	120.50	1063.85	0.001	20.86
	15	365.29	65	**859.58**	–	0.0	1050.22	946.82	0.001	10.15
	15	393.43	22	**804.98**	–	0.0	56.23	986.11	0.001	22.50
	15	396.29	58	**884.69**	–	0.0	28.75	1085.67	0.001	22.72
	15	402.14	47	**870.94**	–	0.0	370.20	1039.74	0.001	19.38
	Avg.			730.23	–	0.0	138.04	824.22	0.001	11.99

6.4 Assessing the Efficiency and Impact of GVNS Metaheuristics for Small Problems

The performance evaluation of the GVNS algorithm applied to solving the RSP-PPMA in small-scale instances is presented comprehensively in Table 3.

To quantify the percentage deviation for every problem instance from its optimum, we compute the percentage deviation (*Dev*) using the formula:

$$Dev = 100 \times \left(\frac{C_{max}^{\text{GVNS}} - opt}{opt} \right)$$

Here, C_{max}^{GVNS} represents the Best makespan obtained by the modified algorithm—GVNS (i.e., GVNS with the constructive heuristic as the initial solution), while *opt* denotes the optimum value obtained through the MIP formulation proposed by Benbrik et al. [3]. The objective of this analysis is to gain insights into the efficacy of this algorithm in finding optimal or near-optimal solutions. A detailed examination of the objective function values reveals that the GVNS algorithm demonstrates competitive performance across the considered problem instances. These results underscore the superior effectiveness of GVNS in identifying optimal solutions for the RSP-PPMA in small-scale instances. Additionally, it is crucial to highlight the computational time aspect. While the MIP formulation provides optimal solutions, the associated CPU times are considerably longer compared to the GVNS approaches. For instance, in the case of the problem instance $(n = 15, T = 365.29, \sigma = 65, \mathcal{P} = \mathcal{P}_3)$, the MIP model took 1050.22 s to find the optimal solution, whereas GVNS identified the same solution in 3.67 s. The deviation column (*Dev*) also provides valuable insights into the optimality of the solutions. Notably, the last row of the table indicates the average performance across all problem instances, revealing consistently low deviations from the optimum solution, with an average deviation of 0.13 %. This indicates that the solutions produced by GVNS are highly reliable and close to optimality, further affirming its effectiveness in solving small-scale instances of the BWRS problem.

Table 3. Evaluation of the GVNS algorithm for the RSP-PPMA in small-scale instances for $\mathcal{P} = \mathcal{P}_3$.

Problem Instance				MIP				GVNS				
$\mathcal{P}$	n	T	σ	Objective value		Gap (%)	CPU (sec)	Best.	Max.	Avg.	CPU (sec)	Dev (%)
				Opt	Best integer							
$\mathcal{P}_3$	10	370.00	22	**550.70**	–	0.0	1.36	**550.70**	550.70	550.70	1.63	0.0
	10	378.00	54	**627.90**	–	0.0	2.87	**627.90**	627.90	627.90	0.67	0.0
	10	366.01	31	**557.45**	–	0.0	7.07	**557.45**	557.45	557.45	0.44	0.0
	10	378.61	47	**612.40**	–	0.0	1.40	**612.40**	612.40	612.40	0.14	0.0
	10	295.20	60	**525.10**	–	0.0	0.69	**525.10**	546.10	527.20	0.58	0.0
	10	355.40	87	**672.90**	–	0.0	0.83	**672.90**	674.90	673.30	0.15	0.0
	10	323.80	65	**580.50**	–	0.0	2.02	**580.50**	590.50	585.04	1.14	0.0
	10	339.80	81	**606.80**	–	0.0	1.01	**606.80**	615.90	607.71	0.62	0.0
	10	424.40	22	**650.10**	–	0.0	1.41	**650.10**	665.10	651.60	0.31	0.0
	10	404.00	60	**582.45**	–	0.0	1.27	**582.45**	582.45	582.45	0.67	0.0
	15	321.00	87	**813.00**	–	0.0	3.51	825.20	926.41	903.86	9.54	1.48
	15	395.86	31	**837.11**	–	0.0	87.20	**837.11**	837.11	837.11	4.34	0.0
	15	397.43	22	**920.51**	–	0.0	145.25	**920.51**	923.96	921.20	3.29	0.0
	15	349.71	60	**893.03**	–	0.0	824.38	**893.03**	893.03	893.03	7.22	0.0
	15	416.43	31	**874.23**	–	0.0	54.61	**874.23**	874.23	874.23	5.65	0.0
	15	394.00	58	**880.20**	–	0.0	120.50	**880.20**	891.95	887.10	3.00	0.0
	15	365.29	65	**859.58**	–	0.0	1050.22	859.59	946.82	876.57	3.67	0.0
	15	393.43	22	**804.98**	–	0.0	56.23	**804.98**	817.03	806.89	5.31	0.0
	15	396.29	58	**884.69**	–	0.0	28.75	**884.69**	993.67	921.44	3.66	0.0
	15	402.14	47	**870.94**	–	0.0	370.20	877.24	880.99	879.11	1.98	0.72
	Avg.			730.23	–	0.0	138.04	731.15	750.43	738.81	2.70	0.13

6.5   Evaluating the Enhancement of Solutions from the Proposed Constructive Heuristic with Metaheuristics for Medium-Scale Problems

In our computational analysis, we integrated a greedy constructive heuristic to generate initial solutions for the GVNS algorithm. This integration aimed to explore potential improvements in the heuristic's effectiveness when utilized within the metaheuristic framework.

To quantify the enhancement achieved, we compute the percentage improvement (Imp) using the formula:

$$Imp = 100 \times \left(\frac{C_{max}^{H} - C_{max}^{GVNS}}{C_{max}^{GVNS}} \right)$$

Here, C_{max}^{H} represents the makespan obtained by the constructive heuristic, while C_{max}^{GVNS} denotes the Best makespan achieved by the modified algorithm-GVNS incorporating the constructive heuristic.

The comprehensive results are presented in Table 4. The table illustrates the average values obtained across all problem instances. Based on the observed variation in the average Imp values across different problem instances, ranging from 8.83 % to 14.40 %, it can be concluded that the modified algorithm—GVNS, effectively enhances the solutions derived from the constructive heuristic. Notably, the last row of the table demonstrates an average improvement (Imp) of 12.11 %, providing a benchmark for the effectiveness of GVNS in enhancing solutions derived from the constructive heuristic, especially for medium-sized instances. Additionally, it's noted that the average Imp values tend to decrease as the number of jobs increases. This trend suggests that the proposed heuristic remains quite competitive when compared to the metaheuristic approach.

Table 4. Average Imp values for enhancing solutions from the proposed constructive heuristic with metaheuristic

Problem Instance		Constructive Heuristic		GVNS				
$\mathcal{P}$	n	C_{max}^{H}	CPU (sec)	Best.	Max.	Avg.	CPU (sec)	Imp (%)
$\mathcal{P}_4$	30	1299.35	0.001	1146.44	1180.8	1160.38	14.09	13.38
	40	1828.26	0.001	1599.15	1649.38	1616.26	15.65	14.40
	50	2390.59	0.001	2138.20	2179.92	2153.14	22.50	11.84
	60	2689.82	0.001	2476.98	2536.86	2501.34	29.53	8.83
	Avg.	2052.01	0.001	1840.19	1886.74	1857.78	20.44	12.11

7   Conclusion

In this paper, we addressed the problem of scheduling stockpile reclamation considering the PPMA in bulk ports. The objective function considered was to find a feasible schedule which minimizes the latest completion time (i.e., makespan).

Given the $\mathcal{NP}$-hard nature of the problem, a novel greedy constructive heuristic has been devised. This heuristic relies on iterative job allocation to machines and prioritized sequencing, all while considering the integration of PPMA into the scheduling process. Consequently, it ensures the appropriate scheduling of jobs within batches. The solutions generated through constructive procedures serve as excellent initial foundations for GVNS algorithm, tailored to handle medium-sized instances with up to 60 jobs and 3 machines (i.e., $\mathcal{P} = \mathcal{P}_4$). Computational experiments conducted on 60 new instances demonstrate that for small-sized instances, GVNS algorithm outperform the MIP formulation in terms of the computing time required to find an optimal solution. Furthermore, for medium-sized instances, GVNS consistently yields superior solutions compared to the proposed constructive heuristic.

Potential future research directions could involve several aspects. Firstly, there is a need to develop further metaheuristic algorithms to allow for a comprehensive comparison and evaluation of the proposed GVNS algorithm. Secondly, exploring the stochastic version of the problem would be relevant. Lastly, exploring the integration of advanced optimization techniques, such as multi-objective optimization methods, could offer valuable insights into addressing the complexities of the RSP-PPMA problem. Moreover, it would be beneficial to develop lower and upper bounds for this problem to facilitate a more thorough comparison of the effectiveness of the proposed algorithms.

References

1. Angelelli, E., Kalinowski, T., Kapoor, R., Savelsbergh, M.W.: A reclaimer scheduling problem arising in coal stockyard management. J. Sched. **19**, 563–582 (2016)
2. Belov, G., Boland, N.L., Savelsbergh, M.W., Stuckey, P.J.: Logistics optimization for a coal supply chain. J. Heuristics **26**(2), 269–300 (2020)
3. Benbrik, O., Benmansour, R., Elidrissi, A.: Mathematical programming formulations for the reclaimer scheduling problem with sequence-dependent setup times and availability constraints. Procedia Comput. Sci. **232**, 2959–2972 (2024). https://doi.org/10.1016/j.procs.2024.02.112
4. Benmansour, R., Braun, O., Hanafi, S., Mladenovic, N.: Using a variable neighborhood search to solve the single processor scheduling problem with time restrictions. In: Sifaleras, A., Salhi, S., Brimberg, J. (eds.) ICVNS 2018. LNCS, vol. 11328, pp. 202–215. Springer, Cham (2019). https://doi.org/10.1007/978-3-030-15843-9_16
5. Benmansour, R., Sifaleras, A.: Scheduling in parallel machines with two servers: the restrictive case. In: Mladenovic, N., Sleptchenko, A., Sifaleras, A., Omar, M. (eds.) ICVNS 2021. LNCS, vol. 12559, pp. 71–82. Springer, Cham (2021). https://doi.org/10.1007/978-3-030-69625-2_6
6. Benmansour, R., Todosijević, R., Hanafi, S.: Variable neighborhood search for the single machine scheduling problem to minimize the total early work. Optim. Lett. **17**(9), 2169–2184 (2023)
7. Boland, N.L., Savelsbergh, M.W.P.: Optimizing the Hunter Valley coal chain. In: Gurnani, H., Mehrotra, A., Ray, S. (eds.) Supply Chain Disruptions: Theory and Practice of Managing Risk, pp. 275–302. Springer, London (2011). https://doi.org/10.1007/978-0-85729-778-5_10

308   O. Benbrik et al.

8. Elidrissi, A., Benmansour, R., Sifaleras, A.: General variable neighborhood search for the parallel machine scheduling problem with two common servers. Optim. Lett. **17**(9), 2201–2231 (2023)
9. Hansen, P., Mladenović, N., Todosijević, R., Hanafi, S.: Variable neighborhood search: basics and variants. EURO J. Comput. Optim. **5**(3), 423–454 (2017)
10. Hu, D., Yao, Z.: Stacker-reclaimer scheduling in a dry bulk terminal. Int. J. Comput. Integr. Manuf. **25**(11), 1047–1058 (2012)
11. Kalinowski, T., Kapoor, R., Savelsbergh, M.W.: Scheduling reclaimers serving a stock pad at a coal terminal. J. Sched. **20**, 85–101 (2017)
12. Mladenović, N., Hansen, P.: Variable neighborhood search. Comput. Oper. Res. **24**(11), 1097–1100 (1997)
13. Sánchez-Oro, J., Pantrigo, J.J., Duarte, A.: Combining intensification and diversification strategies in VNS. An application to the vertex separation problem. Comput. Oper. Res. **52**, 209–219 (2014)
14. Sifaleras, A., Konstantaras, I.: A survey on variable neighborhood search methods for supply network inventory. In: Bychkov, I., Kalyagin, V.A., Pardalos, P.M., Prokopyev, O. (eds.) NET 2018. SPMS, vol. 315, pp. 71–82. Springer, Cham (2020). https://doi.org/10.1007/978-3-030-37157-9_5
15. UNCTAD: Review of maritime transport. United Nations Conference on Trade and Development (2022). http://www.unctad.org
16. Ünsal, Ö.: Reclaimer scheduling in dry bulk terminals. IEEE Access **8**, 96294–96303 (2020)

An Efficient Algorithm for the T-Row Facility Layout Problem

Raúl Martín-Santamaría, Alberto Herrán, Abraham Duarte, and J. Manuel Colmenar[✉]

Universidad Rey Juan Carlos, Calle Tulipán s/n, Móstoles, Madrid, Spain
{raul.martin,alberto.herran,abraham.duarte,josemanuel.colmenar}@urjc.es

Abstract. Facility layout problems represent a challenge to the operations research community. These problems are closely related to real-world scenarios in industry and society, such as the design of production factories or the layout of facilities in medical centers, to name a few. These scenarios have been studied from the theoretical point of view as different optimization problems. Among them, we have studied the T-Row Facility Layout Problem, which considers a layout formed by two orthogonal rows where facilities have to be placed minimizing the material handling cost. To efficiently solve this problem we propose a Variable Neighbourhood Search algorithm which is able to reach all the optimal solutions reported in the literature spending a fraction of the execution time of the previous algorithm.

Keywords: Variable Neighborhood Search · Facility Layout Problem · T-row FLP

1 Introduction

One of the families of problems with more real-world applications is the Facility Layout Problems (FLP) [6]. In brief, this family involves the task of arranging a given set of facilities in a particular layout trying to optimize an objective function which is usually related to the pairwise relationship among the facilities. A single row, two rows, cells or multiple bays are some of the possible layouts for the arrangement usually studied in the literature. Material handling cost or closeness ratio are two different objective functions that respectively represent the cost of moving products from one facility to another, and the need to be nearby due to electricity demand or safety reasons. We refer the reader to the literature for an detailed review [8].

In this context, we have tackled the T-Row Facility Layout Problem (TRFLP), which aims to minimize the total material handling cost (MHC),

This work has been partially supported by the Spanish Ministerio de Ciencia e Innovación (MCIN/AEI/10.13039/501100011033) under grant refs. RED2022-134480-T and PID2021-126605NB-I00, and by ERDF A way of making Europe; and Generalitat Valenciana with grant ref. CIAICO/2021/224.

M. Sevaux et al. (Eds.): MIC 2024, LNCS 14753, pp. 309–315, 2024.
https://doi.org/10.1007/978-3-031-62912-9_29

defined as the weighted sum of the center-to-center distances between each pair of facilities in the layout. Specifically, the TRFLP considers a layout with two orthogonal rows, spaced w_{path}^T units, as in the shape of a T letter (see Fig. 1). This problem was introduced in a recent thesis [2], and a mixed-integer linear programming approach is presented to solve this problem in [1]. The layout is inspired by orthogonal aisles in buildings like hospitals, where the sun impacts differently in the two orthogonal aisles with windows in only one side, and the MHC corresponds to the distance to be walked by nurses to attend patients.

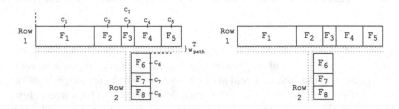

Fig. 1. Example layouts for the TRFLP. On the left, the intersection point c_I is placed at the center of F_3 (c_3), and the discontinuous lines in each row represent the reference point from where distances are calculated. On the right c_I is placed at c_2.

More formally, given a set F of n facilities, $n = |F|$, where each $i \in F$ has an associated length l_i; a weight w_{ij} between each pair of facilities $i, j \in F$; and a layout with two orthogonal rows, row 1 (horizontal) and row 2 (vertical); the TRFLP consists in finding an assignment of facilities to rows $r : F \rightarrow \{1, 2\}$, together with a vector $c \in \mathbb{R}^{(n+1)}$ with the center positions of all the facilities (measured from a fixed left border if $r_i = 1$, or a fixed upper border if $r_i = 2$) plus the position (from the fixed left border) of the vertical row (c_I). Mathematically:

$$\min \ \mathcal{F}(r, c) = \sum_{\substack{i,j \in F \\ i < j}} w_{ij} d_{ij}$$

$$\text{s.t.} \ |c_i - c_j| \geq \frac{l_i + l_j}{2} \qquad i, j \in F, \ i < j, \ r_i = r_j \qquad (1)$$
$$d_{ij} = |c_i - c_j| \qquad\qquad i, j \in F, \ i < j, \ r_i = r_j$$
$$d_{ij} = |c_i - c_I| + c_j + w_{path}^T \qquad i, j \in F, \ r_i = 1, \ r_j = 2$$

The mixed-integer linear programming approach presented in [1] is able to obtain many optimal solutions, but spending long execution times. In this work we propose a Variable Neighborhood Search algorithm able to get all verified optimal solutions, and solve those instances where the exact model reached the time limit. As it will be shown in the experiments, our proposal requires less than 2 s on average to solve any instance, which is several orders of magnitude less computing time than those of the previous work.

2 Variable Neighborhood Search Approach

We decided to tackle this problem using a classical Variable Neighborhood Search (VNS) approach [4]. Therefore, we have defined the three typical elements required by this scheme: construction method, improvement phase and perturbation procedure.

For the construction, we designed a procedure which randomly selects the intersection point and an initial facility, following a Greedy Randomized Adaptive Search Procedure (GRASP) to include new facilities [3]. The greedy function is defined as the contribution to the objective function by any of the non-selected facilities in any position, in either the first or the second row.

Regarding the improvement phase, three neighborhoods have been considered in this work. Two of them are well known in this family of problems, which are those based on the *insert* and *exchange* moves. These neighborhoods are generated by all possible insertions of any facility, and all possible exchanges of two facilities in a given solution, respectively [5]. The third neighborhood corresponds to adjusting the intersection point c_I between the two rows. Although this point can be located at any arbitrary place in the row, the previous work demonstrated that there is always an optimal layout where c_I is located at the center of a facility in the first row. Therefore, this neighborhood will only test the facilities centers in the first row as candidate locations for c_I. Since this is a first approach to the problem, we have considered an improvement process where a *best improvement* local search selects the best move taking into account the three neighborhoods as an *extended neighborhood local search*.

Lastly, our shake procedure randomly applies $k * 5$ moves from either the *insert* or the *exchange* neighborhoods.

Algorithm 1 shows the pseudo-code of this approach, where F is the set of facilities, α is the parameter for the GRASP constructive method and k_{max} is the maximum neighborhood size required by VNS. Notice that a solution S includes the assignment r and the location of the centers c, as defined in Sect. 1.

Algorithm 1: $VNS(F, \alpha, k_{max})$

1: $S \leftarrow GraspConstructive(\alpha, F)$
2: $k \leftarrow 1$
3: **while** $k < k_{max}$ **do**
4:     $S' \leftarrow Shake(F, k, S)$
5:     $S'' \leftarrow ExtendedLocalSearch(S', F)$
6:     **if** $\mathcal{F}(S'') \leq \mathcal{F}(S)$ **then**
7:         $S \leftarrow S', k \leftarrow 1$
8:     **else**
9:         $k \leftarrow k + 1$
10: **return** S

---

3   Computational Experiments

Experiments were run using a single thread on a PC with a Ryzen 1700 (3.7 GHz) CPU, limiting the memory available to the Java Virtual Machine (JVM) to 4 GB. The algorithm has been implemented using Java 21.

In order to perform a fair comparison, we will follow a similar methodology to the state-of-the-art proposal, where instances are divided in two groups. The first group contains instances already existing in the FLP literature, and they will be solved using $w_{path}^T = 3$ and $w_{path}^T = 10$. The second group contains the randomly generated *star instances* from [1], and they will be solved using $w_{path}^T = 0$. All instances have 20 or less facilities.

Since the VNS proposal heavily depends on stochastic factors, in order to guarantee that results are statistically significant, experiments were repeated 100 times, where each experiment consists on 20 iterations of the VNS. Results are summarized using the following metrics. For the VNS, the minimum (*Min*) and average (*Avg*) values of the objective function are reported. Moreover, as the experiment is repeated multiple times, the percentage of executions that reach the best known value is reported (*%best*), and the average time needed for each repetition (*T(s)*). For the state of the art work, labeled as *Model*, the optimal value, if found by the model, is reported (*Min*), along with the execution time needed (*T(s)*).

Tables 1 and 2 compare the performance of the VNS proposal and the state of the art model, using the first set of instances, when $w_{path}^T = 3$ and $w_{path}^T = 10$, respectively. The VNS proposal reaches the best known value for all instances in a fraction of the time required by the state of the art model. Notably, the results are consistent, and not due to random factors. The best value is found in most of the executions, as shown by the *%best* metric, and the average value is extremely close to the best known value. The state of the art model is executed with a time limit of 8 hours, and fails to find the optimal value for instances P18a and P18b, where the VNS obtains solutions quickly.

Table 3 compares the performance of the state-of-the-art model and the VNS proposal using the second set of instances. Again, the VNS reaches all optimal known values, using at most two seconds of computing time on average, while the model requires more than half an hour on average, failing to find the optimal value for the 20b instance in the 8 hours time limit.

Additionally, we have analyzed the performance profile of our VNS algorithm following the methodology proposed in [7]. Specifically, Fig. 2 shows the evolution of the gap to the best known value of the objective function, plotted against the execution time, and averaged for all instances and experiment repetitions. On average, the VNS proposal is able to obtain solutions whose gap is less than 10% to the best known values in less than 10 milliseconds. The gap is further reduced to 1% in 100 milliseconds, and is nearly 0% for execution times longer than 1 s. Therefore, the efficiency of the method is proven. Note the logarithmic scale of the horizontal axis in the figure.

Table 1. Comparison for the first set of instances and $w_{path}^{T} = 3$.

$w_{path}^{T} = 3$	VNS				Model	
Instance	Min	Avg	T(s)	%best	Min	T(s)
Am11a	8902	8904.8	0.09	61	8902	28
Am11b	6118.5	6119.9	0.09	92	6118.5	16
Am12a	2552	2552.0	0.13	100	2552	51
Am12b	2740.5	2741.0	0.14	94	2740.5	40
Am13a	4077	4077.0	0.26	98	4077	121
Am13b	4581.5	4581.7	0.32	98	4581.5	104
Am14_1	4642	4643.2	0.30	54	4642	533
Am14a	4751	4753.3	0.34	90	4751	335
Am14b	4739.5	4739.8	0.39	93	4739.5	407
Am15a	5378	5378.2	0.51	88	5378	953
HK_15b	26446	26448.2	0.50	72	26446	784
P16a	12381	12381.2	0.59	97	12381	4728
P16b	9882.5	9886.7	0.59	46	9882.5	4324
P17a	11956.5	11959.0	0.76	84	11956.5	14250
P17b	12779	12782.1	0.77	70	12779	11200
P17c	7767.5	7767.9	0.84	94	7767.5	8452
P18a	12993.5	12998.5	1.09	38	-	28800
P18b	14542	14551.6	1.05	40	-	28800
P18c	8911.5	8916.9	1.11	36	8911.5	20517

Table 2. Comparison for the first set of instances and $w_{path}^{T} = 10$.

$w_{path}^{T} = 10$	VNS				Model	
Instance	Min	Avg	T(s)	%best	Min	T(s)
Am11a	9852.5	9854.1	0.09	82	9852.5	26
Am11b	6930.5	6930.5	0.08	100	6930.5	12
Am12a	2793.5	2793.5	0.12	99	2793.5	37
Am12b	3081.5	3089.4	0.12	69	3081.5	38
Am13a	4517.5	4517.6	0.28	99	4517.5	124
Am13b	4999	5000.2	0.28	94	4999	114
Am14_1	5169.5	5169.8	0.30	98	5169.5	617
Am14a	5327.5	5329.7	0.29	88	5327.5	310
Am14b	5323	5325.0	0.34	94	5323	465
Am15a	5946.5	5948.0	0.45	96	5946.5	1046
HK_15b	27180	27182.8	0.47	88	27180	724
P16a	13233	13242.8	0.57	90	13233	4401
P16b	10627.5	10644.6	0.57	58	10627.5	3912
P17a	12871	12879.4	0.75	52	12871	13713
P17b	13761	13777.5	0.77	50	13761	18542
P17c	8590	8602.3	0.78	50	8590	6708
P18a	14009.5	14031.1	1.06	54	-	28800
P18b	15616.5	15629.7	1.03	34	-	28800
P18c	9807.5	9811.9	1.12	65	9807.5	20945

Table 3. Comparison for the second set of instances and $w_{path}^{T} = 0$.

$w_{path}^{T} = 0$	VNS				Model	
Instance	Min	Avg	T(s)	%best	Min	T(s)
11a	1702	1702.2	0.16	98	1702	0
11b	2847.5	2847.9	0.11	98	2847.5	2
11c	2301	2302.3	0.10	97	2301	1
11d	2878	2878.2	0.10	99	2878	2
11e	3098.5	3100.7	0.10	91	3098.5	4
12a	5540	5547.1	0.14	86	5540	13
12b	3911	3913.9	0.14	63	3911	6
12c	2529	2531.1	0.15	92	2529	2
12d	4027.5	4028.0	0.14	88	4027.5	8
12e	5583	5583.3	0.14	99	5583	15
13a	3823.5	3824.8	0.19	86	3823.5	11
13b	3290.5	3290.5	0.20	100	3290.5	4
13c	4040.5	4041.0	0.20	98	4040.5	20
13d	4036	4040.4	0.25	68	4036	27
13e	3266	3266.5	0.30	99	3266	8
14a	5276.5	5277.8	0.47	89	5276.5	61
14b	5640	5644.0	0.34	91	5640	48
14c	4150	4155.9	0.33	35	4150	23
14d	4884	4889.7	0.31	91	4884	26
14e	4935.5	4943.6	0.33	54	4935.5	49
15a	5312	5322.8	0.53	61	5312	91
15b	5298.5	5329.3	0.53	47	5298.5	96
15c	4225	4229.8	0.47	72	4225	87
15d	4609.5	4617.5	0.53	65	4609.5	48
15e	4643	4659.5	0.46	33	4643	217
16a	6564	6603.9	0.63	68	6564	234
16b	8356	8361.7	0.66	58	8356	615
16c	8082.5	8104.1	0.63	39	8082.5	387
16d	5521.5	5528.0	0.63	44	5521.5	203
16e	5561	5568.9	0.63	66	5561	289
17a	7853	7869.4	0.81	58	7853	827
17b	9876.5	9915.4	0.80	63	9876.5	2074
17c	7640	7657.4	0.83	73	7640	726
17d	6823	6842.3	0.80	59	6823	599
17e	6736.5	6765.9	0.83	44	6736.5	577
18a	11108.5	11125.3	1.10	36	11108.5	2168
18b	7037.5	7083.8	1.01	17	7037.5	1532
18c	9264.5	9288.2	1.03	27	9264.5	2316
18d	6464	6492.2	1.12	25	6464	1271
18e	8538	8566.0	1.09	34	8538	1936
19a	10046.5	10093.9	1.45	30	10046.5	6054
19b	10203.5	10258.2	1.51	24	10203.5	4755
19c	6637	6666.3	1.52	20	6637	1556
19d	8848	8898.4	1.49	14	8848	7096
19e	9495	9537.1	1.50	24	9495	9430
20a	8257	8300.4	1.84	25	8257	5264
20b	11234.5	11340.7	1.90	22	-	28800
20c	9826	9888.6	1.87	17	9826	12532
20d	11540.5	11578.3	1.90	23	11540.5	13896
20e	8097.5	8157.1	1.77	16	8097.5	9565

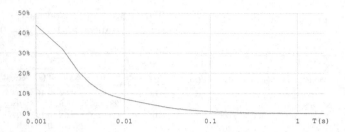

Fig. 2. Runtime behavior of the VNS proposal. The X axis represents the execution time, using a logarithmic scale. The Y axis represents the gap to the best known value, averaged among all instances and all experiment repetitions, for each instant.

4   Conclusions and Future Work

In this work, we have proposed a VNS approach for the T-Row Facility Layout problem. The proposal is based on a GRASP constructive method and a best improvement local search over an extended neighborhood which combines one specific neighborhood for the solved problem, which adjusts the intersection point, and the well-known *insert* and *exchange* neighborhoods. As seen in Sect. 3, combining this simple algorithm components using the VNS metaheuristic allows us to easily outperform the previous approach spending a fraction of its execution time.

The promising results obtained in this work suggest that the existing instances are not challenging enough for an effective comparison, as all optimal values are reached in extremely short computing times. To this end, we will study the application of this ideas to larger instances with up to 120 facilities, as well as different VNS proposals.

References

1. Dahlbeck, M.: A mixed-integer linear programming approach for the T-row and the multi-bay facility layout problem. Eur. J. Oper. Res. **295**(2), 443–462 (2021)
2. Dahlbeck, M.: Solution approaches for facility layout problems. Ph.D. thesis, Dissertation, Göttingen, Georg-August Universität (2021)
3. Feo, T.A., Resende, M.G.: Greedy randomized adaptive search procedures. J. Glob. Optim. **6**, 109–133 (1995)
4. Hansen, P., Mladenović, N., Todosijević, R., Hanafi, S.: Variable neighborhood search: basics and variants. EURO J. Comput. Optim. **5**(3), 423–454 (2017)
5. Herrán, A., Colmenar, J.M., Duarte, A.: An efficient variable neighborhood search for the space-free multi-row facility layout problem. Eur. J. Oper. Res. **295**(3), 893–907 (2021)

6. Hosseini-Nasab, H., Fereidouni, S., Fatemi Ghomi, S.M.T., Fakhrzad, M.B.: Classification of facility layout problems: a review study. Int. J. Adv. Manuf. Technol. **94**, 957–977 (2018)
7. López-Ibáñez, M., Stützle, T.: Automatically improving the anytime behaviour of optimisation algorithms. Eur. J. Oper. Res. **235**(3), 569–582 (2014)
8. Pérez-Gosende, P., Mula, J., Díaz-Madroñero, M.: Facility layout planning. An extended literature review. Int. J. Prod. Res. **59**(12), 3777–3816 (2021)

Interpretability, Adaptability and Scalability of Variable Neighborhood Search

Pierre Hansen[1], Aidan Riordan[2(✉)], and Xavier Hansen[1,2]

[1] GERAD and École des Hautes Études Commerciales, Montreal, QC, Canada
pierre.hansen@gerad.ca
[2] College of Charleston, Charleston, SC, USA
riordanaa@g.cofc.edu

Abstract. Variable Neighborhood Search (VNS) has reached its 25th anniversary as an effective and accessible metaheuristic for combinatorial optimization. This paper explores how VNS's ingenious method to escape local optima exhibits the properties of interpretability, adaptability and scalability, making it well-suited for tackling large and complex real-world problems. We first outline how the simple, modular design of VNS lends itself to insightful problem analysis and systematic formulation of the search space. We then discuss how VNS organically integrates with other methods as a hybrid and readily leverages parallelization and AI/ML capabilities for scalability. Finally, we propose recommendations to further advance VNS through establishing public code repositories and problem libraries, documenting challenges and successes with real-world implementations, actively engaging across metaheuristics, and popularizing VNS as an accessible optimization technique.

Keywords: Variable Neighborhood Search · Metaheuristic · Heuristics

1 Introduction

Variable Neighborhood Search has reached its 25-year milestone as a performant and efficient metaheuristic. Since the initial insights that led to its formulation (Mladenovic, 1995) [47] and since the VNS methodology was first formalized in a general framework (Mladenovic and Hansen, 1997) [48], a generation of researchers explored and developed the field. As we mourn the passing of Nenad Mladenovic, we celebrate his foundational contributions and, as he did throughout his distinguished career, we look ahead to the bright future of VNS and to the next generation of researchers that will continue to advance the field in years to come.

Recent survey papers confirm that VNS is thriving. The number of VNS peer-reviewed publications continues to rise year after year [34] and covers a growing range of disciplines (network design, location theory, chemistry, biology, graph theory, engineering, healthcare management, data mining, cluster analysis,

M. Sevaux et al. (Eds.): MIC 2024, LNCS 14753, pp. 316–330, 2024.
https://doi.org/10.1007/978-3-031-62912-9_30

artificial intelligence) [35,43]. VNS has demonstrated strong performance for classical combinatorial problems (Traveling Salesman Problem, Vehicle Routing Problem, Knapsack Problem) [31,33] and delivered competitive solutions for a wide range of real-world problems [6,19,44,60]. VNS has also spawned many variants and hybrids to solve specific problems [10], with some research reporting competitive results when compared to commercial optimization software (see, for example, a parallel multi-objective VNS hybrid solution compared to commercial optimization packages used for the post-sales network design problem [23] or a parallel VNS algorithm compared to the GAMS software when applied to the dynamic facility layout problem [1]).

This paper explores how the properties of interpretability, adaptability and scalability emerge from VNS's insightful method to escape local optima by systematically shifting local search using neighborhood structures, as shown in the growing number of implementations of VNS for hybrid, parallel and AI/ML optimization.

This paper is organized as follows: 1) Introduction; 2) Interpretability, Adaptability and Scalability of VNS; 3) Hybrid, Parallel and AI/ML VNS implementations; and 4) A proposal for a programme of exploration for the next generation of researchers.

2 Interpretability, Adaptability and Scalability of VNS

VNS's method to avoid the trap of the local optima is simple yet powerful. Through familiarization with a problem, researchers devise and construct neighborhood structures in the solution space to explore a search space systematically. VNS proceeds with a local search in the first neighborhood. If no improvement is found, it performs a perturbation move (the shake) to the next structured neighborhood where the local search process begins anew, starting from the incumbent solution. If a better solution is found, VNS returns to the first neighborhood structure and the entire process loops again, until a defined ending criterion has been met. In General VNS, the neighborhood structure approach is utilized in the local search phase as well.

In their most recent survey paper, Brimberg et al. characterize VNS's innovative approach "for evading the local optimum trap" as its "principal contribution." [10]. We concur and recall from An Introduction to Variable Neighborhood Search (1999) that "[c]ontrary to other metaheuristics based on local search methods, VNS does not follow a trajectory but explores increasingly distant neighborhoods of the current incumbent solution, and jumps from this solution to a new one if and only if an improvement has been made. In this way often favorable characteristics of the incumbent solution, e.g. that many variables are already at their optimal value, will be kept and used to obtain promising neighboring solutions." [30].

VNS can explore the solution space extensively and efficiently because it introduces teleportation (the "shake") as it jumps to intelligently-designed neighborhoods. This teleportation move is not a perturbation of last resort upon reaching a dead end in a trajectory; it is a systematic process for structured exploration via strategically-dispersed local searches. Its efficiency does not come at the cost of severe constraints reducing the search space nor does it require the significant computing resources consumed by population-based metaheuristics.

Further, this systematic and structured approach to the search space affords VNS an additional strength: when a proven global optimum or best solution is known (from prior optimization research), researchers can measure how far the results of their VNS algorithm are from this global optimum for each neighborhood and triangulate the most promising moves to improve their results. This continuous improvement capability is built into the VNS scheme.

The originality of VNS's approach to exploration and exploitation has been noted in the literature by Crainic, Gendreau, Hansen and Mladenovic (2004): "VNS is not a trajectory method (as Simulated Annealing or Tabu Search) and does not specify forbidden moves." [17]; M.A. Akbay et al. (2020): "Unlike most of the local search-based algorithms, VNS uses a set of neighborhood structures instead of a single one. The main idea behind using multiple neighborhood structures is the fact that a local minimum within a neighborhood may not be so for another one. Furthermore, VNS systematically changes the neighborhoods during the search. Thus, it ensures diversification in the search space and overcomes the problem of getting stuck into local optima frequently encountered in local search-based heuristics." [4]; Queiroz dos Santos et al. (2014): "VNS gradually explores neighborhoods more "distant" from the current solution rather than other local search strategies that follow a path." [55]; and Taillard (2023): "Variable neighborhood search (VNS) implements an idea called strategic oscillations. The search alternates intensification and diversification phases." [65].

The essential Variable Neighborhood Search Tutorial (2003) [32] listed eight "desirable properties" of metaheuristics to which VNS could attribute its early success: Simplicity, Precision, Coherence, Efficiency, Effectiveness, Robustness, User-friendliness, and Innovation. The present paper explores three additional emergent properties that VNS has demonstrated empirically over the past two decades and that speak to its present and future value as an integrable metaheuristic, namely: Interpretability, Adaptability and Scalability.

2.1   Interpretability

Each step in VNS can readily be understood as a functional component of a systematic search process that dynamically alternates exploration and exploitation, making VNS highly interpretable. The ease with which users understand the coherence of VNS steps facilitates the diagnosis and resolution of issues in a given implementation and simplifies adapting or altering steps while retaining comprehensibility.

From the outset VNS requires researching and thinking about the problem at hand to define the structure of the search space, i.e. to apprehend the nature of the solution space for the given problem and to design a number of solution neighborhoods of various sizes and types for a systematic search process. As explained in VNS: Principles and Applications, "[t]o construct different neighborhood structures and to perform a systematic search, one needs to have a way for finding the distance between any two solutions, i.e. one needs to supply the solution space with some metric (or quasi-metric) and then induce neighborhoods from it." [31].

VNS anchors the neighborhood structuring process with three facts:

Fact 1. A local minimum with respect to one neighborhood structure is not necessarily a local minimum for another neighborhood structure.

Fact 2. A global minimum is a local minimum with respect to all possible neighborhood structures.

Fact 3. For many problems local minima with respect to one or several neighborhoods are relatively close to each other.

The third fact is an empirical observation. It implies that "often favorable characteristics of the incumbent solution, e.g. that most variables are already at their optimal value, will be kept and used to obtain promising neighboring solutions. Moreover, a local search routine is applied repeatedly to get from these neighboring solutions to local optima." [31].

By asking what and how many neighborhood structures should be used and in what order, as well as what strategy should be used in changing neighborhoods, researchers design a systematic approach to explore and exploit the search space in clearly-defined steps that can readily combine deterministic and stochastic elements. A detailed step-by-step procedure for VNS implementation can be found in VNS: Methods and Applications, which further states "Unlike many other metaheuristics, the basic schemes of VNS and its extensions are simple and require few, and sometimes no parameters. Therefore, in addition to providing very good solutions, often in simpler ways than other methods, VNS gives insight into the reasons for such a performance, which, in turn, can lead to more efficient and sophisticated implementations." [35].

The intuitive modular design of VNS can readily be adjusted based on the landscape of the problem being solved. This facilitates the creation of concisely-defined variants that retain the interpretability of the original VNS scheme while delivering robust results across a wide range of problems. A recent survey of VNS variants can be found in [10]. The interpretability of VNS also encourages collaboration and adoption by other metaheuristic fields, as well as knowledge transfer to similar or even new problems, as discussed in Sect. 3 of the present paper.

2.2 Adaptability

As pointed out by Raidl in his presentation at ICVNS 3 (2014), the modular scheme of VNS facilitates "embedding different improvement methods" as is "explicitly expressed" in the original distinction between Variable Neighborhood Descent, Reduced VNS, VNS and General VNS. With initialization, neighborhood structuring, perturbation and local search selection, VNS presents distinct insertion points for integration and provides a "natural basis for most hybridization patterns", whether using an integrative or collaborative strategy, a sequential or parallel order of execution, weak or strong coupling or more specialized approaches combining problem-specific algorithms or AI/ML techniques [56]. Raidl provides a comprehensive review of the various taxonomies for metaheuristic hybrids in [57].

Section 3 references a variety of VNS applications that underscore the adaptability of VNS for hybrids with other metaheuristics, mathematical programming and AI/ML methods.

2.3   Scalability

The modular scheme of VNS also facilitates parallelization to run tasks concurrently rather than sequentially for added scalability and can readily combine parallelization and hybridization while retaining interpretability. For extremely large and complex real-world problem instances that become intractable for standard metaheuristic approaches, VNS offers a blueprint for problem decomposition [56]. The adaptable scheme of VNS offers multiple paths to approach such large and complex problems and delivers competitive performance at manageable cost and within feasible timeframes. For the largest and most intractable instances, the VNS scheme can be adapted to break down problems into component parts with a "successive approximations decomposition method" [36]. Variable Neighborhood Decomposition Search (VNDS) strategies have been explored for a variety of large-scale real-world problems such as continuous location-allocation [9], power plant cable layout [15] and supply chain management [44]. As with VNS, VNDS has been integrated with hybrid, parallel and/or machine learning in attempts to combine capabilities for improved performance for specific combinatorial problems. Section 3 references a variety of VNS applications that underscore the scalability of VNS with parallelization and decomposition.

   References to the emergent properties of VNS as an interpretable, adaptable and scalable metaheuristic can be found in metaheuristic literature [4, 21, 34, 37] and in papers on VNS implementations in a variety of fields [25, 40, 62, 68].

3   Hybrid, Parallel and AI/ML Implementations of VNS

The VNS Tutorial previously cited [32] identified three sets of promising areas of research: 1) enhancements of the VNS basic scheme (initialization, inventory of neighborhoods, distribution of neighborhoods, ancillary tests); 2) changes to the basic scheme of VNS (use of memory, parallel VNS, hybrids); and 3) new aims ('non-standard uses') for VNS (solutions with bounds on the errors, mixed-integer programming, artificial intelligence). These promising areas of research point to requisite capabilities to take on larger and more difficult challenges representative of the real world which sees an unlimited demand for solving increasingly complex problems. Although much remains to be explored, substantial research over the past two decades has highlighted how VNS is purpose-fit for parallel, hybrid and "non-standard" uses such as integration with AI/ML, showing much promise for scaling optimization methods to solve large and complex combinatorial problems [10, 35].

Hybrid integration, parallelization and AI/ML remain promising areas of research for VNS to augment its capabilities and scale up its performance (including enhancements to the VNS scheme). The interpretability, adaptability and scalability of VNS also position VNS to leverage recent advances in computing to take on larger and more complex real-world problems, including challenging demand-driven, real-time optimization at scale.

3.1 Hybrids

A steadily growing number of published hybrid metaheuristic solutions include a VNS component. The simple modular scheme of VNS facilitates its integration, both for exploration and exploitation of the search space. At its most basic, Reduced VNS delivers strong performance for a modest investment of resources, making it a cost-efficient addition to a hybrid [14,58,73]. The variable neighborhood descent component of VNS, for its part, is an efficient local search that can easily be integrated as a subroutine [29,42,75]. With few parameters to tune and a highly adaptable scheme, VNS has proven a well-suited component in emerging hybrid metaheuristic approaches. Interpretable VNS hybrids can also be enhanced with parallelization and/or with methods from Machine Learning and Artificial Intelligence [28,52,61,74,76].

Figure 1 confirms the growing popularity and utility of VNS as a value-adding component in highly-performant hybrids.

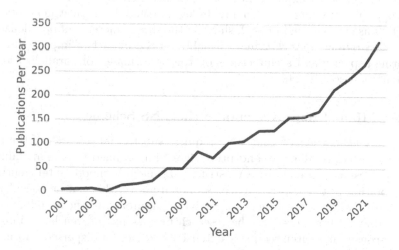

Fig. 1. Publications per year including "Variable Neighborhood Search" and "Hybrid". Made with Scopus Search Tool

Empirical evidence indicates that combining metaheuristics into hybrids to leverage different purpose-fit properties that optimize flexibility and scalability can significantly improve performance for previously intractable real-world problems involving vast quantities of data (including noisy data and data loss)

[26,45,71,72,77]. This is an increasingly important area of research since the recent explosive growth of Big Data opened up new opportunities to take on highly complex real world problems, including a growing set of business and consumer combinatorial problems that require rapid and continuous optimization, such as massive on-demand consumer applications and sophisticated just-in-time industrial production.

3.2 Parallelization

VNS has demonstrated its adaptability to optimize performance for different problem categories while maintaining robust performance as problems get larger and more complex [13,59]. As with all metaheuristics however, performance degrades as the problems get extremely large (the "curse of dimensionality"). The versatility of VNS's modular scheme facilitates parallelization to expand and/or speed up the systematic exploration of the search space [27,46]. VNS can leverage multi-core CPUs (and GPUs) to search neighborhoods concurrently rather than sequentially, with best results thus far provided via asynchronous cooperative sharing of information [38,53]. Such a distributed approach also facilitates the implementation of memory functions to improve solutioning as with reinforcement learning or reactive search [39,55]. For a comprehensive presentation of VNS parallelization approaches, see Crainic et al. [18].

The powerful new capabilities of high-performance computing systems should support further VNS experimentation with parallelization on a larger scale. These opportunities are increasingly being pursued by a new generation of researchers as can be seen in Fig. 2, showing the steady increase of parallelization in new VNS publications. For more insights on the impact of improvements in computing capacity and architectures on the performance of parallelization for metaheuristics, see [16,24].

3.3 AI/ML and Improvements to the VNS Scheme

It is broadly acknowledged that VNS requires substantial research of the problem being solved, its domain, and prior algorithms applied to it. This familiarization process is a prerequisite to structuring neighborhoods intelligently for the exploration of the solution space. As noted above, this can be considered a strength of VNS as it encourages an analytical approach to the search space. It is also a drawback however, as the research process has proven labor-intensive and time-consuming, often requiring extensive searches for dispersed information and sometimes elusive guidance from domain experts.

The search tools recommended 20 years ago in the VNS Tutorial to assist with familiarization, initialization and neighborhood structuring, such as the ISI Web of Knowledge, NEC Research's Citeseer or the Google search engine offered limited capabilities. The new AI tools deployed in the past few years to manage Big Data show transformative potential: massive online databases supply rich structured data, vector databases greatly reduce the effort and time required to locate information, machine learning models and AI algorithms scale

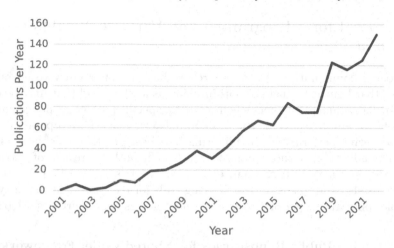

Fig. 2. Publications per year including "Variable Neighborhood Search" and "Parallel". Made with Scopus Search Tool

and accelerate pattern recognition, and the newly-released commercial Large Language Models (LLMs) deliver rapid natural language interactions for initial familiarization with a remarkably wide range of real world phenomena. Big Data's exponential growth has also spurred and enabled the training and leveraging of machine learning to identify complex patterns and relationships, which can be used to enhance optimization. Metaheuristics have in return demonstrated their utility for optimizing machine learning and artificial intelligence capabilities [11,25,50,70].

Although metaheuristics researchers have long experimented with AI and ML approaches, those endeavors required considerable effort, cost, and specialization that limited use and constrained potential. With the drastic reduction in the cost of computational processing and the proliferation of affordable ML models and AI tools in the past few years, metaheuristics can suddenly leverage a range of augmented capabilities to enhance all facets of optimization [8,20,41,66,67].

Pursuing this promising area of research for VNS could involve experimentation with automated, dynamic structuring of neighborhoods, further implementations of learning and memory processes, and the development of real-world problem data repositories with fine-tuned LLMs to transform domain research and problem formulation from laborious isolated efforts into an AI-assisted metaheuristic collaboration at scale. VNS, thanks to its simple, versatile modular scheme and its analytical approach to structuring the search space, is purpose-fit for AI/ML experimentation, as shown by the growing number of published works on VNS implementations including automated programming [22], adaptive memory functions [12], reinforcement learning [5], reactive search [55], automated metaheuristic generation [2], ML-assisted neighborhood selection [54], neural networks [3], as well as VNS applications to improve machine learning [7,64].

4   Proposal for a Programme for Variable Neighborhood Search

The broader community of VNS researchers has an opportunity to encourage practices that benefit the field of optimization as a whole and that may be over-looked by individual researchers when solving a specific problem. The recommen-dations below aim to overcome the tradeoff between general interoperability and problem-specific performance, highlighting how best practices to enhance inter-pretability, adaptability and scalability can benefit both the results of particular implementations and the performance of metaheuristics in general. These recom-mendations also aim to answer the growing calls for increased transparency and verifiable comparative performance assessment across metaheuristics [51,63,69].

4.1   Establish Public Repositories for Shared Code, Frameworks and Libraries

Reusability and replicability of prior implementations are critical to assess the performance of the most promising approaches and to enhance their further development. The field of VNS should encourage all practitioners to share results, code, libraries, frameworks and architectures by establishing and maintaining public repositories easily accessed online. Such repositories could be indexed to facilitate experimentation and exploration of hybrids, parallelization and AI/ML integrations that retain interpretability, adaptability and the ability to scale per-formance for larger and more complex problems. An intentional program to pro-mote transparency and collaborative knowledge-sharing will reduce redundant efforts and costs of implementation while fostering new insights.

4.2   Document the Challenges and Performance of VNS for Solving Large and Complex Real-World Problems and Support Exploration of the Latest Advances in Computing

The VNS field should document the interpretability, adaptability and scalabil-ity of VNS and emphasize demonstrable results for solving large and complex combinatorial problems. This endeavor should explore options to close the gap between many promising theoretical formulations and the still limited number of successful real-world implementations (aka the "Death Valley") via partnerships between academia and business. In collaboration with the broader discipline of optimization, the VNS field should increase experimentation with the evolv-ing capabilities of high-performance computing, Big Data and AI/ML tools to improve performance when taking on previously intractable real-world problems.

4.3   Continue to Engage Proactively with Other Metaheuristics

The VNS field should continue to engage proactively with other metaheuristic fields and with optimization disciplines such as AI/ML to encourage the use of

VNS in hybrid approaches to solve large and complex combinatorial problems. Sponsoring joint conferences and actively seeking the participation of researchers from other metaheuristic fields that have implemented VNS components in their hybrids will benefit the growth of VNS as well as experimentation with enhancements to the VNS scheme. Frequent communication and collaboration between practitioners across a broad range of metaheuristic approaches will benefit the field of optimization with informed combinations of respective strengths required to deliver performant solutions, all while facilitating continued interpretability and adaptability.

4.4 Popularize VNS and Further the Development of Optimization as a Public Utility

In the tradition of the VNS Tutorial and of Nenad Mladenovic's Introduction to VNS at ICVNS 8, [49] the VNS field should continue to popularize the simplicity and efficiency of VNS with the broadest audience, especially with new researchers entering the field of optimization. Such a project should also consider new capabilities delivered by the latest advances in visualization tools, LLMs and low-code applications to develop basic optimization utilities for the general public, especially for small business and nonprofit entities that may be priced out of costly commercial optimization solutions.

References

1. Abedzadeh, M., Mazinani, M., Moradinasab, N., Roghanian, E.: Parallel variable neighborhood search for solving fuzzy multi-objective dynamic facility layout problem. Int. J. Adv. Manuf. Technol. **65**, 197–211 (2013)
2. Adamo, T., Ghiani, G., Guerriero, E., Manni, E.: Automatic instantiation of a variable neighborhood descent from a mixed integer programming model. Oper. Res. Perspect. **4**, 123–135 (2017). https://doi.org/10.1016/j.orp.2017.09.001
3. Adibi, M., Zandieh, M., Amiri, M.: Multi-objective scheduling of dynamic job shop using variable neighborhood search. Expert Syst. Appl. **37**(1), 282–287 (2010). https://doi.org/10.1016/j.eswa.2009.05.001
4. Akbay, M.A., Kalayci, C.B., Polat, O.: A parallel variable neighborhood search algorithm with quadratic programming for cardinality constrained portfolio optimization. Knowl.-Based Syst. **198**, 105944 (2020). https://doi.org/10.1016/j.knosys.2020.105944
5. Alicastro, M., Ferone, D., Festa, P., Fugaro, S., Pastore, T.: A reinforcement learning iterated local search for makespan minimization in additive manufacturing machine scheduling problems. Comput. Oper. Res. **131**, 105272 (2021). https://doi.org/10.1016/j.cor.2021.105272
6. Aloise, D.J., Aloise, D., Rocha, C.T., Ribeiro, C.C., Ribeiro Filho, J.C., Moura, L.S.: Scheduling workover rigs for onshore oil production. Discrete Appl. Math. **154**(5), 695–702 (2006). https://doi.org/10.1016/j.dam.2004.09.021
7. Araújo, T., Aresta, G., Almada-Lobo, B., Mendonça, A.M., Campilho, A.: Improving convolutional neural network design via variable neighborhood search. In: Karray, F., Campilho, A., Cheriet, F. (eds.) ICIAR 2017. LNCS, vol. 10317, pp. 371–379. Springer, Cham (2017). https://doi.org/10.1007/978-3-319-59876-5_41

8. Balas, V.E., Kumar, R., Srivastava, R. (eds.): Recent Trends and Advances in Artificial Intelligence and Internet of Things. ISRL, vol. 172. Springer, Cham (2020). https://doi.org/10.1007/978-3-030-32644-9

9. Brimberg, J., Hansen, P., Mladenović, N.: Decomposition strategies for large-scale continuous location-allocation problems. IMA J. Manag. Math. **17**(4), 307–316 (2006). https://doi.org/10.1093/imaman/dpl002

10. Brimberg, J., Salhi, S., Todosijević, R., Urošević, D.: Variable neighborhood search: the power of change and simplicity. Comput. Oper. Res. **155**, 106221 (2023). https://doi.org/10.1016/j.cor.2023.106221

11. Calvet, L., de Armas, J., Masip, D., Juan, A.A.: Learnheuristics: hybridizing metaheuristics with machine learning for optimization with dynamic inputs. Open Math. **15**(1), 261–280 (2017). https://doi.org/10.1515/math-2017-0029

12. Cao, L., Ye, C.M., Cheng, R., Wang, Z.K.: Memory-based variable neighborhood search for green vehicle routing problem with passing-by drivers: a comprehensive perspective. Complex Intell. Syst. **8**(3), 2507–2525 (2022). https://doi.org/10.1007/s40747-022-00661-5

13. Cazzaro, D., Pisinger, D.: Variable neighborhood search for large offshore wind farm layout optimization. Comput. Oper. Res. **138**, 105588 (2022). https://doi.org/10.1016/j.cor.2021.105588

14. Cheimanoff, N., Fontane, F., Kitri, M.N., Tchernev, N.: A reduced VNS based approach for the dynamic continuous berth allocation problem in bulk terminals with tidal constraints. Expert Syst. Appl. **168**, 114215 (2021). https://doi.org/10.1016/j.eswa.2020.114215

15. Costa, M.C., Monclar, F.R., Zrikem, M.: Variable neighborhood decomposition search for the optimization of power plant cable layout. J. Intell. Manuf. **13**, 353–365 (2002)

16. Crainic, T.: Parallel metaheuristics and cooperative search. In: Gendreau, M., Potvin, J.-Y. (eds.) Handbook of Metaheuristics. ISORMS, vol. 272, pp. 419–451. Springer, Cham (2019). https://doi.org/10.1007/978-3-319-91086-4_13

17. Crainic, T.G., Gendreau, M., Hansen, P., Mladenović, N.: Cooperative parallel variable neighborhood search for the p-median. J. Heuristics **10**(3), 293–314 (2004). https://doi.org/10.1023/B:HEUR.0000026897.40171.1a

18. Davidović, T., Crainic, T.G.: Parallelization strategies for variable neighborhood search. Research report CIRRELT-2013-47, Interuniversity Research Centre on Enterprise Networks, Logistics and Transportation (CIRRELT), Montréal, Canada (2013)

19. De Armas, J., Melián-Batista, B.: Variable neighborhood search for a dynamic rich vehicle routing problem with time windows. Comput. Ind. Eng. **85**, 120–131 (2015). https://doi.org/10.1016/j.cie.2015.03.006

20. De Curtò, J., De Zarzà, I., Roig, G., Cano, J.C., Manzoni, P., Calafate, C.T.: LLM-informed multi-armed bandit strategies for non-stationary environments. Electronics **12**(13), 2814 (2023). https://doi.org/10.3390/electronics12132814

21. Duarte, A., Pantrigo, J.J., Pardo, E.G., Mladenovic, N.: Multi-objective variable neighborhood search: an application to combinatorial optimization problems. J. Global Optim. **63**(3), 515–536 (2015). https://doi.org/10.1007/s10898-014-0213-z

22. Elleuch, S., Jarboui, B., Mladenovic, N.: Variable neighborhood programming - a new automatic programming method in artificial intelligence. Technical report G-2016-21, GERAD, Montreal, Canada (2016)

23. Eskandarpour, M., Zegordi, S.H., Nikbakhsh, E.: A parallel variable neighborhood search for the multi-objective sustainable post-sales network design problem. Int. J. Prod. Econ. **145**(1), 117–131 (2013). https://doi.org/10.1016/j.ijpe.2012.10.013

24. Essaid, M., Idoumghar, L., Lepagnot, J., Brévilliers, M.: GPU parallelization strategies for metaheuristics: a survey. Int. J. Parallel Emergent Distrib. Syst. **34**(5), 497–522 (2019). https://doi.org/10.1080/17445760.2018.1428969
25. Fuksz, L., Pop, P.C.: A hybrid genetic algorithm with variable neighborhood search approach to the number partitioning problem. In: Pan, J.-S., Polycarpou, M.M., Woźniak, M., de Carvalho, A.C.P.L.F., Quintián, H., Corchado, E. (eds.) HAIS 2013. LNCS (LNAI), vol. 8073, pp. 649–658. Springer, Heidelberg (2013). https://doi.org/10.1007/978-3-642-40846-5_65
26. Garcia-Guarin, J., et al.: Smart microgrids operation considering a variable neighborhood search: the differential evolutionary particle swarm optimization algorithm. Energies **12**(16), 3149 (2019). https://doi.org/10.3390/en12163149
27. García-López, F., Melián-Batista, B., Moreno-Pérez, J.A., Moreno-Vega, J.M.: The parallel variable neighborhood search for the p-median problem. J. Heuristics **8**(3), 375–388 (2002). https://doi.org/10.1023/A:1015013919497
28. García-Torres, M., Gómez-Vela, F., Becerra-Alonso, D., Melián-Batista, B., Moreno-Vega, J.M.: Feature grouping and selection on high-dimensional microarray data. In: 2015 International Workshop on Data Mining with Industrial Applications (DMIA), pp. 30–37 (2015). https://doi.org/10.1109/DMIA.2015.18
29. Guo, H., Zhang, L., Ren, Y., Li, Y., Zhou, Z., Wu, J.: Optimizing a stochastic disassembly line balancing problem with task failure via a hybrid variable neighborhood descent artificial bee colony algorithm. Int. J. Prod. Res. **61**(7), 2307–2321 (2023). https://doi.org/10.1080/00207543.2022.2069524
30. Hansen, P., Mladenović, N.: An introduction to variable neighborhood search. In: Voß, S., Martello, S., Osman, I.H., Roucairol, C. (eds.) Meta-Heuristics: Advances and Trends in Local Search Paradigms for Optimization, pp. 433–458. Springer, Boston (1999). https://doi.org/10.1007/978-1-4615-5775-3_30
31. Hansen, P., Mladenović, N.: Variable neighborhood search: principles and applications. Eur. J. Oper. Res. **130**(3), 449–467 (2001). https://doi.org/10.1016/S0377-2217(00)00100-4
32. Hansen, P., Mladenović, N.: A tutorial on variable neighborhood search. Les Cahiers du GERAD G-2003-46, Groupe d'études et de recherche en analyse des décisions, GERAD, Montréal QC H3T 2A7, Canada (2003)
33. Hansen, P., Mladenović, N.: Variable neighborhood search methods. In: Floudas, C.A., Pardalos, P.M. (eds.) Encyclopedia of Optimization, pp. 3975–3989. Springer, Boston (2008). https://doi.org/10.1007/978-0-387-74759-0_694
34. Hansen, P., Mladenović, N., Brimberg, J., Pérez, J.A.M.: Variable neighborhood search. In: Gendreau, M., Potvin, J.-Y. (eds.) Handbook of Metaheuristics. ISORMS, vol. 272, pp. 57–97. Springer, Cham (2019). https://doi.org/10.1007/978-3-319-91086-4_3
35. Hansen, P., Mladenović, N., Moreno Pérez, J.A.: Variable neighbourhood search: methods and applications. Ann. Oper. Res. **175**(1), 367–407 (2010). https://doi.org/10.1007/s10479-009-0657-6
36. Hansen, P., Mladenovic, N., Perez-Brito, D.: Variable neighborhood decomposition search. J. Heuristics **7**, 335–350 (2001)
37. Hansen, P., Mladenović, N., Todosijević, R., Hanafi, S.: Variable neighborhood search: basics and variants. EURO J. Comput. Optim. **5**(3), 423–454 (2017). https://doi.org/10.1007/s13675-016-0075-x
38. Kalatzantonakis, P., Sifaleras, A., Samaras, N.: Cooperative versus non-cooperative parallel variable neighborhood search strategies: a case study on the capacitated vehicle routing problem. J. Glob. Optim. **78**(2), 327–348 (2020). https://doi.org/10.1007/s10898-019-00866-y

39. Kalatzantonakis, P., Sifaleras, A., Samaras, N.: A reinforcement learning-variable neighborhood search method for the capacitated vehicle routing problem. Expert Syst. Appl. **213**, 118812 (2023). https://doi.org/10.1016/j.eswa.2022.118812

40. Karakostas, P., Sifaleras, A.: A double-adaptive general variable neighborhood search algorithm for the solution of the traveling salesman problem. Appl. Soft Comput. **121**, 108746 (2022). https://doi.org/10.1016/j.asoc.2022.108746

41. Karimi-Mamaghan, M., Mohammadi, M., Meyer, P., Karimi-Mamaghan, A.M., Talbi, E.G.: Machine learning at the service of meta-heuristics for solving combinatorial optimization problems: a state-of-the-art. Eur. J. Oper. Res. **296**(2), 393–422 (2022). https://doi.org/10.1016/j.ejor.2021.04.032

42. Kyriakakis, N.A., Sevastopoulos, I., Marinaki, M., Marinakis, Y.: A hybrid Tabu search - variable neighborhood descent algorithm for the cumulative capacitated vehicle routing problem with time windows in humanitarian applications. Comput. Ind. Eng. **164**, 107868 (2022). https://doi.org/10.1016/j.cie.2021.107868

43. Lan, S., Fan, W., Yang, S., Pardalos, P.M., Mladenovic, N.: A survey on the applications of variable neighborhood search algorithm in healthcare management. Ann. Math. Artif. Intell. **89**(8), 741–775 (2021). https://doi.org/10.1007/s10472-021-09727-5

44. Lejeune, M.: A variable neighborhood decomposition search method for supply chain management planning problems. Eur. J. Oper. Res. **175**(2), 959–976 (2006). https://doi.org/10.1016/j.ejor.2005.05.021

45. Liang, Y.L., Kuo, C.C., Lin, C.C.: A hybrid memetic algorithm for simultaneously selecting features and instances in big industrial IoT data for predictive maintenance. In: 2019 IEEE 17th International Conference on Industrial Informatics (INDIN), vol. 1, pp. 1266–1270 (2019). iSSN: 2378-363X. https://doi.org/10.1109/INDIN41052.2019.8972199

46. Menéndez, B., Pardo, E.G., Sánchez-Oro, J., Duarte, A.: Parallel variable neighborhood search for the min-max order batching problem. Int. Trans. Oper. Res. **24**(3), 635–662 (2017). https://doi.org/10.1111/itor.12309

47. Mladenovic, N.: A variable neighborhood algorithm – a new metaheuristic for combinatorial optimization. In: Abstracts of Papers Presented at Optimization Days, p. 112. Montréal (1995). Available as an abstract only

48. Mladenović, N., Hansen, P.: Variable neighborhood search. Comput. Oper. Res. **24**(11), 1097–1100 (1997). https://doi.org/10.1016/S0305-0548(97)00031-2

49. Mladenovic, N.: ICVNS2021: Tutorial by Prof. Mladenovic, Introduction to variable neighborhood search metaheuristic. YouTube Video (2021), tutorial presented at the 8th International Conference on Variable Neighborhood Search (ICVNS 2021), Abu Dhabi, United Arab Emirates

50. Naidu, A., Mittal, A., Kreucher, R., Zhang, A.C., Ortmann, W., Somsel, J.: A systematic approach to develop metaheuristic traffic simulation models from big data analytics on real-world data. Technical report, SAE International (2021). https://doi.org/10.4271/2021-01-0166

51. Osaba, E., et al.: A tutorial on the design, experimentation and application of metaheuristic algorithms to real-world optimization problems. Swarm Evol. Comput. **64**, 100888 (2021). https://doi.org/10.1016/j.swevo.2021.100888

52. Pan, J.-S., Polycarpou, M.M., Woźniak, M., de Carvalho, A.C.P.L.F., Quintián, H., Corchado, E. (eds.): HAIS 2013. LNCS (LNAI), vol. 8073. Springer, Heidelberg (2013). https://doi.org/10.1007/978-3-642-40846-5

53. Polat, O.: A parallel variable neighborhood search for the vehicle routing problem with divisible deliveries and pickups. Comput. Oper. Res. **85**, 71–86 (2017). https://doi.org/10.1016/j.cor.2017.03.009

54. Pugliese, L.D.P., Ferone, D., Festa, P., Guerriero, F., Macrina, G.: Combining variable neighborhood search and machine learning to solve the vehicle routing problem with crowd-shipping. Optim. Lett. (2022). https://doi.org/10.1007/s11590-021-01833-x

55. Queiroz Dos Santos, J.P., De Melo, J.D., Duarte Neto, A.D., Aloise, D.: Reactive search strategies using reinforcement learning, local search algorithms and variable neighborhood search. Expert Syst. Appl. 41(10), 4939–4949 (2014). https://doi.org/10.1016/j.eswa.2014.01.040

56. Raidl, G.R.: Variable neighborhood search hybrids. In: Proceedings of the 3rd International Conference on Variable Neighborhood Search. Vienna University of Technology, Djerba (2014). Presentation

57. Raidl, G.R., Puchinger, J., Blum, C.: Metaheuristic hybrids. In: Gendreau, M., Potvin, J.-Y. (eds.) Handbook of Metaheuristics. ISORMS, vol. 272, pp. 385–417. Springer, Cham (2019). https://doi.org/10.1007/978-3-319-91086-4_12

58. Rettl, M., Pletz, M., Schuecker, C.: Evaluation of combinatorial algorithms for optimizing highly nonlinear structural problems. Mater. Des. 230, 111958 (2023). https://doi.org/10.1016/j.matdes.2023.111958

59. Roshanaei, V., Naderi, B., Jolai, F., Khalili, M.: A variable neighborhood search for job shop scheduling with set-up times to minimize makespan. Future Gener. Comput. Syst. 25(6), 654–661 (2009). https://doi.org/10.1016/j.future.2009.01.004

60. Şevkli, A.Z., Güler, B.: A multi-phase oscillated variable neighbourhood search algorithm for a real-world open vehicle routing problem. Appl. Soft Comput. 58, 128–144 (2017). https://doi.org/10.1016/j.asoc.2017.04.045

61. Shao, Y., Wang, K., Shu, L., Deng, S., Dong, D.J.: Heuristic optimization for reliable data congestion analytics in crowdsourced eHealth networks. IEEE Access 4, 9174–9183 (2016). https://doi.org/10.1100/ACCESS.2016.2646058

62. Sitahong, A., Yuan, Y., Ma, J., Lu, Y., Mo, P.: Effective and interpretable rule mining for dynamic job-shop scheduling via improved gene expression programming with feature selection. Appl. Sci. 13(11), 6631 (2023). https://doi.org/10.3390/app13116631

63. Swan, J., et al.: Metaheuristics "in the large". Eur. J. Oper. Res. 297(2), 393–406 (2022). https://doi.org/10.1016/j.ejor.2021.05.042

64. Syed, M.N.: Feature selection in machine learning via variable neighborhood search. Optim. Lett. 17(9), 2321–2345 (2023). https://doi.org/10.1007/s11590-023-02003-x

65. Taillard, E.D.: Design of Heuristic Algorithms for Hard Optimization: With Python Codes for the Travelling Salesman Problem. GRTOPR, Springer, Cham (2023). https://doi.org/10.1007/978-3-031-13714-3

66. Talbi, E.G.: Machine learning into metaheuristics: a survey and taxonomy. ACM Comput. Surv. 54(6), 129:1–129:32 (2021). https://doi.org/10.1145/3459664

67. Tkatek, S., Bahti, S., Lmzouari, Y., Abouchabaka, J.: Artificial intelligence for improving the optimization of NP-hard problems: a review. Int. J. Adv. Trends Comput. Sci. Eng. 9(5), 7411–7420 (2020). https://doi.org/10.30534/ijatcse/2020/73952020

68. Todosijević, R., Mladenović, M., Hanafi, S., Mladenović, N., Crévits, I.: Adaptive general variable neighborhood search heuristics for solving the unit commitment problem. Int. J. Electr. Power Energy Syst. 78, 873–883 (2016). https://doi.org/10.1016/j.ijepes.2015.12.031

69. Turgut, O.E., Turgut, M.S., Kırtepe, E.: A systematic review of the emerging metaheuristic algorithms on solving complex optimization problems. Neural Comput. Appl. 35(19), 14275–14378 (2023). https://doi.org/10.1007/s00521-023-08481-5

70. Wang, G.G., Tan, Y.: Improving metaheuristic algorithms with information feedback models. IEEE Trans. Cybern. **49**(2), 542–555 (2019). https://doi.org/10.1109/TCYB.2017.2780274

71. Wang, K., Shao, Y., Shu, L., Zhu, C., Zhang, Y.: Mobile big data fault-tolerant processing for ehealth networks. IEEE Netw. **30**(1), 36–42 (2016). https://doi.org/10.1109/MNET.2016.7389829

72. Wang, L., Meng, F., Min, X., Chu, D.: A multi-objective task-driven vehicle routing problem with recirculating delivery and its solution approaches. In: 2021 4th International Conference on Artificial Intelligence and Big Data (ICAIBD), pp. 687–694 (2021). https://doi.org/10.1109/ICAIBD51990.2021.9459022

73. Xiong, F., Xing, K.: Meta-heuristics for the distributed two-stage assembly scheduling problem with bi-criteria of makespan and mean completion time. Int. J. Prod. Res. **52**(9), 2743–2766 (2014). https://doi.org/10.1080/00207543.2014.884290

74. Yuan, Z., Gao, J.: Dynamic uncertainty study of multi-center location and route optimization for medicine logistics company. Mathematics **10**(6), 953 (2022). https://doi.org/10.3390/math10060953

75. Zhang, B., Pan, Q.K., Meng, L.L., Zhang, X.L., Jiang, X.C.: A decomposition-based multi-objective evolutionary algorithm for hybrid flowshop rescheduling problem with consistent sublots. Int. J. Prod. Res. **61**(3), 1013–1038 (2023). https://doi.org/10.1080/00207543.2022.2093680

76. Zhang, H., He, Z., Man, Y., Li, J., Hong, M., Tran, K.P.: Multi-objective optimization of flexible flow-shop intelligent scheduling based on a hybrid intelligent algorithm. In: Tran, K.P. (ed.) Artificial Intelligence for Smart Manufacturing: Methods, Applications, and Challenges. RELIABILITY, pp. 97–119. Springer, Cham (2023). https://doi.org/10.1007/978-3-031-30510-8_6

77. Zhao, F., Qin, S., Zhang, Y., Ma, W., Zhang, C., Song, H.: A hybrid biogeography-based optimization with variable neighborhood search mechanism for no-wait flow shop scheduling problem. Expert Syst. Appl. **126**, 321–339 (2019). https://doi.org/10.1016/j.eswa.2019.02.023

Exploring the Integration of General Variable Neighborhood Search with Exact Procedures for the Optimization of the Order Batching Problem

Sergio Gil-Borrás[1]([✉]) and Eduardo G. Pardo[2]

[1] Dept. Sistemas Informáticos, Universidad Politécnica de Madrid, Madrid, Spain
sergio.gil@upm.es
[2] Dept. Computer Science, Universidad Rey Juan Carlos, Madrid, Spain
eduardo.pardo@urjc.es

Abstract. This paper studies the influence of combining exact algorithms with heuristic procedures for the optimization of the Order Batching Problem. The problem looks to minimize the time that an operator needs to pick all items from a set of orders within a warehouse, when all orders are known before starting the picking process. It involves solving different tasks, such as batching the orders in groups or routing the picker through the warehouse. In this proposal a previous General Variable Neighborhood Search proposal for the batching task has been integrated with exact algorithms for the routing task. Several experiments have been designed to test the performance of the constructed algorithms.

Keywords: Order Batching Problem · General Variable Neighborhood Search · Exact algorithms

1 Introduction

The growth of e-Commerce and online sales has led companies to develop and improve their supply chain management processes. The evolution of supply chain models, which can be traced in the literature from the early 1980s to the present day, has contributed to increased productivity in supply chain companies. In the last decade, there has been a significant increase in the number of articles related to supply chain management.

This article focuses on the activities that occur within logistics warehouses, specifically the item picking process. Multiple factors can affect the picking process [29]. Different metrics can be used to analyze the quality and productivity of a service, such as balancing operator workload, meeting delivery deadlines, saving energy, or reducing picker travel time. Particularly, we study the Order Batching Problem, which consists of optimizing the picking process when the picking policy is based on the concept of batching. This means that a group of orders are packed together before the picking starts. Then, all items in the same

M. Sevaux et al. (Eds.): MIC 2024, LNCS 14753, pp. 331–343, 2024.
https://doi.org/10.1007/978-3-031-62912-9_31

batch are picked on a single route. Orders cannot be divided into more than one batch, and batches cannot exceed a predefined maximum capacity, which may be based on weight and/or volume restrictions. The picking strategy discussed in this paper falls into the picker-to-part category. The influence of other factors, such as storage policy, batch size, time window, or batch selection, is not studied here.

This article addresses the Order Batching Problem with a single order picker in a rectangular shape, one-block warehouse. To solve this variant of the Order Batching Problem, it is necessary to solve two main tasks: The Batching Task and the Routing Task. The batching task consists of pre-grouping the orders into batches for the picker to pick. The routing task consists of defining a route for the picker to collect the order items that make up each generated batch. The objective is to minimize the time required to pick up all items from the orders received. This objective function is also known as minimizing the picking time.

The warehouse studied here is a rectangular layout. It consists of a variable number of parallel aisles, crossed by two perpendicular aisles, one front cross aisle and one back cross aisle. Additionally, the warehouse has only one input/output point, denoted as depot. In Fig. 1, we show an example of the layout of the studied warehouse.

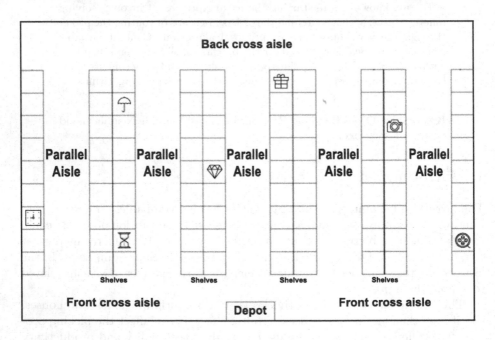

Fig. 1. Warehouse layout.

In order to evaluate the performance of the combination of GVNS with exact procedures, exact algorithms are applied to specific tasks parts of the problem.

Particularly, we compare the use of exact and heuristic algorithms to construct the initial solution provided to the GVNS. Also, we evaluate the performance of the GVNS when using exact and heuristic algorithms for the routing task. It is important to note that the routing task is used to assess the objective function of the solutions obtained within the search process performed by the GVNS, thus highlighting the close relationship between them. Through the experiments performed, we have observed that the use of exact algorithms for either constructing and routing, in combination with GVNS for batching, produces better results than previous approaches in both cases.

The main contribution of this work is to determine the most effective combination of algorithms that work best with GVNS, when it is used as a batching algorithm to solve the OBP problem. To achieve this task we evaluated both, exact and heuristic algorithms, for solving other tasks within the OBP that are different from the batching task. Our findings demonstrate that the use of exact algorithms yields the best results for these tasks. To achieve our objective, we conducted two experiments. The first experiment aims to determine the most effective constructive algorithm to provide initial solutions to the GVNS for the batching task. The second experiment aims to identify the most effective routing algorithm to be combined with GNVS to evaluate the solutions provided.

The rest of the paper is organized as follows. In Sect. 2, we present the state of the art of the order batching family of problems, and focus our attention on the offline variant of the OBP with a single picker. Here, we also review the most outstanding heuristic routing procedures in the literature. In Sect. 3, we present the algorithms for dealing with the batching task of the problem considered. Section 4 compiles the computational results obtained with the proposed algorithms in some well-known data sets. Finally, our conclusions are presented in Sect. 5.

2 State of the Art

OBP is a family of optimization problems related to the picking of goods from a warehouse. It uses a picking policy based on the order batching strategy. Within this family of problems, we review the literature on the classical Order Batching Problem (OBP), the most well-known and simple variant of the family. Theoretical studies of the OBP indicate that the problem is $\mathcal{NP}$-difficult. Following the need to find solutions to the problem in a short time, heuristics and metaheuristics have been applied to tackle the problem, although we can also find exact solutions in the literature [25]. The First-Come-First-Served (FCFS) strategy was one of the first heuristic strategies proposed and used in practice to assign orders to batches in a warehouse. This strategy has been widely used for its simplicity. The first approach to OBP was proposed in [7]. In this case, the routing task was handled by an automated storage and retrieval system (AS/RS), while the batching task was handled by simple heuristics. Other widely used heuristic methods in this problem are SEED methods [8] and Saving methods [34]. A study in which the authors compare the efficiency of several of these first methods can be found in [3]. The first metaheuristic algorithm applied to classical OBP was based on a Genetic Algorithm (GA) and was proposed in [19]. Subsequently, in [1] a method based on the Variable Neighborhood Search (VNS) methodology was presented; in [17] an Iterated Local Search (ILS) and in [18] a Tabu Search (TS). In [27] the authors proposed a new ILS algorithm with a Tabu Threshold (TT). Since 2017, an explosion of papers have been published dealing with different variants of the problem, among which we highlight those dealing with the classical OBP. In [21] a Multi-Start VNS is presented. In [26] several heuristics and metaheuristics are compared, among which GA, TS, and Simulated Annealing (SA) stand out. Hybrid metaheuristics have also been used for the problem, among which the Hybrid Genetic Algorithm (HGA) in [36]; and the Adaptive Large Neighborhood Search (ALNS) hybridized with TS in [37].

3   Algorithmic Proposal

This section presents the algorithmic proposal used for each task of the OBP tackled in the article. In Sect. 3.1 we present a GVNS algorithm for the batching task. In Sect. 3.2 several methods are introduced as an initial solution of the GVNS algorithm. Finally, in Sect. 3.3, different routing methods are proposed.

3.1   Batching Algorithms

This section introduces the General Variable Neighborhood Search framework [14,24] as the fundamental algorithm used in our experimental approach to address the batching task within the Order Batching Problem. GVNS, an off-shoot of the Variable Neighborhood Search (VNS) methodology, was introduced by Mladenovic and Hansen [24]. It represents a versatile approach that aims to solve complex optimization challenges. Central to its operation is the principle of dynamically changing the neighborhood structure to navigate away from local optima. The VNS paradigm encompasses a spectrum of variations that differ in their use of stochastic or deterministic exploration techniques, or a combination of both. For an in-depth exploration of VNS and its variants, interested readers are referred to seminal works such as [13,15,24]. Notable among the alternative adaptations of the VNS methodology are the Variable Formulation Search (VFS) [28], the Multi-objective Variable Neighborhood Search [4], and the Parallel Variable Neighborhood Search [5,23], each of which offers a different perspective on addressing optimization challenges.

In Algorithm 1 we present a schema of the GVNS method proposed in this paper. The method receives three input parameters: i) an initial solution ($solution$); ii) the largest neighborhood to be explored ($k_{\max}$); and iii) the maximum time ($t_{\max}$). The initial solution will be calculated using an external method. One of the purposes of this work is to study the best algorithm to generate this initial solution by comparing exact and heuristic methods, which are described in Sect. 3.2.

Algorithm 1. General Variable Neighborhood Search

1: **function** GVNS($solution, k_{\max}, t_{\max}$)
2:     **repeat**
3:         $k \leftarrow 1$
4:         **while** $k \leq k_{\max}$ **do**
5:             $solution' \leftarrow$ Shake($solution, k$)
6:             $solution'' \leftarrow$ VND($solution'$)
7:             $k \leftarrow$ NeighborhoodChange($solution, solution'', k$)
8:         **end while**
9:     **until** $t < t_{max}$
10:    **return** $solution$
11: **end function**

GVNS includes three different procedures to explore the current neighborhood and determine whether there has been improvement. First, the method **Shake**, performs a perturbation of the solution to escape from local optima. In particular, the proposed shake procedure is based on random swaps of orders between different batches. The method **VND** runs an improvement procedure based on the exploration of several neighborhoods through one or more local search procedures, the result of which is a local optimum with respect to all the neighborhoods explored. This method is detailed in Algorithm 2. Third, the procedure **NeighborhoodChange**, determines if there has been any improvement in the solution.

Algorithm 2. Variable Neighborhood Descent

1: **function** VND($solution$)
2:     $[\mathcal{N}_1, \mathcal{N}_2, \mathcal{N}_3]$
3:     $k \leftarrow 1$
4:     $k_{max} \leftarrow 3$
5:     $best \leftarrow solution$
6:     **repeat**
7:         $solution' \leftarrow$ LocalSearch($best, \mathcal{N}_k$)
8:         **if** eval($solution'$) < eval($best$) **then**
9:             $best \leftarrow solution'$
10:          $k = 1$
11:         **else**
12:             $k = k + 1$
13:         **end if**
14:     **until** $k > k_{max}$
15:     **return** $best$
16: **end function**

The proposed VND procedure [6] includes three neighborhood structures $[\mathcal{N}_1, \mathcal{N}_2, \mathcal{N}_3]$: a swap move that exchanges two orders belonging to different batches (denoted as $\mathcal{N}_1$), a swap move that exchanges two orders belonging to one batch with one order belonging to a different batch (denoted as $\mathcal{N}_2$), and finally a shift move that removes an order from a batch and inserts it into another batch (denoted as $\mathcal{N}_3$).

3.2 Constructive Procedure

The constructive method provides an initial solution to the GVNS algorithm. We have compared several algorithms as a constructive method. The algorithms compared are two First Come First Served (FCFS) methods, a SEED aisle-based algorithm, a classical Mixed-Integer Linear Programming (MILP) model where the objective function is to minimize the number of batches of the solution, and a combination of FCFS with a new GVNS (denoted as FCFS+GVNS) which tries to minimize the number of batches of the solution. The solution provided by FCFS+GVNS algorithm is then utilized as the initial solution for the main GVNS studied in this paper.

The MILP algorithm is executed using Gurobi, a well-known mathematical optimization solver. The MILP model used minimizes the number of batches in the solution. The following model is described in detail below. The necessary parameters and variables for the definition of the problem are introduced in Table 1.

Table 1. Parameters and variables for the OBP.

Parameters	
$n \rightarrow$ Number of customer orders available at the system	
$m \rightarrow$ Upper bound of the number of batches (a straightforward value is $m = n$)	
$w_i \rightarrow$ Number of items of order o_i for $1 \leq i \leq n$	
$W \rightarrow$ Maximum number of articles that can be included in a batch (device capacity)	
Variables	
$x_{ji} \rightarrow$	$\begin{cases} 1, & \text{if order } o_i \text{ is assigned to batch } b_j, \\ 0, & \text{otherwise.} \end{cases}$
$y_j \rightarrow$	$\begin{cases} 1, & \text{if batch } b_j \text{ contains some order } o, \\ 0, & \text{otherwise.} \end{cases}$

The objective function of the model is to minimize the number of batches of the solution, and it is given by:

$$\min \sum_{j=1}^{m} y_j \tag{1}$$

The set of feasible solutions, in both cases, is given by the following constraints:

- Constraints in (2) guarantee that each order is assigned only to one batch:

$$\sum_{j=1}^{m} x_{ji} = 1, \quad \forall i \in \{1, \ldots, n\}. \tag{2}$$

- Constraints in (3) guarantee that the maximum capacity of each batch is not exceeded:

$$\sum_{i=1}^{n} w_i x_{ji} \leq W, \quad \forall j \in \{1, \ldots, m\}. \tag{3}$$

- Constraints in (4) guarantee that y_j is activated if $batch_j$ contains some order o:

$$y_j \geq \sum_{i=1}^{n} x_{ji}, \quad \forall j \in \{1, \ldots, m\}. \tag{4}$$

- Constraints in (5) state that the variables x_{ji} are binary:

$$x_{ji} \in \{0, 1\}, \quad \forall j \in \{1, \ldots, m\} \text{ and } \forall i \in \{1, \ldots, n\}. \tag{5}$$

- Finally, the constraints in (6) state that the variables y_j are binary:

$$y_j \in \{0, 1\}, \quad \forall j \in \{1, \ldots, m\}. \tag{6}$$

3.3 Routing Algorithms

The picking operation is a critical and resource-intensive process in warehouses. Optimizing both batching and picking tasks together can reduce picking time, up to 35% [2].

In this paper, we study rectangular warehouses which have been widely studied in the literature. In particular, many algorithms have been developed to design efficient routes in this type of warehouse. Several routing heuristics have been introduced in the literature [12,30,31,33], among which the S-Shape, Largest Gap, and Combined algorithms stand out as particularly significant and widely used. Some exact methods have also been published. An exact method based on dynamic programming [32] accurately generates the optimal path in this context. In addition, a MILP model implemented in Cplex was presented in [35]. Among the previous algorithms, we compare 3 routing algorithms. The Combined heuristic, and two exact methods, one based on dynamic programming [32], and the other on the MILP model implemented in Cplex [35].

4   Experiments

This work presents two experiments that evaluate the use of exact and heuristic algorithms in various OBP tasks. The first experiment focuses on batch generation and compares different constructive algorithms to provide an initial solution to the GVNS metaheuristic. The second experiment compares various exact and heuristic algorithms for route generation.

All heuristic methods compared in this section, including those of the state of the art, were coded in Java 8 and run on an Intel (R) Core (TM) 2 Quad CPU Q6600 2.4 Ghz computer, with 4 GB DDR2 RAM memory and Ubuntu 20.04.1 64 bit LTS operating system.

In Sect. 4.1, we present the sets of instances used in both experiments. In Sects. 4.2 and 4.3 we present the experiments conducted and analyze the results obtained.

4.1   Instances

For this study, we have selected two datasets commonly used in the state of the art to study different variants of order batching problems. These datasets are available at https://grafo.etsii.urjc.es/optsicom/oobp/. Both datasets consist of instances of rectangular single-block warehouses with two cross-aisles and a single depot. The Dataset #1 includes 80 instances corresponding to four distinct warehouses (labeled W1, W2, W3, and W4). This dataset was first introduced in [1] and has since been utilized in various related works, such as [9–11,21–23]. The Dataset #2 consists of 64 instances that correspond to a single warehouse (labeled W5). It was originally proposed in [16] and has also been used in several related publications, including [11,20,21].

4.2   Experiment #1

The objective of this initial experiment is to evaluate the effect of the constructive procedure on the performance of GVNS + S-Shape. The results presented in this experiment are based on the final solution provided by the whole method, including GVNS for the batching algorithm and S-Shape as the routing algorithm, rather than just the initial solution generated by the constructive algorithms. Specifically, in this experiment, we evaluate two constructive algorithms that attempt to minimize the number of solution batches to provide a good initial solution. Two algorithms were tested: a MILP model run on Gurobi and a new GVNS that works as a constructive algorithm. Additionally, we included in the comparison several simple heuristics, such as First Come First Served, and aisle-based SEED algorithms.

It is important to note that the goal of this experiment is to find the most effective constructive algorithm to reduce picking time when combined with GVNS in the batching task. However, it should be noted that when the constructive algorithm selected is executed in isolation, it may not achieve a good picking time itself.

Table 2 reports the results of Experiment #1. As it is possible to observe, the exact MILP model achieved the best deviation, improving with the next algorithm compared to SEED by 0.23%. Furthermore, the exact MILP model achieved the best average time, improving on the next algorithm compared to SEED by 92 s. However, the SEED algorithm achieved a better number of best-known values.

Table 2. Comparison based on the picking time of the influence of the constructive procedures when combined with GVNS for batching task, and S-Shape for the routing task.

	FCFS	FCFS C.	SEED	F+GVNS	MILP
AVG (s)	31,374	31,194	30,967	31,133	**30,875**
Dev. (%)	1.63%	1.31%	0.90%	0.94%	**0.67%**
CPU time (s)	**191**	**191**	**191**	195	194
# Best	24/144	17/144	**50/144**	24/144	42/144

4.3 Experiment #2

The objective of this group of experiments is to evaluate the effect of different routing strategies and their influence on the performance of the GVNS used for the batching.

In the first experiment, reported in Table 3, we compare the influence of three routing algorithms: an MILP model run in Cplex [35] (Valle), a Dynamic Programming algorithm [32] (Ratliff), and a popular heuristic algorithm called Combined (Combined), when using a simple heuristic such as FCFS for the batching task.

Table 3. Comparison of several algorithms for the routing task when using a simple heuristic as FCFS for batching task.

	Valle	Ratliff	Combined
AVG (s)	**31,247**	**31,247**	32,051
Dev. (%)	**0.00%**	**0.00%**	2.41%
CPU t. (s)	118.462	0.005	**0.001**
# Best	**144/144**	**144/144**	0/144

In the next two experiments, we consider the previous algorithms: an MILP model running in Cplex [35] (Valle), a Dynamic Programming algorithm [32] (Ratliff), and a well-known heuristic algorithm called "Combined" (Combined), by comparing six routing algorithms. Particularly, we proposed some hybrid combinations among them, including two hybrid algorithms of a Dynamic Programming with Combined denoted as "Combined+Ratliff" and "Combined>Ratliff", and two hybrid algorithms of an MILP with "Combined" denoted as "Combined+Valle" and "Combined>Valle". In these cases where the "+" symbol appears, it means that the "Combined" algorithm is utilized in the local search procedure, while the Neighborhoodchange procedure uses either Ratliff or Valle. Additionally, when the ">" symbol is present, the "Combined" algorithm is used to evaluate the solutions provided by the GVNS during the search process, while Ratliff or Valle is used only to evaluate the best solution found by GVNS at the end of the process, in search for a further improvement.

Table 3 reports first part of the results of Experiment #2, which used a simple heuristic as the batching method. Here, we observe that Valle and Ratliff obtained the best deviation and average time. In addition, they certify an exact solution. The main difference here is the CPU time. The Combined algorithm obtained the best time with 1 ms, after which we can find Ratliff with 5ms, and finally Valle with almost 2 min. It is worth mentioning that having a short CPU time is crucial when using the routing method within the GVNS metaheuristic, since many evaluations are performed.

Table 4. A comparison study of several routing algorithms when using the FCFS as constructive algorithm and GVNS metaheuristic for batching task without time limit.

	Ratliff	Combined + Ratliff	Combined > Ratliff	Combined > Valle	Combined	Combined + Valle
AVG (s)	**28,170**	28,234	28,312	28,312	28,709	31,198
Dev. (%)	**0.31%**	0.84%	1.08%	1.08%	2.06%	13.21%
CPU t. (s)	430	**123**	124	461	124	461
# Best	**95/144**	33/144	11/144	11/144	0/144	0/144

Tables 4 and 5 present the second part of the results of Experiment #2. In this case GVNS metaheuristic was used as the batching method. The difference between the results reported in both tables is the time limit. Particularly, in Table 4 there is not time limit for the batching task, while in Table 5 the execution of the VND algorithm within the GVNS used for the batching task is truncated after 120 s.

In Table 4, it can be observed that Ratliff achieved the best deviation by improving the Combined+Ratliff algorithm in 0.53% and reducing the average time in 64 s. However, it should be noted that the Ratliff algorithm takes 307 s longer than the Combined+Ratliff algorithm in terms of CPU time. In these results, the execution of the VND algorithm used in the batching method was not truncated, then VND algorithm does not have a limit on the execution time.

Table 5. A comparison study of several routing algorithms when using the FCFS as constructive algorithm and GVNS metaheuristic for batching task when limiting the batching task to 120 s.

	Ratliff	Combined + Ratliff	Combined > Ratliff	Combined > Valle	Valle	Combined + Valle	Combined
AVG (s)	**31,131**	31,132	31,141	31,145	31,185	31,198	31,921
Dev. (%)	0.31%	**0.30%**	0.32%	0.32%	0.45%	0.51%	2.63%
CPU t. (s)	**120**	**120**	**120**	333	350	461	120
# Best	57/144	**68/144**	36/144	48/144	34/144	30/144	0/144

In Table 5, it can be seen that Ratliff improves the average time of the "Combined+Ratliff" algorithm in only 1 s. However, the "Combined+Ratliff" algorithm achieves better deviation scores in 0.01%, and the number of best scores is improved by 11 best scores compared to the "Ratliff" algorithm. For these values, the "Ratliff" and "Combined+Ratliff" algorithms produce very similar results with no significant differences. In these results, the execution of the VND algorithm used in the batching method was truncated after 120 s.

5 Conclusions

This paper studies the combination of a GVNS with several exact algorithms to solve the Order Batching Problem. As a first conclusion, we observed that the combination of exact algorithms for tasks such as construction of the initial solution or routing, with heuristic algorithms for the batching task, can substantially improve current state-of-the-art results.

Particularly, we propose to use a GVNS method in the state of the art for the batching task, together with two exact procedures. The first exact procedure is to construct the solution provided to the GVNS as a starting point. In this case, the use of a MILP model, which tries to minimize the number of batches of the solutions, performed the best. As far as the second exact procedure is concerned, it is devoted to the routing task. In this case, the best combination includes the use of an exact algorithm based on dynamic programming.

Future research lines can be derived from this work. The proposed configuration of exact and heuristic procedures can be applied to other proposals, substituting the GVNS used for the batching task. Furthermore, the next step in this investigation is to compare our proposed GVNS configuration with the current state-of-the-art algorithms for the problem.

Acknowledgments. This research has been partially supported by grants PID2021-125709OA-C22 and RED2022-134480-T, funded by MCIN/AEI/10.13039/501100011033 and by "ERDF A way of making Europe"; grant CIAICO/2021/224 funded by Generalitat Valenciana; grant M2988 funded by "Proyectos Impulso de la Universidad Rey Juan Carlos 2022"; and "Cátedra de Innovación y Digitalización Empresarial entre Universidad Rey Juan Carlos y Second Episode" (Ref. ID MCA06).

References

1. Albareda-Sambola, M., Alonso-Ayuso, A., Molina, E., De Blas, C.S.: Variable neighborhood search for order batching in a warehouse. Asia-Pac. J. Oper. Res. **26**(5), 655–683 (2009)
2. De Koster, R.B.M., Roodbergen, K.J., Van Voorden, R.: Reduction of walking time in the distribution center of De Bijenkorf. In: Speranza, M.G., Stähly, P. (eds.) New Trends in Distribution Logistics. LNE, vol. 480, pp. 215–234. Springer, Heidelberg (1999). https://doi.org/10.1007/978-3-642-58568-5_11
3. De Koster, R.B.M., Van Der Poort, E.S., Wolters, M.: Efficient order batching methods in warehouses. Int. J. Prod. Res. **37**(7), 1479–1504 (1999)
4. Duarte, A., Pantrigo, J.J., Pardo, E.G., Mladenović, N.: Multi-objective variable neighborhood search: an application to combinatorial optimization problems. J. Glob. Optim. **63**(3), 515–536 (2015)
5. Duarte, A., Pantrigo, J.J., Pardo, E.G., Sánchez-Oro, J.: Parallel variable neighbourhood search strategies for the cutwidth minimization problem. IMA J. Manag. Math. **27**(1), 55–73 (2013)
6. Duarte, A., Mladenovic, N., Sánchez-Oro, J., Todosijević, R.: Variable neighborhood descent (2018)
7. Elsayed, E.A.: Algorithms for optimal material handling in automatic warehousing systems. Int. J. Prod. Res. **19**(5), 525–535 (1981)
8. Gibson, D.R., Sharp, G.P.: Order batching procedures. Eur. J. Oper. Res. **58**(1), 57–67 (1992)
9. Gil-Borrás, S., Pardo, E.G., Alonso-Ayuso, A., Duarte, A.: New VNS variants for the online order batching problem. In: Sifaleras, A., Salhi, S., Brimberg, J. (eds.) ICVNS 2018. LNCS, vol. 11328, pp. 89–100. Springer, Cham (2019). https://doi.org/10.1007/978-3-030-15843-9_8
10. Gil-Borrás, S., Pardo, E.G., Alonso-Ayuso, A., Duarte, A.: Basic VNS for a variant of the online order batching problem. In: Benmansour, R., Sifaleras, A., Mladenović, N. (eds.) ICVNS 2019. LNCS, vol. 12010, pp. 17–36. Springer, Cham (2020). https://doi.org/10.1007/978-3-030-44932-2_2
11. Gil-Borrás, S., Pardo, E.G., Alonso-Ayuso, A., Duarte, A.: GRASP with variable neighborhood descent for the online order batching problem. J. Glob. Optim. **78**(2), 295–325 (2020)
12. Hall, R.W.: Distance approximations for routing manual pickers in a warehouse. IIE Trans. **25**(4), 76–87 (1993)
13. Hansen, P., Mladenović, N.: Variable neighborhood search: principles and applications. Eur. J. Oper. Res. **130**(3), 449–467 (2001)
14. Hansen, P., Mladenović, N., Brimberg, J., Pérez, J.A.M.: Variable neighborhood search. In: Gendreau, M., Potvin, J.-Y. (eds.) Handbook of Metaheuristics. ISORMS, vol. 272, pp. 57–97. Springer, Cham (2019). https://doi.org/10.1007/978-3-319-91086-4_3
15. Hansen, P., Mladenović, N., Moreno-Pérez, J.A.: Variable neighbourhood search: methods and applications. Ann. Oper. Res. **175**(1), 367–407 (2010)
16. Henn, S.: Algorithms for on-line order batching in an order picking warehouse. Comput. Oper. Res. **39**(11), 2549–2563 (2012)
17. Henn, S., Koch, S., Doerner, K.F., Strauss, C., Wäscher, G.: Metaheuristics for the order batching problem in manual order picking systems. Bus. Res. **3**(1), 82–105 (2010)

18. Henn, S., Wäscher, G.: Tabu search heuristics for the order batching problem in manual order picking systems. Eur. J. Oper. Res. **222**(3), 484–494 (2012)
19. Hsu, C.M., Chen, K.Y., Chen, M.C.: Batching orders in warehouses by minimizing travel distance with genetic algorithms. Comput. Ind. **56**(2), 169–178 (2005)
20. Menéndez, B., Bustillo, M., Pardo, E.G., Duarte, A.: General variable neighborhood search for the order batching and sequencing problem. Eur. J. Oper. Res. **263**(1), 82–93 (2017)
21. Menéndez, B., Pardo, E.G., Alonso-Ayuso, A., Molina, E., Duarte, A.: Variable neighborhood search strategies for the order batching problem. Comput. Oper. Res. **78**, 500–512 (2017)
22. Menéndez, B., Pardo, E.G., Duarte, A., Alonso-Ayuso, A., Molina, E.: General variable neighborhood search applied to the picking process in a warehouse. Electron. Notes Discrete Math. **47**, 77–84 (2015)
23. Menéndez, B., Pardo, E.G., Sánchez-Oro, J., Duarte, A.: Parallel variable neighborhood search for the min-max order batching problem. Int. Trans. Oper. Res. **24**(3), 635–662 (2017)
24. Mladenović, N., Hansen, P.: Variable neighborhood search. Comput. Oper. Res. **24**(11), 1097–1100 (1997)
25. Muter, I., Öncan, T.: An exact solution approach for the order batching problem. IIE Trans. **47**(7), 728–738 (2015)
26. Nicolas, L., Yannick, F., Ramzi, H.: Order batching in an automated warehouse with several vertical lift modules: optimization and experiments with real data. Eur. J. Oper. Res. **267**(3), 958–976 (2018)
27. Öncan, T.: MILP formulations and an iterated local search algorithm with tabu thresholding for the order batching problem. Eur. J. Oper. Res. **243**(1), 142–155 (2015)
28. Pardo, E.G., Mladenović, N., Pantrigo, J.J., Duarte, A.: Variable formulation search for the cutwidth minimization problem. Appl. Soft Comput. **13**(5), 2242–2252 (2013)
29. Pardo, E.G., Gil-Borrás, S., Alonso-Ayuso, A., Duarte, A.: Order batching problems: taxonomy and literature review. Eur. J. Oper. Res. **313**(1), 1–24 (2023). https://doi.org/10.1016/j.ejor.2023.02.019
30. Petersen, C.G.: Routeing and storage policy interaction in order picking operations. Decis. Sci. Inst. Proc. **31**(3), 1614–1616 (1995)
31. Petersen, C.G.: An evaluation of order picking routeing policies. Int. J. Oper. Prod. Manag. **17**(11), 1098–1111 (1997)
32. Ratliff, H.D., Rosenthal, A.S.: Order-picking in a rectangular warehouse: a solvable case of the traveling salesman problem. Oper. Res. **31**(3), 507–521 (1983)
33. Roodbergen, K.J., Petersen, C.G.: How to improve order picking efficiency with routing and storage policies. In: Progress in Material Handling Practice, vol. 1, pp. 107–124 (1999)
34. Rosenwein, M.B.: A comparison of heuristics for the problem of batching orders for warehouse selection. Int. J. Prod. Res. **34**(3), 657–664 (1996)
35. Valle, C.A., Beasley, J.E., Da Cunha, A.S.: Optimally solving the joint order batching and picker routing problem. Eur. J. Oper. Res. **262**(3), 817–834 (2017)
36. Yang, J., Zhou, L., Liu, H.: Hybrid genetic algorithm-based optimisation of the batch order picking in a dense mobile rack warehouse. PLoS ONE **16**(4), e0249543 (2021)
37. Žulj, I., Kramer, S., Schneider, M.: A hybrid of adaptive large neighborhood search and tabu search for the order-batching problem. Eur. J. Oper. Res. **264**(2), 653–664 (2018)

VNS-Based Matheuristic Approach to Group Steiner Tree with Problem-Specific Node Release Strategy

Tatjana Davidović[1]($\boxtimes$) (iD) and Slobodan Jelić[2] (iD)

[1] Mathematical Institute of the Serbian Academy of Sciences and Arts,
Kneza Mihaila 36, 11000 Belgrade, Serbia
`tanjad@mi.sanu.ac.rs`
[2] Faculty of Civil Engineering, University of Belgrade, Bulevar kralja Aleksandra 73,
11000 Belgrade, Serbia
`sjelic@grf.bg.ac.rs`

Abstract. For a given undirected graph $G = (V, E)$ with a non-negative weight function $w : E \to \mathbb{R}_+$ and subsets $G_1, \ldots, G_k$ of V, the Group Steiner Tree (GST) problem consists of constructing a tree $T = (V_T, E_T)$ with minimal cost, where $V_T \subseteq V$, $E_T \subseteq E$, and T spans at least one node from each of the groups. We develop a VNS-based metaheuristics approach for solving the GST problem. Our main contribution is that we propose a new problem-specific node release strategy that mimics the steps of a VNS-based heuristic. Instead of exploring different neighborhoods by combinatorially enumerating neighboring solutions, as in classical local search, we use a provably good Integer Linear Programming (ILP) formulation to solve a sequence of subproblems of the original problem. Our approach leads to an improvement over the state-of-the-art Gurobi solver both in terms of quality and runtime of the instances available in the literature.

Keywords: integer programming formulation · subtour elimination · metaheuristic methods · hybrid heuristic · decomposition strategy

1 Introduction

The Group Steiner Tree (GST) is a well-known NP-hard combinatorial optimization problem that was introduced in [26] and is studied in both theoretical and applied computing societies. Given an undirected graph $G = (V, E)$, $|V| = n$,

---

This research was partially supported by Serbian Ministry of Science, Technological Development, and Innovations, Contract No. 451-03-66/2024-03/200029. The funds were also provided by the Science Fund of the Republic of Serbia, Green program of cooperation between science and industry, project: EO and in-situ based information framework to support generating Carbon Credits in forestry.

M. Sevaux et al. (Eds.): MIC 2024, LNCS 14753, pp. 344–358, 2024.
https://doi.org/10.1007/978-3-031-62912-9_32

$|E| = m$, with edge-weight function $w : E \to \mathbb{R}_+$, and a family of subsets of V, $\mathcal{G} = \{G_1, \ldots, G_k\}$, $k \in \mathbb{N}$, $G_i \neq \emptyset$, which are called *groups*, the problem is to find such a tree $T = (V_T, E_T)$ that $V_T \cap G_i \neq \emptyset$ for each $i = 1, 2, \ldots, k$ and

$$\sum_{e \in E_T} w(e)$$

is minimized.

The first application of GST comes from the problem of routing in VLSI design [26, 30]. Interesting applications also arose in the team formation problem in social networks [20], where the goal is to find a cohesive subnetwork of experts who collectively have all the necessary skills to complete a predefined project represented as a set of skills. Motivated by the team formation problem, GST can also be applied to other problems in the social sciences, e.g., the government formation problem [16]. The database and data mining communities are also interested in GST as there are interesting applications for keyword search in relational databases [4, 5].

Due to its NP-hardness, GST is also very popular in the society of approximation algorithms and heuristics. In the seminal paper by Garg et al. [9] a polylogarithmic approximation ratio algorithm with randomized rounding technique is presented. Duin et al. [7] presented heuristic algorithms based on the reduction of GST to the Steiner tree problem on graphs. Ant colony and genetic algorithms for heuristic solutions are also studied [25]. The latest result involving the general VNS algorithm for the GST problem is proposed by Matijević et al. in [22]. There are also experimental studies on exact and approximation algorithms for the node-weighted and edge-weighted group Steiner tree problem [28]. However, this problem variant is out of the scope of this paper.

The transformation from GST to Steiner tree described in [7] can be used to obtain a Steiner tree instance that is solvable with the exact solver [19]. Unfortunately, this transformation uses a big-M strategy that introduces a terminal connected by expensive edges to all nodes in a given group. The LP relaxation proposed in [19] for such Steiner tree instances has a large integrality gap, implying a large gap between lower and upper bounds during the branch-and-cut process. We are motivated to give stronger formulations that are extended by additional isolated node inequalities and work well in practice. In this paper, we use the subtour elimination formulation as it has stronger relaxation than the natural cuts-based formulation [24]. It is not a rooted formulation, but has an exponential number of constraints. Therefore, to improve its efficiency, we use a simple constraint generation procedure that successively creates only violated constraints.

In this paper, we focus on a matheuristic algorithm with a node release strategy that mimics the VNS concept of searching over different neighborhood structures. One of the most important aspects of our approach is the Integer Linear Programming (ILP) formulation for GST. Starting from a polyhedral approach, different ILP formulations for GST have been investigated in several papers [8, 10, 24]. In [24] an overview of all GST formulations is given together with the comparisons of the strength of their relaxations. We compare our solutions on three types of randomly generated GST instances available in the literature [7,22], with solutions obtained by Gurobi [11] when solving subtour elimination formulations with and without a given initial feasible solution. One of the most specific parts of our algorithm is the problem-specific node release strategy, which follows the VNS-based rules to define the number and order of node release with respect to the current solution.

Having in mind that all the methods considered in this paper are based on an ILP formulation of the GST problem, we present the subtour elimination formulation with a brief literature review. We propose a matheuristic algorithm, named SUBTVNS, that uses the subtour elimination formulation with a distance-based node release strategy. Our method is compared with the branch-an-cut algorithm applied to the subtour elimination formulation. Specifically, we consider the variant with and without initial solution. The first variant, called SUBTINIT, uses an initial feasible solution found with the algorithm by Ihler in [15]. The second variant, called SUBT, does not use an initial feasible solution, i.e., it is not given any potential help in the initialization phase.

Contributions. The main contributions of this work can be summarized as follows:

- a matheuristic algorithm for GST with problem-specific node release strategy (SUBTVNS) is developed;
- the number of instances solved to optimality with SUBTVNS is improved compared to SUBTINIT and SUBT;
- the quality of the feasible solution provided by SUBTVNS is improved compared to the solution found by the other two methods under the same settings;
- SUBTVNS required significantly less runtime for instances solved to optimality with both SUBTVNS and SUBTINIT.

All three methods use the branch-and-cut procedure implemented in Gurobi [11] and the subtour elimination formulation [10,24]. Therefore, SUBT refers to the Gurobi solver running branch-and-cut procedure with default parameters, while SUBINIT additionally uses an initial feasible solution computed with the procedure from [15]. On the other hand, SUBTVNS uses VNS-based rules to generate subproblems of the original problem, which are solved by Gurobi within a given time limit.

Notation. Here we provide a brief overview of notations used in the paper. For $n \in \mathbb{N}$, $[n] = \{1, 2, \ldots, n\}$. Let $G = (V, E)$, $H \subseteq V$, then $E(H) = \{e = \{s, t\} \in E : s \in H, t \in H\}$. For $v \in V$, $\delta(v) = \{u \in V : \{u, v\} \in E\}$ and $\Delta(v) = \{e \in E : e = \{v, u\}, u \in V\}$ denote the node-based and edge-based neighborhoods of v in the graph G, respectively. For the tree $T = (V_T, E_T)$ in the graph G, $v \in T$ means $v \in V_T$. If $w : E \to \mathbb{R}_+$ is a non-negative weight function given as part of the GST instance and $T = (V_T, E_T)$ is a tree with $V_T \subseteq V$, $E_T \subseteq E$, then $c(T) = \sum_{e \in E_T} w(e)$.

Paper Organization. Our paper is structured as follows. In Sect. 2 we describe matheuristic algorithms in general, especially the ones related to the VNS metaheuristic, together with a relevant literature overview. The explored mathematical model in the form of the integer programming formulation is described in Sect. 3. In Sect. 4, the main part of the paper, the proposed matheuristic (SUBTVNS) with problem-specific node release strategy is presented. Section 5 describes the experimental environment, the instances, and the results of the experiments performed with the considered methods. Summary of the presented results and the possible directions of future work are provided in Sect. 6.

2 Matheuristics

Matheuristics [29] are optimization methods obtained by hybridizing exact solvers with metaheuristics. The main idea is to use metaheuristic rules for creating subproblems to be treated by the exact solvers. The popularity of matheuristics constantly grows during the last decade [2,3]. In the majority of the cases, matheuristics are general-purpose optimization algorithms, i.e., they do not explore *a priori* knowledge about the considered problem. That enables their application to the wide spectrum of problems.

The decomposition of the original problem into subproblems can be done by fixing values of some (binary) variables using metaheuristic rules [17,21]. Then, exact solver is invoked to determine values of the remaining variables. More precisely, the definition of subproblems refers to the exploration of variable states. Each (binary) variable can be in one of the two possible states: fixed or released. The fixed state means that the exact solver is not allowed to change the value of this variable during the optimization process. Therefore, the main goal is that the solver finds the best value of the variables in released state.

An alternative way to create subproblems controls the number of variables allowed to change the value [14]. Namely, it is not important which variables would change values, it is just specified how many of them could be modified. There are also some other approaches as can be seen from [2,3]. However, it is important to note that sometimes, matheuristics can be considered as exact algorithms, i.e., given enough resources (unlimited memory and running time) they can provide optimal solution.

Usually, the exact solvers work on mathematical programming formulation of the considered problem. More precisely they explore Mixed Integer Linear Programming, (MILP) formulations. The logical question that arises is how the mathematical model influences the performance of the resulting matheuristic method and it is investigated in [1]. However, there are also matheuristic that operate solely on the combinatorial formulations, for example, the Fixed Set Search method applied in [18].

We are particularly interested in the neighborhood based metaheuristics combined with mathematical programming optimization techniques to design effective matheuristics. We review three state-of-the-art matheuristics based on Variable Neighborhood Search (VNS) [12,13,23]: Variable Neighborhood Branching (VNB) [14], Variable Neighborhood Decomposition Search for 0–1 MIP problems (VNDS-MIP) [21], and Variable Intensity Neighborhood Search (VINS)

[17]. These matheuristics rely on Mixed Integer Linear Programming, (MILP) formulation and operate exclusively on binary variables. This fact should not be considered as a major drawback as the reformulation of a model containing integer variables into a model with binary variables only is straightforward.

When employing the binary variables, the Hamming distance can be used to measure the diversity of the solutions and to introduce neighborhood structures. For two solutions $x = (x_1, \ldots, x_n)$ and $x' = (x'_1, \ldots, x'_n)$, Hamming distance is defined as

$$d_H(x, x') = \sum_{i \in B} | x_i - x'_i |, \tag{1}$$

where B denotes the subset of binary variables. Now, the neighborhood structures can be defined as:

$$\mathcal{N}_k(x) = \{x' \in X \mid d_H(x, x') \leq k\}. \tag{2}$$

More precisely, neighborhood $\mathcal{N}_k(x)$ contains all the solutions that have at most k different values for binary variables. It is worth noting that $\mathcal{N}_k(x) \subset \mathcal{N}_{k+1}(x)$, which means that if neighborhood $\mathcal{N}_{k+1}(x)$ is completely explored, there is no need to search in neighborhood $\mathcal{N}_k(x)$.

Local search in the space of binary variables applied in the VNB method [14] is realized by invoking exact solver on a subproblem obtained by adding the set of neighborhood-defining constraints to the original model. At the same time, the constraints are imposed to the objective function value such that only solutions improving it become feasible.

The other two methods, VNDS-MIP [21] and VINS [17], use the more general distance function defined between a feasible solution x and the solution of the linear relaxation y. The difference between each pair of corresponding variables can take any value from the interval $[0, 1]$. The variables are sorted in the non-decreasing order according this distance and various criteria are used to fix/release a subset of them. Although the reviewed matheuristic are general-purpose methods, they were inspiration for the development of the proposed SUBTVNS method, a matheuristic with problem-specific node release strategy that we describe in more detail in Sect. 4.

3   Integer Programming Formulations

There are three integer programming formulations for the GST problem that have been studied in the literature [24]. The cut-based formulation is a natural ILP formulation based on the idea of covering all cuts separating a group from a given root to satisfy connectivity requirements. As a solution of the relaxation of the natural cut-based formulation is the weakest lower bound of the optimal solution [24], it is reasonable to use the other two formulations. The flow-based formulation is a compact version of the cut-based formulation, where at least one unit of flow must be sent from each group to the root. Although the flow-based formulation has polynomial number of constraints, it is impractical to solve instances with a large number of groups with any modern ILP solver. On the other hand, both the cut-based and flow-based formulations are root formulations, which means that we need to solve one instance of the problem for each possible choice of designated root. Therefore, we consider the subtour elimination formulation because it is not a root formulation, it has a better lower bound for the optimal solution compared to the cut-based formulation, and does not require a list of explicitly created constraints. This formulation requires a constraint generation procedure that creates violating constraints and iteratively adds them to the constraint pool. The idea is to eliminate subtours by removing cycle edges, which ensures that the generated subgraph is a tree.

In addition to the edge variables $x_e \in \{0,1\}$, $e \in E$ that decide which edges are included in the solution, we introduce vertex variables $y_v \in \{0,1\}$, $v \in V$. If it is decided that the vertex $v \in V$ is included in the solution, y_v is set to 1, otherwise to 0. We start with the subtour elimination formulation that is presented in [24].

$$\min \sum_{e \in E} w(e) x_e$$

$$\sum_{v \in G_i} y_v \geq 1, \quad i \in [k], \tag{3}$$

$$\sum_{e \in E} x_e = \sum_{v \in V} y_v - 1, \tag{4}$$

$$\sum_{e \in E(W)} x_e \leq \sum_{v \in W} y_v - y_u, \quad W \subseteq V : G_i \subsetneq W, \forall i \in [k], u \in W \tag{5}$$

$$\sum_{e \in E(W)} x_e \leq \sum_{v \in W} y_v - 1, \quad W \subseteq V : G_i \subseteq W, \text{ for some } i \in [k], \tag{6}$$

$$x_e, y_v \in \{0,1\} \quad e \in E, v \in V. \tag{7}$$

Constraints (3) ensure that each group is covered by at least one node in the solution tree. The solution tree is characterized as an acyclic graph whose number of edges is equal to the number of nodes minus one. The constraints (4) and (5) require that the number of edges is (at most) one less than the number of nodes for each subset that is selected as part of the solution. The constraints (6) are stronger versions of the constraints (5) for all cases where $W \subset V$ contains a group G_i, $i \in [k]$. In these cases, we know from the constraints (3) that at least one of the nodes in W must be part of the solution. The constraints (4), (5) and (6) are also called subtour elimination constraints, because they prevent the existence of cycles in any subgraph of the solution graph [10,27]. We consider some additional cut inequalities that could strengthen the relaxation of this ILP obtained by omitting the integrality in constraints (7).

To improve the quality of the lower bounds in the branch-and-cut process, we add a set of cutting planes given by inequalities which remove isolated nodes during the branch-and-cut process. The formulation from [10,24,27] starts with the empty pool of constraints (5) and (6). During the branching process, some of the solutions that violate (5) and (6) may contain nodes that are not incident to any of the currently selected edges. To strengthen the formulation, we add the following inequality constraints:

$$\sum_{e \in \Delta(v)} x_e \geq y_v, \quad v \in V. \tag{8}$$

This formulation is used in all the methods presented in this paper within the Gurobi framework.

4   The Proposed Matheuristic Algorithm SUBTVNS

The main idea of our algorithm is to mimic VNS-based metaheuristic methods, where the enumerative traversal over a given neighborhood is replaced by an ILP solver (using a specific ILP formulation of the problem) that finds the best possible solution in a given time. The size of the neighborhood is controlled by the release factor ρ, which determines the percentage of the total number of nodes whose model variables are released, i.e., these nodes can be added to or removed from the solution. A pseudocode of the algorithm is shown in Algorithm 1. The input arguments are as follows:

- G - input graph,
- $\mathcal{G}$ - a family of groups,
- α - step size of the release factor, typically $\alpha = 0.1$,
- t_{max} - global time limit for the solver,
- t_s - time limit for the subproblem (typically $t_s = t_{max}/10$).

The total execution time is measured and saved as the value of the variable t. It is used to check the stopping criterion for SUBTVNS. The variable *start* is initialized by calling TIME() in the line 4. To measure the total elapsed time at any execution point, the value of *end* is updated at that point and the difference *end* − *start* is calculated. In the line 5, the subroutine ALLPAIRSSHORTEST-PATHS is called to calculate the distance matrix D using Dijkstra's algorithm. The matrix D contains the length of the shortest paths between the individual node pairs. As the distance matrix D is used in the two subroutines IHLER and NODEPRIORITY, it is calculated once in the preprocessing phase. In addition to the entries of the matrix D, the value δ, a diameter of the graph G that is calculated as the maximum of the distances between any two nodes in G, is also computed in the preprocessing phase.

The initialization of SUBTVNS algorithm is performed in lines 7 to 12. First, the initial feasible solution is computed using the subroutine IHLER that implements the algorithm proposed in [15]. The best found solution S_B is initialized with the initial feasible solution S_C, the node priorities with respect to S_C are calculated and nodes are sorted in ascending order by priorities p, and the release factor ρ is set to 0, because all node variables y_v, $v \in V$, are set to $1/0$, depending on whether v is spanned by the initial solution or not. In line 12, the Gurobi model is initialized based on the solution found in step 7.

Algorithm 1. SUBTVNS matheuristic

1: **procedure** SUBTVNS($G, \mathcal{G}, t_s, t_{max}, \alpha$)
2:     $t_{best} \leftarrow 0$
3:     $solved \leftarrow False$
4:     $start \leftarrow \text{TIME}()$
5:     $D \leftarrow \text{ALLPAIRSSHORTESTPATHS}(G, w)$
6:     $\delta \leftarrow \max_{u,v} D[u, v]$
7:     $S_C \leftarrow \text{IHLER}(G, \mathcal{G}, D)$
8:     $S_B \leftarrow S_C$
9:     $p \leftarrow \text{NODEPRIORITIES}(G, \mathcal{G}, D, S_B, \delta)$
10:    sort nodes in ascending order by priorities p
11:    $\rho \leftarrow 0$
12:    initialize model based on S_C
13:    $end \leftarrow \text{TIME}()$
14:    $t \leftarrow end - start$
15:    $t_{best} \leftarrow t$
16:    **while** $(t < t_{max}) \wedge (\neg solved)$ **do**
17:        **if** $\rho < 1$ **then**
18:            $\rho \leftarrow \rho + \alpha$
19:            $t_s \leftarrow \min(t_{max} - t, t_s)$
20:        **else**
21:            $t_s \leftarrow t_{max} - t$
22:        release last $\rho \cdot n$ variables in the model
23:        optimize model to find S_C with time limit t_s
24:        $end \leftarrow \text{TIME}()$
25:        $t \leftarrow end - start$
26:        **if** S_C is optimal to current subproblem and $\rho = 1$ **then**
27:            $S_B \leftarrow S_C$
28:            $solved \leftarrow True$
29:            $t_{best} \leftarrow t$
30:        **else**
31:            **if** $c(S_C) < c(S_B)$ **then**
32:                $S_B \leftarrow S_C$
33:                $t_{best} \leftarrow t$
34:                $p \leftarrow \text{NODEPRIORITIES}(G, \mathcal{G}, D, S_B)$
35:                add constraint $c(S) \leq c(S_B)$
36:                $\rho \leftarrow 0$
37:                sort nodes in ascending order by priorities p
38:        $end \leftarrow \text{TIME}()$
39:        $t \leftarrow end - start$
40:    **return** S_B, t_{best}

The algorithm for calculating the initial solution [15] is not presented as a separate subroutine as it would go beyond the scope of this paper. It is a greedy algorithm that iteratively expands the solution tree until all groups are covered. Although this greedy approach leads to a solution whose cost is at most k times optimal, it is easy to implement and produces high-quality solutions in most cases. From line 16 to 39 we improve the current solution S_C while total execution time is not exceeded and the optimal solution is not found. If the release factor ρ is less than 1, it is increased by α and the time limit for the subproblem is updated. Otherwise, the solver uses the time limit t_s, which is set to the remaining time to solve the original problem. After updating ρ and t_s, the last $\rho \cdot n$ variables in the model are released and S_C is recalculated by running the solver with the time limit t_s. The current solution is optimal if, after the optimization of a current subproblem (line 23), the status of the Gurobi solver is OPTIMAL and the release factor is equal to 1. Otherwise, if the current solution S_C is better (i.e. with lower cost) than S_B, it becomes the newly found best solution and the node priorities must be recalculated. In addition, the release factor ρ is reset to the initial value 0 and the nodes are sorted in ascending order with respect to the priorities $p_S(v)$, $v \in V$. The while loop is terminated when the total time t has exceeded t_{max} or the optimal solution has been found. In the former case, the algorithm returns the best feasible solution found.

4.1 Node Priorities

Here we explain the calculation of node priorities, used to determine the order of node release in Algorithm 1. By incrementally release nodes, the SUBTVNS algorithm searches in different neighborhoods of the current solution. The proposed strategy to determine the priority of a node in the list, takes into account two properties of that node with respect to the current solution: **proximity** and **coverage**. We believe that a node should be a part of the improved solution if it is close to the current solution and covers as many groups as possible. We introduce the proximity of node v to the current solution S as follows:

$$\mathsf{prox}(v, S) = \frac{d(v, S)}{\mathsf{diam}(G)}, \tag{9}$$

where $d(v, S)$ is the length of the shortest path from v to the current solution S (i.e., length of the shortest path from v to the closest node in S), and $\mathsf{diam}(G)$ is a diameter of the graph G. Diameter of G is used as a normalizing term.

The value for coverage of node v with respect to solution S, denoted by $\mathsf{cov}(v, S)$ depends on the fact whether v is in S or not. If $v \in S$, then the coverage is the number of groups exclusively covered by this node (i.e., it does not count the groups covered by other nodes in the solution S) divided by the total number of groups k. Otherwise, if $v \notin S$, then the coverage is calculated as the total number of groups covered by v divided by k. Finally, the priority of v with respect to S is calculated as follows:

$$p_S(v) = \mathsf{prox}(v, S) - \mathsf{cov}(v, S). \tag{10}$$

From the Eq. (10) it follows that $p_S(v)$ can take values from the interval $[-1, 1]$. The node v has a higher priority if it is closed to S and covers a large number of groups. The procedure to calculate priority vector p_S is summarized in Algorithm 2.

Algorithm 2. Node priorities calculation

1: **procedure** NODEPRIORITIES(G, $\mathcal{G}$, D, S, δ)
2:     **for** $v \in V$ **do**
3:         $d(v, S) \leftarrow \min_{u \in S} D[v, u]$
4:         $\mathsf{prox}(v, S) \leftarrow \frac{d(v,S)}{\delta}$
5:         **if** $v \in S$ **then**
6:             $\mathsf{cov}(v, S) \leftarrow \frac{|\{g \in \mathcal{G}: v \in g \wedge \exists u \in S, u \in g\}|}{k}$
7:         **else**
8:             $\mathsf{cov}(v, S) \leftarrow \frac{|\{g \in \mathcal{G}: v \in g\}|}{k}$
9:         $p_S(v) \leftarrow \mathsf{prox}(v, S) - \mathsf{cov}(v, S)$
10:     **return** p_S

5 Experiments

5.1 Implementation

All methods are implemented with the programming language C++ and Gurobi v9.5.0 C++ API. The experiments are carried out on a workstation with an Intel Xeon E7-4850 v3 processor at a standard frequency of 2.2 GHz, 8 GB RAM and the operating system Ubuntu 16.04.4 LTS.

The formulation used has an exponential number of constraints, and it is impractical to generate the entire pool of constraints in advance. Instead, we have implemented a constraint generation procedure to find the set W that violates one of the constraints in (5) and (6). This procedure is implemented using breath-first search traversal to find a cycle with minimum length. At each node of the branch-and-cut procedure, the values of the variables x_e, $e \in E$ and y_v, $v \in V$ induce a subgraph H of G. An inequality in (5) or (6) is violated if and only if there is a $W \subseteq V(H)$ such that $E(W)$ contains a cycle in H. From this characterization it follows that it is sufficient to find a cycle of minimal length and construct a violated inequality in (5) and (6) by using the set of nodes and edges in this cycle. If there are no cycles, then there are no violated inequalities. The procedure for calculating the minimum length cycle at the branch-and-cut node can be efficiently implemented by inheriting the C++ Gurobi API class GRBCallback. We have implemented this procedure in such a way that violating inequalities are included in the pool of lazy constraints.

5.2 Instances

The procedure for generating instances is described in [22] and is presented in the study [7]. Compared to [22], we considered a larger set of instances by extending the range of the number of nodes (n). We generated instances with 100, 200, 300, 400, 500, 600, 700, 800, 900, 1000 and 2000 nodes, where the parameters m and k were determined using the same procedure as in [22]. We generated a total of 990 instances with three types of distances between nodes. In all experiments, t_{max} is set to 600 s, $t_s = 60$ s, and $\alpha = 0.1$.

5.3   Results

In Table 1 we have indicated the number of instances in which an optimal solution was found in the column four, the number of instances in which a feasible solution was found (whose optimality is not proven) in the column 5, and the number of instances in which the method found no feasible solutions at all (in the last column). SUBTVNS found the largest number of optimal solutions for all instance types compared to SUBTINIT and SUBT. Approximately 90% out of 330 instances for each instance type were solved optimally. SUBT had poor performance because it needed to spend a considerable amount of time finding the initial feasible solution that was not given. In some cases, SUBT was not even able to find the first feasible solution.

Table 1. Method - three methods considered: SUBTVNS - our matheuristic method, SUBTINIT - branch-and-cut algorithm applied to the subtour formulation with a given initial solution, SUBT - branch-and-cut algorithm applied to the subtour formulation without an initial solution, **Instance type** - one of three instance groups based on the type of distance between nodes, **Total** - number of instances in a group, **# Optimal** - number of instances where an optimal solution was found, **# Feasible** - number of instances where a feasible solution was found, **# Feasible not found** - number of instances where no feasible solution was found.

Method	Instance type	Total	# Optimal	# Feasible	# Feasible not Found
SUBTVNS	EUCLID	330	294 (89.09%)	36 (10.91%)	0 (0.00%)
	GRID	330	296 (89.70%)	34 (10.30%)	0 (0.00%)
	RANDOM	330	302 (91.52%)	28 (8.48%)	0 (0.00%)
SUBTINIT	EUCLID	330	225 (68.18%)	105 (31.82%)	0 (0.00%)
	GRID	330	234 (70.91%)	96 (29.09%)	0 (0.00%)
	RANDOM	330	266 (80.61%)	64 (19.39%)	0 (0.00%)
SUBT	EUCLID	330	209 (63.33%)	111 (33.64%)	10 (3.03%)
	GRID	330	213 (64.55%)	106 (32.12%)	11 (3.33%)
	RANDOM	330	257 (77.88%)	70 (21.21%)	3 (0.91%)

As SUBTVNS outperforms SUBTINIT and SUBT in terms of the number of instances in which an optimal solution was found, in Table 2 we have included only instances in which both SUBTVNS and SUBTINIT found feasible solutions that are not proved to be optimal. As can be seen in Table 1, out of the three instance types, EUCLID instances had the lowest number of optimal solutions found. However, looking at the improvement of the objective function value for non-optimal solutions (Table 2), the improvement is achieved in the majority of EUCLID instances, i.e., in 87.5% of the considered instances. In addition, the average relative improvement of 3.89% is the largest also for EUCLID instances. This means that SUBTVNS reduced the objective function value of the feasible solutions found by 3.89% on average (considering only the instances where the solution was improved).

Table 2. Instance type - one of three instance groups based on the type of distance between the nodes, **# of improvements** - total number of instances in which SUBTVNS found a feasible solution with objective function values smaller than SUBTINIT, **Total** - total number of instances in which both SUBTVNS and SUBTINIT found a feasible solution, **% of improvements** - *# of improvements* in percent of *Total*, **Average relative improvement %** - average ratio between the difference of the objective function values found by SUBTINIT and SUBTVNS for all instances in which SUBTVNS found a better solution.

Instance type	# of improvements	Total	% of improvements	Average relative improvement (%)
GRID	12	16	75.00 %	2.71 %
RANDOM	9	16	56.25 %	1.98 %
EUCLID	14	16	87.50 %	3.89 %

If both SUBTVNS and SUBTINIT have found optimal solutions, it makes sense to compare the total time it took to find these optimal solutions. Based on previous experimental studies in [22], it has been shown that the number of groups k is a parameter that significantly affects the problem hardness. Even basic theoretical work on the Steiner tree problem [6] can be easily applied to the group Steiner tree problem and it leads to the conclusion that the group Steiner tree problem becomes solvable in polynomial time as soon as the number of groups is fixed to a constant. Figure 1 shows the dependence of speedup factor on the number of groups k. The speedup factor is calculated as the ratio of the running times of SUBTINIT and SUBTVNS for the instances in which both algorithms provide optimal solutions. The average speedup factors calculated for all instances with the same number of groups k are presented and compared in Fig. 1. It can be seen that, on average, the running time on EUCLID instances shows the greatest improvement as the value of k increases.

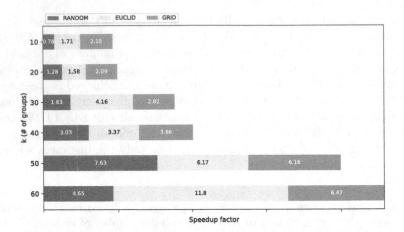

Fig. 1. Dependence of speed-up factor on the number of groups. Speed-up factor is the ratio between SUBTINIT and SUBTVNS. The numbers in the horizontal bars indicate the average speedup of SUBTVNS compared to SUBTINIT, calculated for instances of certain types for which both methods have found optimal solutions. The Instance types, based on the type of distances, are indicated by different colors. (Color figure online)

6   Conclusion

The biggest challenge in solving GST with approximation algorithms and heuristics is dealing with instances containing a large number of groups. This is the critical parameter for the GST problem, which significantly affects both the theoretical and practical hardness of this problem. In this work, we have proposed a VNS-based matheuristic method (SUBTVNS) that is able to provide solutions of higher quality than exact Gurobi solver for the same mathematical programming formulation and within the same global time limit. The most interesting result was the increase of the speed-up factor with increasing number of groups for all instance types considered. The remaining challenges for future work go in two directions. First, it is interesting to investigate and possibly improve the asymptotic behavior of the proposed SUBTVNS when solving even larger instances. Along this path, future work will involve the implementation of metaheuristics approach where the size of the VNS-based neighborhood is controlled by adding constraints that limit the number of nodes added to and/or removed from the current feasible solution instead of specifying nodes themselves. Second, it is also interesting to extend this approach to more general versions of the GST problem, e.g., node-weighted and prize-collecting versions. Because the node-weighted version has important practical applications, it will be possible to test our approach on real-world instances.

References

1. Ahmed, M.B., Hvattum, L.M., Agra, A.: The effect of different mathematical formulations on a matheuristic algorithm for the production routing problem. Comput. Oper. Res. **155**, 106232:1–106232:19 (2023). https://doi.org/10.1016/j.cor.2023.106232
2. Boschetti, M.A., Letchford, A.N., Maniezzo, V.: Matheuristics: survey and synthesis. Int. Trans. Oper. Res. **30**(6), 2840–2866 (2023). https://doi.org/10.1111/itor.13301
3. Boschetti, M.A., Maniezzo, V.: Matheuristics: using mathematics for heuristic design. 4OR **20**(2), 173–208 (2022). https://doi.org/10.1007/s10288-022-00510-8
4. Coffman, J., Weaver, A.C.: An empirical performance evaluation of relational keyword search techniques. IEEE Trans. Knowl. Data Eng. **26**(1), 30–42 (2014). https://doi.org/10.1109/TKDE.2012.228
5. Ding, B., Yu, J.X., Wang, S., Qin, L., Zhang, X., Lin, X.: Finding top-k min-cost connected trees in databases. In: 2007 IEEE 23rd International Conference on Data Engineering, pp. 836–845 (2007). https://doi.org/10.1109/ICDE.2007.367929
6. Dreyfus, S.E., Wagner, R.A.: The Steiner problem in graphs. Networks **1**(3), 195–207 (1971). https://doi.org/10.1002/net.3230010302
7. Duin, C.W., Volgenant, A., Voß, S.: Solving group Steiner problems as Steiner problems. Eur. J. Oper. Res. **154**(1), 323–329 (2004). https://doi.org/10.1016/S0377-2217(02)00707-5
8. Ferreira, C.E., de Oliveira Filho, F.M.: Some formulations for the group Steiner tree problem. Discrete Appl. Math. **154**(13), 1877–1884 (2006). https://doi.org/10.1016/j.dam.2006.03.028
9. Garg, N., Konjevod, G., Ravi, R.: A polylogarithmic approximation algorithm for the group Steiner tree problem. J. Algorithms **37**(1), 66–84 (2000). https://doi.org/10.1006/jagm.2000.1096
10. Goemans, M.X.: The Steiner tree polytope and related polyhedra. Math. Program. **63**(1), 157–182 (1994). https://doi.org/10.1007/BF01582064
11. Gurobi Optimization, LLC: Gurobi Optimizer Reference Manual (2023). https://www.gurobi.com
12. Hansen, P., Mladenović, N., Brimberg, J., Pérez, J.A.M.: Variable neighborhood search. In: Gendreau, M., Potvin, J.-Y. (eds.) Handbook of Metaheuristics. ISORMS, vol. 272, pp. 57–97. Springer, Cham (2019). https://doi.org/10.1007/978-3-319-91086-4_3
13. Hansen, P., Mladenović, N., Todosijević, R., Hanafi, S.: Variable neighborhood search: basics and variants. EURO J. Comput. Optim. **5**(3), 423–454 (2017). https://doi.org/10.1007/s13675-016-0075-x
14. Hansen, P., Mladenović, N., Urošević, D.: Variable neighbourhood search and local branching. Comput. Oper. Res. **33**(10), 3034–3045 (2006). https://doi.org/10.1016/j.cor.2005.02.033
15. Ihler, E.: Bounds on the quality of approximate solutions to the group Steiner problem. In: Möhring, R.H. (ed.) WG 1990. LNCS, vol. 484, pp. 109–118. Springer, Heidelberg (1991). https://doi.org/10.1007/3-540-53832-1_36
16. Jelić, S., Ševerdija, D.: Government formation problem. CEJOR **26**(3), 659–672 (2018). https://doi.org/10.1007/s10100-017-0505-8
17. Jovanović, P., Davidović, T., Lazić, J., Mitrović Minić, S.: The variable intensity neighborhood search for 0-1 MIP. In: Proceedings of the 42nd Symposium on Operations Research, SYM-OP-IS 2015, Srebrno jezero, Serbia, pp. 229–232 (2015)

18. Jovanovic, R., Voß, S.: Matheuristic fixed set search applied to the multidimensional knapsack problem and the knapsack problem with forfeit sets. OR Spectrum (2024, in press). https://doi.org/10.1007/s00291-024-00746-2
19. Koch, T., Martin, A.: Solving Steiner tree problems in graphs to optimality. Networks **32**(3), 207–232 (1998). https://doi.org/10.1002/(SICI)1097-0037(199810)32:3⟨207::AID-NET5⟩3.0.CO;2-O
20. Lappas, T., Liu, K., Terzi, E.: Finding a team of experts in social networks. In: Proceedings of the 15th ACM SIGKDD International Conference on Knowledge Discovery and Data Mining, KDD 2009, pp. 467–476. Association for Computing Machinery, New York (2009). https://doi.org/10.1145/1557019.1557074
21. Lazić, J., Hanafi, S., Mladenović, N., Urošević, D.: Variable neighbourhood decomposition search for 0–1 mixed integer programs. Comput. Oper. Res. **37**(6), 1055–1067 (2010). https://doi.org/10.1016/j.cor.2009.09.010
22. Matijević, L., Jelić, S., Davidović, T.: General variable neighborhood search approach to group Steiner tree problem. Optim. Lett. **17**(9), 2087–2111 (2023). https://doi.org/10.1007/s11590-022-01904-7
23. Mladenović, N., Hansen, P.: Variable neighborhood search. Comput. Oper. Res. **24**(11), 1097–1100 (1997). https://doi.org/10.1016/S0305-0548(97)00031-2
24. Myung, Y.S.: A comparison of group Steiner tree formulations. J. Korean Inst. Ind. Eng. **37**(3), 191–197 (2011). https://doi.org/10.7232/JKIIE.2011.37.3.191
25. Nguyen, T.D.: A fast approximation algorithm for solving the complete set packing problem. Eur. J. Oper. Res. **237**(1), 62–70 (2014). https://doi.org/10.1016/j.ejor.2014.01.024
26. Reich, G., Widmayer, P.: Beyond Steiner's problem: a VLSI oriented generalization. In: Nagl, M. (ed.) WG 1989. LNCS, vol. 411, pp. 196–210. Springer, Heidelberg (1990). https://doi.org/10.1007/3-540-52292-1_14
27. Salazar, J.J.: A note on the generalized Steiner tree polytope. Discrete Appl. Math. **100**(1), 137–144 (2000). https://doi.org/10.1016/S0166-218X(99)00200-0
28. Sun, Y., Xiao, X., Cui, B., Halgamuge, S., Lappas, T., Luo, J.: Finding group Steiner trees in graphs with both vertex and edge weights. Proc. VLDB Endow. **14**(7), 1137–1149 (2021). https://doi.org/10.14778/3450980.3450982
29. Voss, S., Stutzle, T., Maniezzo, V.: MATHEURISTICS: Hybridizing Metaheuristics and Mathematical Programming. Springer, New York (2009). https://doi.org/10.1007/978-1-4419-1306-7
30. Zachariasen, M., Rohe, A.: Rectilinear group Steiner trees and applications in VLSI design. Math. Program. **94**(2), 407–433 (2003). https://doi.org/10.1007/s10107-002-0326-x

A Basic Variable Neighborhood Search for the Planar Obnoxious Facility Location Problem

Sergio Salazar(ID), Abraham Duarte(ID), and J. Manuel Colmenar(✉)(ID)

Universidad Rey Juan Carlos, Calle Tulipán s/n, Móstoles, Madrid, Spain
{sergio.salazar,abraham.duarte,josemanuel.colmenar}@urjc.es

Abstract. Obnoxious facility location problems are devoted to choose the best location for a given set of facilities considering that, despite they should not be close to population communities, their service is needed, like the case of airports, paper factories or nuclear plants. In this paper we study the planar multiple obnoxious facility location problem. Our approach is based on a first discretization of the instance where a Basic Variable Neighborhood Search algorithm is applied. Our results improve the state of the art spending less than a third of the execution time of the second best algorithm.

Keywords: Basic Variable Neighborhood Search · Obnoxious Facility Location Problem · Greedy Randomized Adaptive Search Procedure

1 Introduction

The family of problems devoted to facility location is wide, and their applications range from the different variants of the distribution problem [2] to the facility location considering capacity planning for pandemics [7]. In this family we can also find problems where the facilities to be located generate a negative impact around them, maybe due to noise, pollution or associated heavy traffic. These obnoxious facilities like landfills, smelly factories or airports are needed by the society, since they provide service to the population, but they should be located not too close to residential neighborhoods, according to different objective functions and requirements. These are the obnoxious facility location problems [1].

One of the problems that belongs to this family is the planar multiple obnoxious facility location problem which, as defined in [3], consists in locating a given number of facilities in the plane, considering the continuous space. The objective in this problem is to maximize the shortest distance between facilities and communities, taking into account that the minimum distance between two facilities must be greater or equal to a given value D.

This work has been partially supported by the Spanish Ministerio de Ciencia e Innovación (MCIN/AEI/10.13039/501100011033) under grant refs. RED2022-134480-T, PID2021-126605NB-I00 and by ERDF A way of making Europe; and Generalitat Valenciana with grant ref. CIAICO/2021/224.

M. Sevaux et al. (Eds.): MIC 2024, LNCS 14753, pp. 359–364, 2024.
https://doi.org/10.1007/978-3-031-62912-9_33

More formally, let S be a solution formed by p points a_i in the continuous space limited by the unit square, $a_i \in [0,1] \times [0,1]$, let C be the set of n communities, and let $d(a_i, a_j)$ be the distance between points a_i and a_j. The objective function is defined in Eq. (1), where D is the minimum separation distance, which, as in the case of p and C, are given as data of the instance.

$$\max \mathcal{F}(S) \quad = \min_{\substack{a_i \in S \\ x_j \in C}} d(a_i, x_j)$$
$$\text{s.t.} \quad d(a_i, a_j) \geq D \quad \forall a_i, a_j \in S, i \neq j \tag{1}$$

This problem has been previously studied, proposing a mathematical model and some heuristics based on Voronoi diagrams [6], representing the state of the art of the problem previous to this paper. In this work, we propose a Basic Variable Neighborhood Search (BVNS) approach which, supported by a first step which performs a discretization of the instance, is able to improve the previous results spending less than a third of the execution time of the previous proposals.

2 Algorithmic Proposal

In this work, we propose an algorithmic approach based on three main steps: an initial discretization of the instance, moving the problem from the continuous to the discrete domain, a simple process to create new solutions, and a BVNS process which improves this initial solution. Next, these phases are described.

2.1 Discretization of the Instance

As stated before, this problem is defined in the continuous space. Therefore, in order to apply our metaheuristic proposal, we have first defined a discretization of the instance. This process will be the first step of the algorithm, and it will determine the candidate positions where the facilities can be located.

To this aim, we follow a strategy already studied in the literature, where the Voronoi points are calculated for the given communities [3]. However, the Voronoi edges still belong to the continuous domain. Therefore, we propose to create a grid of points belonging to the Voronoi edges, separated by a certain distance β. This way, the number of initial candidate points is greatly reduced by both the Voronoi edges and the separation distance. Figure 1 shows an example with $n = 15$ and $\beta = 0.1$. As seen, the number of candidate locations, shown as blue points, is very small in relation to the continuous space.

2.2 Construction of Solutions

Any Variable Neighborhood Search proposal requires an initial solution to begin the exploration [8]. In this work, we decided to build initial solutions following a Greedy Randomized Adaptive Search Procedure (GRASP) strategy [4]. This process uses as greedy function the distance from each point to the closest community. The GRASP selection, using $\alpha = 0.5$, is repeated until p points are selected. This initial solution is sent to the BVNS process.

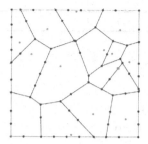

Fig. 1. Example of Voronoi discretization with $n = 15$ and $\beta = 0.1$. (Color figure online)

2.3   Basic Variable Neighborhood Search

Since this is a first approach to the problem using the VNS metaheuristic, we have selected one of the simplest implementations, which is the Basic VNS [5].

Algorithm 1 shows the pseudo-code of our method, which receives an initial solution S_0 provided by the constructive method, a set of candidate points as a result of the discretization of the instance, I_D, and a maximum value for the k parameter, k_{max}. The algorithm begins setting the value of k to 1 and starting the main loop. A new solution S' is generated in step 3 by the *Shake* process. In this case, we implemented a perturbation which removes the worst point, that is, the one with the shortest distance to a community and, if k is greater than 1, randomly removes $k - 1$ different points. The improving phase begins in step 5, where a *LocalSearch* process obtains a new solution S'' from S' in step 6. This method is a typical best improvement local search based on an exchange move of a selected point and a non-selected point in I_D. The rest of the code is the customary BVNS implementation of the acceptance of better solutions and management of k depending on the improvements. Finally, the best solution is returned in step 20.

3   Computational Experiments

Following the previous work, we have dealt with two sets of communities, with sizes $n = 100$ and $n = 1000$ [6]. Considering two different distances, $D = 1/\sqrt{p}$ and $D = 1/\sqrt{2p}$, and 19 values for p, from 2 to 20, a total number of 76 instances are studied. For each instance, we compare the results from our BVNS proposal with the three algorithms described in [6]. These algorithms are denoted in the original work as *Algorithm 1*, *Algorithm 2* and *Algorithm 3*. We denote them as A1, A2 and A3 in our results. For each instance, the final BVNS algorithm has been executed 10 times for the $n = 100$ instances and 100 times for $n = 1000$. The accumulated results are next reported.

Table 1 shows the average values obtained for the objective function for each value of n and D. As seen, the average values reached by BVNS are the best in 3 out of the 4 groups of instances, while algorithm A2 reaches the best value in the resulting one. The best values are highlighted in bold font. Therefore, the

---

Algorithm 1: $BVNS(S_0, I_D, k_{max})$

1: $k \leftarrow 1$
2: **while** $k < k_{max}$ **do**
3:   $S' \leftarrow Shake(I_D, k, S)$
4:   $improve \leftarrow true$
5:   **while** $improve$ **do**
6:     $S'' \leftarrow LocalSearch(S', I_D)$
7:     **if** $F(S'') > F(S')$ **then**
8:       $S' \leftarrow S''$
9:     **else**
10:       $improve \leftarrow false$
11:     **end if**
12:   **end while**
13:   **if** $F(S') \geq F(S)$ **then**
14:     $S \leftarrow S'$
15:     $k \leftarrow 1$
16:   **else**
17:     $k \leftarrow k + 1$
18:   **end if**
19: **end while**
20: **return** S

---

average behavior of the BVNS algorithm is better than the previous proposals, followed by algorithm A3. Notice that, despite the improvement of the values is achieved after the third decimal figure, this is not an unusual result since the space defined by the instance is the unit square.

Table 1. Average objective function values for the algorithms under study.

n	D	BVNS	A1	A2	A3
100	$1/\sqrt{p}$	**0.101978**	0.099460	0.101149	0.101475
	$1/\sqrt{2p}$	**0.111155**	0.110095	0.110907	0.110937
1000	$1/\sqrt{p}$	0.040822	0.040862	**0.040989**	0.041000
	$1/\sqrt{2p}$	**0.044043**	0.043939	0.044004	0.044010
	Average	**0.074499**	0.073589	0.074262	0.074355

Table 2 shows the number of instances where each algorithm reached the best value in the experiment. As seen, BVNS reaches 57 out of the 76 instances, being dominant in the instances with $n = 100$. Notice that A3 obtains 56 best values, reaching the highest results in the larger instances where $n = 1000$.

Table 3 shows the average deviation for each group of instances in relation to the best result. In this case, our BVNS proposal reaches the best values in 3 out of the 4 groups of instances. These results prove that our algorithm is the most stable, followed again by A3.

Table 2. Number of instances each studied algorithm reached the best solution.

n	D	BVNS	A1	A2	A3
100	$1/\sqrt{p}$	**14**	8	11	12
	$1/\sqrt{2p}$	**18**	11	12	12
1000	$1/\sqrt{p}$	11	13	**17**	**17**
	$1/\sqrt{2p}$	14	14	13	**15**
	Sum	**57**	46	53	56

Table 3. Average deviation values for the algorithms under study.

n	D	BVNS	A1	A2	A3
100	$1/\sqrt{p}$	**0.008238**	0.036808	0.019194	0.016187
	$1/\sqrt{2p}$	**0.000169**	0.010142	0.002681	0.002391
1000	$1/\sqrt{p}$	0.006140	0.005024	0.001682	**0.001326**
	$1/\sqrt{2p}$	**0.000702**	0.002823	0.001681	0.001527
	Average	**0.003812**	0.013699	0.006309	0.005358

Finally, Table 4 shows the average execution time, in seconds, for each algorithm. Notice that the execution time of BVNS is calculated as the average of the sum of the execution time of each experiment run. The execution times for algorithms A1, A2 and A3 were obtained from the previous paper [6]. As seen, A1 obtains the best values, showing impressive execution times, although the algorithm is not getting the best results in the previously studied metrics. Our BVNS proposal is the second best performer, spending less than a third of the time compared to the second best performer in terms of objective function and number of best values, which is A3.

It is worth noticing that our BVNS proposal was run on a laptop computer with an Intel i7 7500 processor with 16 Gb of RAM on a Manjaro Linux operating system. According to the authors, the executions of A1, A2 and A3 algorithms were run on a PowerEdge R720 server machine provided with an Intel E5 2650 processor with 128 Gb of RAM, which is a much more powerful environment than our laptop computer.

Table 4. Average execution times (in seconds) for the algorithms under study.

n	D	BVNS	A1	A2	A3
100	$1/\sqrt{p}$	0.571053	**0.000155**	0.588421	0.925263
	$1/\sqrt{2p}$	0.625158	**0.000088**	1.860000	2.594737
1000	$1/\sqrt{p}$	6.126316	**0.003648**	1.412105	2.197895
	$1/\sqrt{2p}$	6.123316	**0.001288**	27.015263	34.679474
	Average	3.361461	**0.001295**	7.718947	10.099342

As seen, our BVNS proposal is able to obtain the best average objective function value, the highest number of best results and the lowest deviation, spending a very short execution time. Considering that this is an initial work where the BVNS proposal is simple, we are confident on obtaining better results in the future using more complex VNS variants.

4 Conclusions and Future Work

In this work we propose a Basic Variable Neighborhood Search approach for the planar multiple obnoxious location problem. Since the algorithm is based on discrete solutions, we propose a discretization of the instance following a grid approach guided by Voronoi points. From this point on, new solutions are constructed using a Greedy Randomized Adaptive Search Procedure, which are improved by means of the BVNS using a shake method which randomly removes selected points, and a best improvement local search based on an exchange move.

Our results show that the BVNS proposal is able to get a better average value for the objective function than the previous method, obtaining the best result in 57 out of the 76 studied instances. In addition, our proposal reaches these figures being the most stable proposal, and spending a third of the computation time of the second best performer in terms of quality.

Currently, we are working on different local search strategies that will take advantage of the continuous space, being more flexible than the proposed discretization. In addition, some other VNS variants are studied.

References

1. Church, R.L., Drezner, Z.: Review of obnoxious facilities location problems. Comput. Oper. Res. **138**, 105468 (2022)
2. Davoodi, M., Rezaei, J.: Bi-sided facility location problems: an efficient algorithm for k-centre, k-median, and travelling salesman problems. Int. J. Syst. Sci. Oper. Logist. **10**(1), 2235814 (2023)
3. Drezner, Z., Kalczynski, P., Salhi, S.: The planar multiple obnoxious facilities location problem: a Voronoi based heuristic. Omega **87**, 105–116 (2019)
4. Feo, T.A., Resende, M.G.: Greedy randomized adaptive search procedures. J. Glob. Optim. **6**, 109–133 (1995)
5. Hansen, P., Mladenović, N., Todosijević, R., Hanafi, S.: Variable neighborhood search: basics and variants. EURO J. Comput. Optim. **5**(3), 423–454 (2017)
6. Kalczynski, P., Drezner, Z.: Extremely non-convex optimization problems: the case of the multiple obnoxious facilities location. Optim. Lett. **16**, 1153–1166 (2022)
7. Liu, K., Liu, C., Xiang, X., Tian, Z.: Testing facility location and dynamic capacity planning for pandemics with demand uncertainty. Eur. J. Oper. Res. **304**(1), 150–168 (2023)
8. Mladenović, N., Hansen, P.: Variable neighborhood search. Comput. Oper. Res. **24**(11), 1097–1100 (1997)

Temporal Action Analysis in Metaheuristics: A Machine Learning Approach

Panagiotis Kalatzantonakis[ID], Angelo Sifaleras[✉][ID], and Nikolaos Samaras[ID]

Department of Applied Informatics, School of Information Sciences,
University of Macedonia, 156 Egnatia Str., 54636 Thessaloniki, Greece
pkalatzantonakis@uom.edu.gr, {sifalera,samaras}@uom.gr

Abstract. This study explores the use of Autoregressive Integrated Moving Average (ARIMA) and Long Short-Term Memory (LSTM) machine learning models in metaheuristic algorithms, with a focus on a modified General Variable Neighborhood Search (GVNS) for the Capacitated Vehicle Routing Problem (CVRP). We analyze the historical chain of actions in GVNS to demonstrate the predictive potential of these models for guiding future heuristic applications or parameter settings in metaheuristics such as Genetic Algorithms (GA) or Simulated Annealing (SA). This "optimizing the optimizer" approach reveals that, the history of actions in metaheuristics provides valuable insights for predicting and enhancing heuristic selections. Our preliminary findings suggest that machine learning models, using historical data, offer a pathway to more intelligent and data-driven optimization strategies in complex scenarios, marking a significant advancement in the field of combinatorial optimization.

Keywords: Intelligent Heuristic Decision-Making · Data-Driven Metaheuristic Strategies · Machine Learning Enhanced Combinatorial Optimization · Offline Metaheuristic Algorithm Configuration

1 Introduction

1.1 Metaheuristics in Combinatorial Optimization

Metaheuristic algorithms have evolved significantly to address complex and NP-hard challenges in combinatorial optimization [12,13], such as the CVRP [4]. Historically, these algorithms have evolved from simple solution-seeking methods to sophisticated adaptive frameworks capable of intelligently navigating complex solution spaces. This evolution reflects a continuous effort to enhance efficiency and effectiveness in finding near-optimal solutions, especially in computationally demanding scenarios.

M. Sevaux et al. (Eds.): MIC 2024, LNCS 14753, pp. 365–370, 2024.
https://doi.org/10.1007/978-3-031-62912-9_34

1.2 Machine Learning Integration in Metaheuristics

The integration of machine learning into metaheuristics marks the latest advancement in this field, representing a significant leap in computational intelligence. Building upon historical progress, our research incorporates ARIMA [5] and LSTM models [7] into the GVNS for the CVRP, aiming to capture and leverage temporal dynamics in optimization processes. This integration is not just an innovation, but a response to the growing need for more precise predictive capabilities in dynamic environments. Based on recent studies in parallel execution [1,8,9], learning-based neighborhood search [10,14], and large neighborhood search adaptations [6], our approach seeks to harness the potential of machine learning to further refine and improve metaheuristic strategies.

2 Methodological and Experimental Setup

In this section, we outline our comprehensive methodological framework, which begins with the innovative reconfiguration of the GVNS [2,11] for the CVRP. This approach is crucial for generating a robust dataset, essential for the subsequent training and optimization of ARIMA and LSTM models.

2.1 GVNS-Driven Data Collection and Analysis

The proposed modification of the GVNS metaheuristic consists of a different neighborhood selection step, and it is tailored towards generating unbiased data across CVRP instances. Through multithreaded data collection and extensive preprocessing, including normalization and structuring, we prepare the data set for pattern analysis using ARIMA and LSTM models. This strategic approach seeks to evolve traditional metaheuristic algorithms into intelligent, adaptive systems.

Data collection during GVNS iterations involves tracking each heuristic's application and outcome, quantified as binary values (success or failure) and continuous values (degree of solution improvement). This detailed data collection is crucial for a robust analysis. The data set is then preprocessed for analysis. ARIMA models are used for regression analysis to identify linear trends, while LSTM networks address classification issues, adept in processing sequential data. This combination allows for a comprehensive analysis of patterns in decision making. We used CVRP instances from CVRPLib [3] (sets A, B, and X), which offer various complexities, to validate our methodology in a structured environment.

2.2 Model and Parameter Optimization

In our study, both ARIMA and LSTM models underwent meticulous optimization processes to enhance their predictive accuracy for the GVNS algorithm.

ARIMA's role was to forecast the "reward" value, with our analysis revealing some seasonality potentially influenced by GVNS's cyclic phases. The Augmented Dickey-Fuller and KPSS tests, combined with Fourier Transform and Seasonal Decomposition, confirmed the time series' stationarity, leading us to favor ARIMA over SARIMA. We explored a range of parameters, evaluated the models on Akaike Information Criterion (AIC) and Mean Squared Error (MSE), and settled on the ARIMA (5, 1, 5) model for its optimal balance of AIC and high PRAUC, indicating its effectiveness in predicting heuristic improvements. In optimizing the ARIMA model, specific parameters are pivotal: "p" representing the order of autoregression, "d" the degree of differencing, and "q" the moving average window, that together define the model's structure. The performance of ARIMA (5, 1, 5) was evaluated using the Mean Squared Error (MSE) and Akaike Information Criterion (AIC), with lower values in both indicating better model fit. AIC was particularly crucial for comparing the quality of different models. Additionally, the Precision-Recall Area Under Curve (PRAUC) metric was utilized to assess the binary classification effectiveness of ARIMA, an important aspect given the imbalanced nature of our dataset.

Hyperparameter tuning of the LSTM model was conducted using the Hyperband method, targeting key parameters such as LSTM units, dropout rate, and learning rate, which unveiled a preference for a BiLSTM structure to better capture temporal dependencies. The optimization involved systematic exploration of critical hyperparameters. LSTM units affect the model's complexity and its ability to discern data patterns, with higher units offering greater complexity at the cost of computational resources. Dropout rate mitigates overfitting by omitting units during training, while the learning rate is vital for effective model training, avoiding minima overshoots. The choice of loss function (MSE, MAE, Binary Cross-Entropy) influences error quantification, and activation functions (sigmoid, ReLU, tanh) affect data signal processing, crucial for learning. BiLSTM's bidirectional approach improves predictive accuracy by utilizing past and future data. Batch size and epochs set the training sample size and cycles, and the optimizer (SGD, RMSprop, Adam) impacts learning speed and efficiency. The attention mechanism further refines the model by concentrating on particular input sequence segments, boosting performance on complex time-series tasks.

Table 1 details the models that perform the best. Also, the final hyperparameter configuration for the BiLSTM model, as detailed in Table 2, was strategically chosen to strike a balance between computational resources and predictive accuracy. This resulted in an optimized BiLSTM model. Both the ARIMA and LSTM models were meticulously fine-tuned to complement each other, thus providing comprehensive predictive insights within the GVNS framework.

Table 1. Top 3 ARIMA Models

Rank	p	d	q	AIC	MSE
Top 3 Models Based on AIC					
1	5	0	5	470.7783	2.0643
2	5	1	5	471.7483	1.8388
3	5	1	6	472.6345	1.8477
Top 3 MSE-based models					
1	3	1	6	495.4785	1.8168
2	5	1	2	487.4079	1.8381
3	6	1	6	475.7171	1.8388

Table 2. LSTM Model Hyperparameters

Hyperparameter	Range	Best Value
LSTM Units	32 to 512 (step: 32)	256
Dropout Rate	0.0 to 0.5 (step: 0.1)	0.1
Learning Rate	1e-4 to 1e-2 (sampling="log")	0.0079
Loss Functions	MSE, MAE, Binary Cross-Entropy	Mean Squared Error
Activation Functions	sigmoid, relu, tanh	tanh
Bidirectional setting	True/False	True
Batch Size	32 to 512	256
Epochs	10 to 100	100
Optimizers	SGD, RMSprop, Adam	Adam
Attention Mechanism	True/False	True

3 Results and Analysis

Our study showcases the potential of machine learning, particularly ARIMA and LSTM, in interpreting the sequence of actions in metaheuristic algorithms such as GVNS, GA and SA. By analyzing historical data from heuristic applications, we demonstrate how these models can predict and influence future heuristic choices, thus optimizing the decision-making process within these algorithms.

The selection of ARIMA and LSTM models in our study illustrates the complexity of decision-making in forecasting actions. ARIMA effectively predicts continuous outcomes such as "reward", providing linear insights, while LSTM excels in binary classification, crucial for different decision-making scenarios. The ARIMA(5, 0, 5) model, with its high AIC, accurately predicts "reward" values. This suggests that maintaining a focus on recent historical actions, up to five steps back, could be crucial to accurately forecasting outcomes in metaheuristic processes. Despite its limitations in accuracy and PRAUC, its ability to capture short-term historical trends is notable.

Conversely, the BiLSTM model significantly surpasses ARIMA in both accuracy and PRAUC, demonstrating its superior capability in binary classification and effective handling of sequential data. This highlights its potential as a robust tool for guiding heuristic decisions in metaheuristic algorithms. For a detailed comparison of the performance of the models, particularly highlighting their respective strengths in predictive accuracy, readers are encouraged to refer to Table 3, which presents a comprehensive overview of the performance metrics of the top models.

In conclusion, the findings of this study have far-reaching implications for the broader field of optimization and algorithm design. The successful integration of ARIMA and LSTM models within metaheuristics such as GVNS, GA, and SA demonstrates a promising path toward more intelligent, data-driven decision-making processes. This approach can be extended to other complex optimization

Table 3. Top Models Performance (ARIMA & LSTM)

Metric	Best Model based on AIC (5, 0, 5)	Best Model based on MSE (3, 1, 6)	LSTM Model
MSE	2.06431	1.81686	-
AIC	470.7783	495.4785	-
Accuracy	0.52658	0.53291	0.788956
PRAUC	0.57666	0.48160	0.673414

scenarios, opening up new avenues for research in algorithm efficiency and effectiveness. Future studies might explore the integration of different machine learning models or delve into real-time data adaptation, further advancing the field of combinatorial optimization. By leveraging historical data to inform heuristic choices, this research contributes to the ongoing evolution of metaheuristic algorithms, moving them toward more adaptive, predictive, and efficient frameworks.

4 Exploring the Future of Machine Learning in Metaheuristics

Our study, focused on analyzing data generated by a modified VNS approach for CVRP, indicates the potential of machine learning models like ARIMA and LSTM in enhancing metaheuristic algorithms. While our research is specific to VNS, the principle can be extended to other metaheuristics such as Genetic Algorithms and Simulated Annealing. For example, in GA, the history of genetic operations could be analyzed to predict their effectiveness, while in SA, the sequence of temperature adjustments and their outcomes could inform future adjustments. Integrating ML into these algorithms involves challenges such as adapting to unique operational frameworks, ensuring data quality, and managing computational demands. Future research should explore the broad application of machine learning models in various optimization contexts, integrate real-time data for adaptive strategies, and investigate advanced machine learning methodologies. This trajectory aims to significantly enhance the problem-solving capabilities of metaheuristics, leading to more optimized solutions in diverse and complex optimization scenarios.

5 Conclusions

This study represents a pioneering effort to blend machine learning with metaheuristics, specifically through the lens of time-series analysis. Integrating ARIMA and LSTM models into the VNS framework for the CVRP demonstrated the potential to significantly enhance the algorithm's decision-making process. Our preliminary findings pave the way for future research in this direction, promising more efficient and effective solutions in combinatorial optimization's vast and challenging domain. The generalization of this approach to other metaheuristics holds substantial promise, heralding a new era in the development of optimization strategies.

References

1. Abdelhafez, A., Luque, G., Alba, E.: Parallel execution combinatorics with meta-heuristics: comparative study. Swarm Evol. Comput. **55**, 100692 (2020)
2. Brimberg, J., Salhi, S., Todosijević, R., Urošević, D.: Variable neighborhood search: the power of change and simplicity. Comput. Oper. Res. **155**, 106221 (2023)
3. CVRPLIB - all instances. http://vrp.atd-lab.inf.puc-rio.br/index.php/en. Accessed 01 Feb 2024
4. Dantzig, G.B., Ramser, J.H.: The truck dispatching problem. Manag. Sci. **6**(1), 80–91 (1959)
5. Dickey, D.A., Fuller, W.A.: Distribution of the estimators for autoregressive time series with a unit root. J. Am. Stat. Assoc. **74**(366a), 427–431 (1979)
6. Hendel, G.: Adaptive large neighborhood search for mixed integer programming. Math. Program. Comput. **14**(2), 185–221 (2022)
7. Hochreiter, S., Schmidhuber, J.: Long short-term memory. Neural Comput. **9**(8), 1735–1780 (1997)
8. Kalatzantonakis, P., Sifaleras, A., Samaras, N.: Cooperative versus non-cooperative parallel variable neighborhood search strategies: a case study on the capacitated vehicle routing problem. J. Glob. Optim. **78**(2), 327–348 (2020)
9. Kalatzantonakis, P., Sifaleras, A., Samaras, N.: On a cooperative VNS parallelization strategy for the capacitated vehicle routing problem. In: Matsatsinis, N.F., Marinakis, Y., Pardalos, P. (eds.) LION 2019. LNCS, vol. 11968, pp. 231–239. Springer, Cham (2020). https://doi.org/10.1007/978-3-030-38629-0_19
10. Kalatzantonakis, P., Sifaleras, A., Samaras, N.: A reinforcement learning-variable neighborhood search method for the capacitated vehicle routing problem. Expert Syst. Appl. **213**, 118812 (2023)
11. Mladenović, N., Hansen, P.: Variable neighborhood search. Comput. Oper. Res. **24**(11), 1097–1100 (1997)
12. Monteiro, A.C.B., França, R.P., Arthur, R., Iano, Y.: The fundamentals and potential of heuristics and metaheuristics for multiobjective combinatorial optimization problems and solution methods. In: Multi-Objective Combinatorial Optimization Problems and Solution Methods, pp. 9–29. Academic Press (2022)
13. Talbi, E.G.: Machine learning into metaheuristics: a survey and taxonomy. ACM Comput. Surv. (CSUR) **54**(6), 1–32 (2021)
14. Thevenin, S., Zufferey, N.: Learning variable neighborhood search for a scheduling problem with time windows and rejections. Discrete Appl. Math. **261**, 344–353 (2019)

A Variable Neighborhood Search Approach for the S-labeling Problem

Marcos Robles[✉] , Sergio Cavero , and Eduardo G. Pardo

Universidad Rey Juan Carlos, Madrid, Spain
{marcos.robles,sergio.cavero,eduardo.pardo}@urjc.es

Abstract. The S-labeling problem is a graph layout problem that assigns numeric labels to the vertices of a graph. It aims to minimize the sum of the minimum numeric label assigned to each pair of adjacent vertices. In this preliminary work, we propose the use of the Variable Neighborhood Search (VNS) framework to test different Shake procedures and Local Search methods for the problem. We compare our VNS variants with the state-of-the-art Population-based Iterated Greedy algorithm on a set of benchmark instances. The results show that our VNS methods can obtain competitive solutions with a low deviation, but they are not able to improve the best-known values. We discuss the strengths and weaknesses of our proposal and suggest some future research directions. This work lays the groundwork for future research into the S-Labeling problem using Variable Neighborhood Search.

Keywords: Graph labeling · Variable Neighborhood Search · S-labeling

1 Introduction

The S-labeling problem is a Graph Layout Problem (GLP) [2] originally related to the problem of packaging (0,1)-matrices [8]. This problem has been studied from practical and theoretical perspectives. Particularly, several lower bounds and intrinsic properties of optimal labelings have been proposed [3]. Additionally, we can find in the literature exact solutions based on mixed integer programming [7], and approximate methods like the Population-based Iterated Greedy (PIG) algorithm proposed in [6], which is currently the state of the art of the problem.

Formally, given an undirected graph $G = (V, E)$ where V and E are the set of vertices and edges respectively, we define a labeling ϕ by assigning a unique label (i.e., a number) $l \in \mathbb{Z}$, s.t. $1 \leq l \leq |V|$, to each vertex. This relationship is done by the bijective function ϕ which, for a given vertex v, returns its corresponding label. Given embedding ϕ of a graph G, to evaluate the objective function of the solution, denoted as S-labeling number (SL), we compute for each edge e, the minimum label associated to the pair of vertices which are endpoints of e. Then we sum all the minimum labels previously computed. Formally, the objective function is defined as $SL(\phi, G) = \sum_{(u,v) \in E} \min\{\phi(u), \phi(v)\}$.

© The Author(s), under exclusive license to Springer Nature Switzerland AG 2024
M. Sevaux et al. (Eds.): MIC 2024, LNCS 14753, pp. 371–376, 2024.
https://doi.org/10.1007/978-3-031-62912-9_35

For instance, let us consider the graph $G_1(V_1, E_1)$ depicted in Fig. 1 with vertices $V_1 = \{A, B, C, D, E\}$ and edges $E_1 = \{(A, B), (A, C), (B, C), (C, D), (D, E)\}$. Additionally, let ϕ_1 be a labeling where $\phi_1(A) = 2$, $\phi_1(B) = 3$, $\phi_1(C) = 1$, $\phi_1(D) = 4$, and $\phi_1(E) = 5$. For this labeling, the S-labeling number is computed as follows:

$$SL(\phi_1, G_1) = \min(\phi_1(A), \phi_1(B)) + \min(\phi_1(A), \phi_1(C)) + \min(\phi_1(B), \phi_1(C))$$
$$+ \min(\phi_1(C), \phi_1(D)) + \min(\phi_1(D), \phi_1(E))$$
$$= \min(2, 3) + \min(2, 1) + \min(3, 1) + \min(1, 4) + \min(4, 5)$$
$$= 2 + 1 + 1 + 1 + 4 = 9$$

The goal of this problem is to identify the embedding ϕ^* that minimizes the S-labeling number for the given graph.

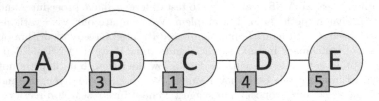

Fig. 1. Example of the labeling of a graph.

In our pursuit of a novel perspective on the S-labeling problem, we advocate the adoption of trajectory-based metaheuristics. Specifically, in this paper we propose multiple variants of Variable Neighborhood Search (VNS) [1,4]. To evaluate our proposal, we conduct a comparative analysis with respect to the best solutions found for the S-labeling in [6].

2   Algorithmic Proposal

VNS is a metaheuristic approach that consists of three key steps: Shake, Local Search, and NeighborhoodChange [4]. The Local Search is an intensification strategy that explores neighborhoods looking for better solutions. The Shake is a diversification strategy commonly used to escape from local optima. Finally, the NeighborhoodChange is a mechanism to explore different solution spaces. Among the different variants of VNS, Basic Variable Neighborhood Search (BVNS) involves a Local Search and a Shake procedure, Variable Neighborhood Descent (VND) entails the iteration of different Local Search procedures, and General Variable Neighborhood Search (GVNS) is a modification of BVNS where the Local Search is substituted with a VND. See [4,5] for a detailed description of the methods.

In this proposal, we set the NeighborhoodChange to be a Sequential NeighborhoodChange [5]. Therefore, our focus turns to exploring variations in the

Shake and Local Search phases. Specifically, we present different strategies for three VNS variants: BVNS, VND, and GVNS.

We propose the definition of two neighborhoods based on two classic moves in the context of combinatorial optimization problems. The first one is the swap movement. This operation involves two vertices, u, and v, with labels l_1 and l_2 respectively. The labels are exchanged, resulting in a new labeling ϕ' such that $\phi'(u) = l_2$ and $\phi'(v) = l_1$. The second movement studied is the insert move, which can be viewed as a series of consecutive swap moves to the closest label in the solution. Given a vertex u and a label l such that $\phi(u) \neq l$, the vertex u is exchanged with the vertex $\phi(u) + 1$ if $l > \phi(u)$ or $\phi(u) - 1$ if $\phi(u) > l$, for a number of iterations until the label l is assigned to u obtaining a new labeling.

To explore these two neighborhoods, we use a Local Search procedure. A Local Search can follow a first- or best-improvement strategy. In this proposal, the swap neighborhood is explored following a first improvement strategy, i.e., the first exchange that improves the current solution, is performed. On the other hand, the insert Local Search, uses the best insertion for the first vertex that enhances the current solution quality. This method follows a first-improvement strategy for the selected vertex and a best-improvement strategy for determining the position.

Moving to the Shake step, we propose three strategies. All of them are influenced by a diversification parameter k that determines the number of labels that are affected by the shake procedure. The first variant, named **ShuffleShake**, consists of randomly shuffling the label of a subset of vertices. The subset of affected vertices is formed by those whose label is in the range $[1, k]$. The second variant, denoted as **NeighborhoodShake**, executes k random movements from both the neighborhood swap and the insertion neighborhood. If k is odd, a move is chosen from the swap neighborhood, otherwise from the insert neighborhood. The third variant, named **InverseShake**, consists of selecting a subset of k vertices from the range $[1, k]$ and assigning the highest labels to the vertices with the lowest initial labels, and vice versa. For example, given a value $k = 4$ and the example depicted in Fig. 1, this strategy would result in the labeling $\phi''(A) = 3$, $\phi''(B) = 2$, $\phi''(C) = 4$, $\phi''(D) = 1$, and $\phi''(E) = 5$.

3   Preliminar Experimentation

In this section, we conduct a comprehensive experimentation to identify the most effective combination of Shake and Local Search for each VNS variant in the context of S-labeling. We establish a time limit of 300 s as the stopping criterion for each method. New iterations are initiated continuously starting from a random initial solution, which is then improved with a VNS variant, until the time limit is reached. We test our proposal over a subset of 20 instances, drawn from the state-of-the-art paper for the problem [6]. The algorithms were coded in Java 17 and all experiments were conducted on a system equipped with an Intel Xeon Gold 6226R processor running at 2.90 GHz and supported by 119 GB of RAM. We used several metrics to evaluate the performance, including the average objective function (O.F.), the computation time (CPU.T.) measured in seconds, the deviation (Dev.) to the optimal/best-known solutions, and the percentage of best solutions discovered.

In Table 1 we introduce the different VNS configurations tested, obtained as the result of combining the previously proposed Shake and Local Search (LS) methods presented in the previous section.

Table 1. Configurations of VNS tested.

VNS Variant	Shake	LS
$BVNS_1$	InverseShake	Insert
$BVNS_2$	NeighborhoodShake	
$BVNS_3$	ShuffleShake	
$BVNS_4$	InverseShake	Swap
$BVNS_5$	NeighborhoodShake	
$BVNS_6$	ShuffleShake	
VND_1	–	Insert+Swap
VND_2	–	Swap+Insert
$GVNS_1$	InverseShake	VND_1
$GVNS_2$	NeighborhoodShake	
$GVNS_3$	ShuffleShake	
$GVNS_4$	InverseShake	VND_2
$GVNS_5$	NeighborhoodShake	
$GVNS_6$	ShuffleShake	

In Table 2 we report the results of testing the BVNS variants. Generally, using a swap-based Local Search leads to superior results. This could be attributed to the more aggressive nature of swap operations, enabling substantial changes to the solution within a shorter timeframe. In terms of Shakes, it is evident that BVNS variants that use NeighborhoodShake produce inferior solutions compared to other variants. On the contrary, ShuffleShake gives slightly improved results. Consequently, $BVNS_6$ emerges as the most effective variant in this experiment.

Table 2. Comparision of the BVNS variants.

	$BVNS_1$	$BVNS_2$	$BVNS_3$	$BVNS_4$	$BVNS_5$	$BVNS_6$
O.F.	1539906.25	1687174.65	1548403.20	1489568.65	1576198.70	**1489481.00**
CPU.T. (s)	300.00	300.00	300.00	300.00	300.00	300.00
Dev. (%)	1.77	10.58	1.95	**0.04**	4.48	**0.04**
% Best	0	0	0	45	0	**55**

Next, we examine the VND variants. As shown in Table 3, VND_1, which begins the exploration with the insertion neighborhood, consistently produces superior solutions in all the instances. Based on these observations, we conclude that VND_1 is the most effective variant among those studied.

Table 3. Comparison of the VND variants.

	VND_1	VND_2
O.F.	**1606570.90**	1629959.80
CPU.T. (s)	303.30	301.20
Dev. (%)	**0.00**	1.15
% Best	**100**	0

Finally, we perform an evaluation of the GVNS variants. The results are shown in Table 4. As we can observe, $GVNS_2$ and $GVNS_5$ consistently have a higher deviation than the others. $GVNS_6$ demonstrates a better result in the percentage of best solutions found, deviation, and the average objective function. This remarkable performance positions $GVNS_6$ as the most effective variant among the methods studied.

Table 4. Comparison of the GVNS variants.

	$GVNS_1$	$GVNS_2$	$GVNS_3$	$GVNS_4$	$GVNS_5$	$GVNS_6$
O.F.	1505096.60	1524217.00	1496944.45	1508176.85	1531479.30	**1488297.65**
CPU.T. (s)	300.05	300.25	300.15	300.00	300.00	300.00
Dev. (%)	0.67	1.57	0.25	0.74	1.93	**0.02**
% Best	10	0	15	0	0	**75**

4 Final Results

In our study, we found that although the state-of-the-art method outperformed others in the comparison, the performance of our proposed VNS variants was remarkably close. A statistical analysis using the Wilcoxon test was conducted to compare the VNS variants against the state-of-the-art method. At a significance level of 0.05, we obtained p-values of 0.0001, 0.00008, and 0.00012 for $BVNS_6$, VND_1 and $GVNS_6$, respectively. These p-values indicate statistically significant differences between the performances of the VNS variants and the state-of-the-art method.

Furthermore, within the VNS framework, VND_1 was the least effective, with $BVNS_6$ and $GVNS_6$ emerging as stronger alternatives. This observation leads us to propose that the development of a more sophisticated BVNS or GVNS algorithm could be beneficial. Specifically, innovative approaches to constructing and perturbing the initial solution may prove to be more critical than the mere presence of multiple neighborhood structures for diversification (Table 5).

Table 5. Comparison of the best VNS variants with the state-of-the-art results in [6].

	PIG [6]	BVNS 6	VND 1	GVNS 6
O.F.	**1477107.65**	1489481.00	1606570.90	1488297.65
CPU.T.(s)	**200.00**	300.00	303.30	300.00
Dev. (%)	**0.00**	0.72	6.17	0.60
% Best	**95**	5	0	0

5   Conclusions

In this research, we present 14 different variants of VNS tailored for the S-labeling problem. Each variant starts from a random solution and incorporates a unique combination of Shake and Neighborhood exploration in a VNS configuration. Among the evaluated variants, the GVNS approach emerged as the most effective one. The solutions obtained by our BVNS and GVNS variants resulted in a deviation under 1% with respect to those obtained with the best state-of-the-art method. Therefore, a proposal based on VNS could be competitive with further improvements.

To improve the effectiveness of VNS for S-labeling, several promising avenues for future research are proposed. First, exploring novel neighborhoods, especially for the GVNS variant, could lead to significant improvements. Second, the introduction of memory-based mechanisms, could further improve the performance of the algorithm. Finally, the investigation of alternative construction methods might be crucial for further progress.

References

1. Cavero, S., Pardo, E.G., Duarte, A.: A general variable neighborhood search for the cyclic antibandwidth problem. Comput. Optim. Appl. **81**(2), 657–687 (2022)
2. Díaz, J., Petit, J., Serna, M.: A survey of graph layout problems. ACM Comput. Surv. (CSUR) **34**(3), 313–356 (2002)
3. Fertin, G., Rusu, I., Vialette, S.: The S-labeling problem: an algorithmic tour. Discret. Appl. Math. **246**, 49–61 (2018)
4. Hansen, P., Mladenović, N., Brimberg, J., Pérez, J.A.M.: Variable Neighborhood Search. Springer, Cham (2019)
5. Hansen, P., Mladenović, N., Todosijević, R., Hanafi, S.: Variable neighborhood search: basics and variants. EURO J. Comput. Optim. **5**(3), 423–454 (2017)
6. Lozano, M., Rodriguez-Tello, E.: Population-based iterated greedy algorithm for the S-labeling problem. Comput. Oper. Res. **155**, 106224 (2023)
7. Sinnl, M.: Algorithmic expedients for the S-labeling problem. Comput. Oper. Res. **108**, 201–212 (2019)
8. Vialette, S.: Packing of (0, 1)-matrices. RAIRO-Theor. Inf. Appl. **40**(4), 519–535 (2006)

Improving Biased Random Key Genetic Algorithm with Variable Neighborhood Search for the Weighted Total Domination Problem

Alejandra Casado[1] , Jesús Sánchez-Oro[1](✉) , Anna Martínez-Gavara[2] ,
and Abraham Duarte[1]

[1] Universidad Rey Juan Carlos, 28933 Móstoles, Madrid, Spain
{alejandra.casado,jesus.sanchezoro,abraham.duarte}@urjc.es
[2] Universitat de València, 46100 Burjassot, Valencia, Spain
gavara@uv.es
https://grafo.etsii.urjc.es/en/

Abstract. The Weighted Total Domination Problem (WTDP) belongs to the family of dominating set problems. Given a weighted graph, the WTDP consists in selecting a total domination set D such that the sum of vertices and edges weights of the subgraph induced by D plus, for each vertex not in D, the minimum weight of its edge to a vertex in D is minimized. A total domination set D is a subset of vertices such that every vertex, is at least adjacent to one vertex in D. This problem arises in many real-life applications closely related to covering and independent set problems, however it remains computationally challenging due to its $\mathcal{NP}$-hardness. This work presents a Variable Neighborhood Search procedure to tackle the WTDP. In addition, we develop a Biased Greedy Randomized Adaptive Search Procedure that keeps adding elements once a feasible solution is found in order to produce high-quality initial solutions. We perform extensive numerical analysis to look into the influence of the algorithmic components and to disclose the contribution of the elements and strategies of our method. Finally, the empirical analysis shows that our proposal outperforms the state-of-art results, supported by an statistical analysis.

Keywords: Weighted Total Domination Problem · Graph Domination · Metaheuristics · Variable Neighborhood Search

1 Introduction

The Weighted Total Domination Problem belongs to the family of graph domination problems. Domination problems aims to select a subset of nodes from a

---

This work has been partially supported by the "Ministerio de Ciencia e Innovación" under grant ref. PID2021-125709OA-C22 and PID2021-126605NB-I00.

given input graph in such a way that all the nodes in the graph are dominated, where the domination criterion usually depends on the problem variant.

Given a weighted and undirected graph $G = (V, E)$, where V is the set of vertices and E is the set of edges, let us define the neighborhood $N(v)$ of a vertex v as the adjacent vertices to v, i.e., $N(v) = \{u \in V : (u, v) \in E\}$. Similarly, the closed neighborhood of a vertex v is defined as $N[v] = N(v) \cup \{v\}$. As a weighted graph, each vertex and edge have an associated weight, which is represented by $\omega(v)$ and $\omega(u, v)$, respectively.

A dominating set S over a graph $G = (V, E)$ is a subset $S \subseteq V$ where every vertex $v \in \{V \setminus S\}$ is adjacent to a vertex in S. In WTDP, the word *total* indicates that not only vertices in $\{V \setminus S\}$ must be adjacent to a vertex in S, but all the vertices in V must be adjacent to at least one vertex in S. Then, the WTDP consists of selecting a total dominating set which minimizes the following objective function:

$$f(S) = \sum_{s \in S} w(s) + \sum_{e \in E(S)} w(e) + \sum_{v \in V \setminus S} \min\{w(s, v) : u \in N(v) \cap S\}$$

where $E(S)$ represents the set of edges in which both endpoints belongs to S.

WTDP has not been widely studied in the literature. The problem was originally proposed in [4], together with three integer linear programming formulations which are able to solve instances up to 50 vertices within a 1800 s time horizon. Then, [1] proposed two new mixed-integer programming models, as well as a genetic algorithm and a greedy randomized adaptive search procedure, solving instances up to 125 vertices. It is interesting to propose new methods for dealing with more complex instances which are closer to real-life scenarios.

2   Variable Neighborhood Search

In this research, an algorithm based on Variable Neighborhood Search (VNS) is proposed for providing high-quality solutions for the WTDP. VNS is a metaheuristic [5] designed for considering neighborhood changes for avoiding getting stuck in local optima.

Among the wide variety of VNS schemes, this work is focused on Basic VNS (BVNS), which balances diversification and intensification during the neighborhood changes. The scheme of BVNS is presented in Algorithm 1.

BVNS requires from two input parameters: $k_{\max}$, the maximum neighborhood to be explored, and S, the initial solution,. The former is an input parameter that must be configured, while the latter is the solution used as starting point of the search. Although VNS methodology suggests that a random initial solution lead to high quality solutions, it has been experimentally tested that a good starting point usually reduces the computational effort of the complete algorithm [8,9]. The proposed method for generating initial solutions is depicted in Sect. 3.

Algorithm 1. VNS($k_{\max}, S$)

1: $k \leftarrow 1$
2: **while** $k \leq k_{\max}$ **do**
3:      $S' \leftarrow Shake(S, k)$
4:      $S'' \leftarrow Improve(S')$
5:      $k \leftarrow NeighborhoodChange(S, S'', k)$
6: **end while return** S

The method starts in the first neighborhood (step 1), and iterates until reaching the maximum predefined neighborhood $k_{\max}$ (steps 2–6). In each iteration, three phases are performed: *Shake* (step 3), responsible for diversification; *Improve* (step 4), focused on intensification; and *NeighborhoodChange* (step 5), which selects the next neighborhood to explore.

The shake procedure perturbs the incumbent solution S by randomly selecting a solution in the neighborhood under exploration. In the context of WTDP, this solution is selected by removing k vertices from S and then incorporating new ones until it becomes feasible again. In order to favor diversity, both the removed and added nodes are selected at random.

The resulting solution S' is not a local optimum due to the random selection of the vertices involved in the process. Therefore, a local improvement method is applied, based on the exchange move operator. In particular, this move operator consists of removing one or more nodes from the incumbent solution, replacing them with new ones. In the context of WTDP, a single exchange is considered, which removes a vertex and replace it with another one. Then, the local improvement explores the neighborhood conformed by all the solutions that can be reached by a single exchange move. This neighborhood is explored following a first improvement strategy to reduce the computational effort, which has been successfully applied in graph domination problems [2]. Only feasible moves are accepted, so in order to try a movement, it is necessary that all the vertices that are dominated only by the removed node are also dominated by the newly included one. Including this constraint in the local search procedure drastically reduces the number of evaluations of the objective function, thus reducing the computational effort of the procedure. The method stops when no improvement is found when exploring the neighborhood of the incumbent solution.

3   Biased Greedy Randomized Adaptive Search Procedure

Biased Greedy Randomized Adaptive Search Procedure (Biased GRASP) differs from traditional GRASP scheme in the function for selecting the next vertex to be included in the solution. Traditional GRASP schemes randomly select the vertex from a restricted candidate list conformed with the most promising vertices which have not been selected yet. Instead of this random function, Biased GRASP consider other probability functions that incorporate bias in the construction of the solutions [7], which will eventually lead to increase the quality of the constructed solutions.

In this research, the biased randomization of GRASP proposed in [3] is considered, which used an empirical non-uniform probability distribution. The original biased randomization are memory-less techniques which do not consider information from previous iterations. The proposed constructive procedure adds a frequency memory function for leveraging the information generated in previous iterations [6]. The associated greedy function for selecting the next vertex is then evaluated as a combination of quality (Q) and diversity (D), weighted by a $\beta \in [0,1]$ parameter which indicates the relevance of each component:

$$g(u) = \beta \cdot D + (1 - \beta) \cdot Q$$

In the context of WTDP, the quality metric is evaluated as the first part of the objective function evaluation, i.e., $\sum_{s \in S} w(s) + \sum_{e \in E(S)} w(e)$. The second part of the objective function evaluation is not considered since it is the most computationally demanding part.

The diversity metric is based on evaluating the number of solutions in which the node has not appeared, prioritizing the selection of nodes which have not been included in the solutions for a large number of iterations.

The selection of the next candidate differs from the traditional GRASP in the set of available candidates. In Biased GRASP, the restricted candidate list is not considered, since the method selects the next vertex using a discrete non-uniform distribution based on the probabilities evaluated for each element in the candidate list. The roulette selection method is implemented to increase the efficiency of the procedure, using the accumulated probabilities to randomly select the next vertex.

Since the randomness increases diversity, it is interesting to generate more than a single solution to leverage the potential of the Biased GRASP constructive procedure. As it is customary in the literature, 100 independent constructions are considered, using the best one as initial solution for BVNS.

4   Experiments and Results

The proposed algorithm has been developed using Java 17 and all the experiments have been performed in an AMD Ryzen 9 5950x (3.4 GHz) with 128 GB RAM. The dataset used is the one proposed in the best work found in the literature, conformed with 135 instances with vertices ranging from 100 to 125.

All the tables reports the following metrics: Avg., the average objective function value obtained by the algorithm; Time (s), the computing time in seconds required by the algorithm; Dev. (%), the average deviation with respect to the best solution found in the experiment; and # Best, times that the algorithm reaches the best solution in the experiment.

The experiments are designed to compare the proposed algorithm with the best method found in the state of the art [1]. In particular, it proposes four mathematical models which are able to optimally solve the smallest instances and a genetic algorithm named GA1 where the initial population is generated using a traditional GRASP algorithm.

The first experiment evaluates BVNS when comparing it with GA1 over the set of instances considered. Table 1 shows the performance of BVNS in this set of instances with the results grouped by number of nodes.

Table 1. Comparison of BVNS and GA1 over the set of large instances presented in the original research.

n	Algorithm	Avg.	Time (s)	Dev. (%)	#Best
75	BVNS	421.78	4.43	0.02	43
	GA1	423.87	9.91	0.34	39
100	BVNS	525.53	11.14	0.05	42
	GA1	526.51	23.13	0.23	36
125	BVNS	611.31	22.83	0.12	39
	GA1	611.80	47.76	0.20	37

The results show how BVNS is able to outperform GA1 in every group of instances, requiring half of the computing time on average. The deviation of BVNS is close to zero, so it is able to reach high-quality solutions in those cases in which it does not match the best one. In order to confirm that there are statistically significant differences between the results of both algorithms, we have conducted the well-known pairwise Wilcoxon statistical test, obtaining a p-value smaller than 0.01. This result indicates that the differences between BVNS and GA1 are statistically significant, emerging BVNS as a competitive algorithm for solving the WTDP

Finally, we have extended the set of instances with a set of more challenging ones conformed with graphs where nodes are ranging from 200 to 500. The results obtained with both BVNS and GA1 are depicted in Table 2.

Table 2. Comparison of BVNS and GA1 in the set of more challenging instances.

n	Algorithm	Avg.	Time (s)	Dev. (%)	#Best
200	BVNS	871.40	100.73	0.00	45
	GA1	937.08	962.77	6.72	1
350	BVNS	1332.04	616.34	0.00	45
	GA1	1794.06	1824.58	29.79	0
500	BVNS	1776.40	1164.68	0.00	45
	GA1	2539.57	1900.48	37.56	0

The results show that the proposed algorithm maintains the trend of requiring half of the computing time than GA1, probing the scalability of BVNS. In terms of quality, BVNS continues obtaining consistently better results than

GA1. In this case, GA1 is not able to reach any best solution when considering instances with 350 and 500 nodes. Therefore, these size of instances appears to be the limit of the previous proposal.

5   Conclusions and Future Work

This research presents a BVNS algorithm for solving the Weighted Total Domination Problem. The initial solution for BVNS is constructed using Biased GRASP, a combination which is not common in the associated literature, resulting in an effective but efficient algorithm for solving this hard combinatorial optimization problem.

The algorithm is tested over a set of instances where the optimal value is known, showing its efficacy and, then, over a set of more challenging instances comparing it with the best proposal found in the literature. Again, BVNS is able to reach the best solutions requiring half of the computing time, emerging as a competitive method for solving the WTDP.

References

1. Álvarez-Miranda, E., Sinnl, M.: Exact and heuristic algorithms for the weighted total domination problem. Comput. Oper. Res. **127**, 105157 (2021)
2. Casado, A., et al.: An iterated greedy algorithm for finding the minimum dominating set in graphs. Math. Comput. Simul. **207**, 41–58 (2023)
3. Ferone, D., et al.: Enhancing and extending the classical GRASP framework with biased randomisation and simulation. J. Oper. Res. Soc. **70**(8), 1362–1375 (2019)
4. Ma, Y., Cai, Q., Yao, S.: Integer linear programming models for the weighted total domination problem. Appl. Math. Comput. **358**, 146–150 (2019)
5. Mladenović, N., Hansen, P.: Variable neighborhood search. Comput. Oper. Res. **24**(11), 1097–1100 (1997)
6. Napoletano, A., et al.: Heuristics for the constrained incremental graph drawing problem. Eur. J. Oper. Res. **274**(2), 710–729 (2019)
7. Mauricio, G.C.R., Celso, C.R.: Optimization by GRASP - Greedy Randomized Adaptive Search Procedures. Springer Nature, Cham (2016)
8. Sánchez-Oro, J., Mladenović, N., Duarte, A.: General variable neighborhood search for computing graph separators. Optim. Lett. **11**, 1069–1089 (2017)
9. Sánchez-Oro, J., et al.: Variable neighborhood scatter search for the incremental graph drawing problem. Comput. Optim. Appl. **68**, 775–797 (2017)

Optimization of Fairness and Accuracy on Logistic Regression Models

Javier Yuste$^{(\boxtimes)}$ ⓘ, Eduardo G. Pardo ⓘ, and Abraham Duarte ⓘ

Universidad Rey Juan Carlos, Móstoles, 28933 Madrid, Spain
{javier.yuste,eduardo.pardo,abraham.duarte}@urjc.es

Abstract. Decision-making software is used to automatically make informed decisions by leveraging large amounts of data. Advances in machine learning have extended the implementation of these systems to processes that have a significant impact on the lives of people, such as credit scoring, employment applications, or insurance rates. Due to their impact, these systems must guarantee fairness from social and legal points of view, operating in a non-discriminatory manner. Several methods have been studied in the literature to improve the fairness of these systems, but often at the cost of accuracy. In this work, we propose two methods based on the Variable Neighborhood Search scheme to optimize the fairness of machine learning models after the training phase. In particular, we apply the proposed approaches to optimize Linear Regression models, which are frequently used in decision-making software. The proposed methods are competitive with a state-of-the-art Hill Climbing algorithm, using a set of publicly available instances.

Keywords: Search-Based Software Engineering · Fairness optimization · Variable Neighborhood Search

1 Introduction

Decision-making software is increasingly being used to automatically make decisions in systems that have a significant impact on the lives of people [6]. Due to the importance of such decisions, these systems must guarantee fairness from both a social and a legal point of view [10]. That is, these systems must operate in a non-discriminatory manner [6]. Discrimination is defined as "treating a person or particular group of people differently, especially in a worse way from the way in which other people are treated, because of their race, gender, sexuality, etc." [1]. Due to the importance of software fairness in decision-making software systems, several researchers have investigated this issue [3,4,8].

In recent times, advances made in the area of Machine Learning (ML) have extended the use of ML models for decision-making systems. In this context, perhaps the simplest approach to mitigate biased decisions is the removal of sensitive attributes from the training data. However, this method is not effective in mitigating bias, due to indirect relations between different attributes [9]. This

M. Sevaux et al. (Eds.): MIC 2024, LNCS 14753, pp. 383–389, 2024.
https://doi.org/10.1007/978-3-031-62912-9_37

J. Yuste et al.

phenomenon is known as the red-lining effect [5] or indirect discrimination [13]. As an alternative, other more sophisticated methods have been proposed to apply bias mitigation at different stages: before training ML models (pre-processing), during the training process (in-processing), or after the training phase (post-processing) [8]. Regardless of the stage at which bias mitigation methods are applied, most of the proposed approaches reduce bias at the cost of accuracy, a trade-off known as the price of fairness [4].

In this work, we propose an approach based on the Variable Neighborhood Search (VNS) scheme to optimize the fairness of ML models after the training phase. In particular, we apply the proposed approach to Linear Regression models. These models are widely used in decision-making processes due to their explainability, which is critical in these scenarios. We favorably compare the proposed approach with a recent method based on Hill Climbing [8] over six different instances that are publicly available [3].

2 Problem Definition

Linear Regression (LR) is a statistical model used for classification that estimates the probability that an event occurs on the basis of a set of variables. An LR model is made up of n variables or coefficients, represented by a vector of dimension n $(b_0, b_1, \ldots, b_{n-1})$. In Eq. (1), we represent the computation of the prediction of an LR model with five variables that receives four values as input (x_1, x_2, x_3, x_4).

$$Linear(x_1, x_2, x_3, x_4) = b_0 + b_1 \cdot x_1 + b_2 \cdot x_2 + b_3 \cdot x_3 + b_4 \cdot x_4 \qquad (1)$$

The result obtained from Eq. (1), which will be denoted as Y, is then used to make a prediction using a sigmoid function, represented in Eq. (2). In a binary classification context, predictions greater than or equal to 0.5 are labeled 1, while predictions less than 0.5 are labeled 0.

$$P(Y) = \frac{1}{1 + e^{-Y}} \qquad (2)$$

To measure the performance of the model, we use two metrics: Accuracy and Statistical Parity Difference (SPD). Accuracy is a metric that evaluates the frequency with which a model makes a correct classification. It is calculated as the number of correct predictions divided by the total number of predictions. The higher its value, the better the model. Therefore, the accuracy should be maximized. SPD analyzes the independence of protected attributes from predictions. That is, the fairness of the model. The calculation of this metric is shown in Eq. (3), where $P(Y)$ represents the prediction of the LR model (see Eq. (2)) and D denotes a privileged or unprivileged group. This metric computes the difference between the rate of favorable outcomes received by the unprivileged group and the favorable outcomes received by the privileged group. The best value for this metric is zero, where there is no bias. The higher the absolute value of the metric, the greater the bias towards one of the groups (and the less fair the model). Therefore, the absolute value of SPD should be minimized.

$$\begin{aligned} \text{SPD} =& \text{Probability}(P(Y) \geq 0.5 \mid D = \text{unprivileged}) \\ &- \text{Probability}(P(Y) \geq 0.5 \mid D = \text{privileged}) \end{aligned} \quad (3)$$

3 Algorithmic Proposal

In this work, we propose two different methods based on the VNS scheme [11]. In particular, we propose an approach based on Basic Variable Neighborhood Search (BVNS) and an approach based on Variable Neighborhood Descent (VND). In the BVNS approach, three input parameters are received: an initial solution x, a maximum perturbation size k_{max}, and a maximum time t_{max}. The method tries to improve the solution iteratively until the maximum computational time t_{max} is reached. At each iteration, the solution x is first perturbed by a shake procedure and then improved by a local search. The size of the perturbation introduced by the shake procedure is guided by the value of a variable k, which is initially set to one. If the resulting solution x', obtained at the end of an iteration, is better than x, then x' is saved as the new best solution x and k is reset to one. Otherwise, k is incremented by one unit. If k is greater than k_{max}, then its value is reset to one. Once the maximum time has been reached, the method returns the best solution found during the search process.

In the VND approach, two input parameters are received: an initial solution x and a set of neighborhood structures N. The method tries to improve the solution iteratively by exploring the different neighborhood structures in N until the current solution is a local optimum within all the neighborhoods. First, a variable l is set to one. At each iteration, the solution x is improved by performing a local search within the neighborhood structure N_l. If the resulting solution x' is better than x, then x' is saved as the new best solution x and l is reset to one. Otherwise, l is incremented by one unit. When all neighborhood structures have been explored without finding a better solution than the current best solution x ($l \geq |N|$), then the method returns the best solution found during the search.

For the aforementioned methods, we use three neighborhood structures previously proposed by Hort et al. [8]. The first neighborhood structure, named Reduction, consists of multiplying a coefficient of the LR model by a random value within the range [-0.1, 0.1]. The second neighborhood structure, named Adjustment, consists of multiplying a coefficient of the LR model by a random value within the range [0.9, 1.1]. The third neighborhood structure, named Vector, consists of multiplying all coefficients of the LR model by a random value within the range [0.9, 1.1]. For the BVNS method, the Reduction neighborhood structure is used both as the shake procedure and within the local search. For the VND method, all neighborhood structures are explored. In particular, they are explored in the following order: Reduction, Adjustment, and Vector. All local search procedures follow a first improvement approach. For the BVNS method, the parameters t_{max} and k_{max} have been experimentally set to 10.

4 Experimental Results

For the experiments, we used four publicly available real-world datasets previously used [8]: Adult Census Income (adult) [2], Bank Marketing (bank) [12], Correctional Offender Management Profiling for Alternative Sanctions (COMPAS) [14], and Medical Expenditure Panel Survey (MEPS19) [7]. In total, these datasets contain six protected attributes. We compared the performance of the proposed methods with a Hill-Climbing (HC) approach recently proposed by Hort et al. [8]. First, we split the data into three sets (training, validation, and test) as described in [8] and train the LR model on the training data. The trained model is then optimized using the methods under comparison. To account for the stochastic behavior of the optimization methods under comparison, we run them 30 times for each instance. Importantly, the same initial trained model is used for every method. Therefore, the differences obtained are only due to the optimization process carried out by each method, not to the training process of the LR model. All experiments have been performed in the same computing environment, an Ubuntu 20.04.1 LTS with an AMD EPYC 7643 CPU with 16 cores and 32 GB RAM, using Python 3.7.17. In the case of the HC method mentioned above, we use the original implementation crafted by the authors [8].

In Table 1, we present the results obtained. For each method and instance of the comparison, we report the average improvement in accuracy ($\overline{\Delta\text{ Accuracy}}$) and the average reduction in SPD ($\overline{\nabla\text{ SPD}}$), which represents the fairness of the resulting model. For each metric, we highlight the best result in bold font with gray background and the second-best result in bold font. As it can be observed, the BVNS and VND approaches are able to simultaneously improve both accuracy and fairness in all cases. Regarding the accuracy metric, both the BVNS and VND methods achieve better or at least equal values than HC in all cases. With respect to SPD, the VND method is able to obtain better results for all instances. In Table 2, we present the average CPU time consumed ($\overline{\text{CPUt (s)}}$). Again, we highlight the best result in bold font with gray background and the second-best result in bold font. As it can be observed, the VND method is the fastest in four out of six cases, while the HC approach is the fastest in the other two cases.

Table 1. Comparison of the results obtained with an HC [8] method, a BVNS approach, and a VND procedure.

Dataset	Attribute	Δ Accuracy			∇ SPD		
		HC [8]	BVNS	VND	HC [8]	BVNS	VND
adult	race	0.0010	0.0010	0.0012	0.0073	0.0073	0.0081
	sex	0.0012	0.0009	0.0016	0.0027	0.0022	0.0054
bank	age	0.0015	0.0044	0.0037	0.0038	0.0146	0.0153
COMPAS	race	0.0141	0.0141	0.0141	0.0256	0.0256	0.0256
	sex	0.0151	0.0130	0.0357	0.0350	0.0431	0.2071
MEPS19	race	0.0098	0.0124	0.0103	0.0276	0.0323	0.0357

Table 2. Comparison of the computational time consumed by an HC [8] method, a BVNS approach, and a VND procedure.

Dataset	Attribute	CPUt (s)		
		HC [8]	BVNS	VND
adult	race	10.23	11.56	5.69
	sex	10.46	10.19	7.26
bank	age	7.18	14.37	1.95
COMPAS	race	4.53	19.76	9.29
	sex	4.57	17.43	21.95
MEPS19	race	5.41	13.12	3.58

5 Conclusions

In this paper, two different methods, based on the BVNS and VND schemes, have been proposed to optimize the fairness (in terms of SPD) and accuracy of LR models. In both cases, the same neighborhood structures proposed in a recent state-of-the-art HC method have been explored. As it has been shown in the experimental results, the BVNS and VND methods are able to obtain better overall results than the state-of-the-art HC algorithm, thanks to the diversification introduced by a shake procedure in the case of BVNS and the systematic exploration of different neighborhood structures in the case of VND.

The problem described in this work needs to be further explored to find better optimization strategies. First, additional neighborhood structures must be studied that better explore the continuous values of the decision variables of the problem. Additionally, we believe that this problem should be tackled in a multi-objective approach by using dominance criteria to build non-dominated sets of solutions. Following this line of research, the problem could then easily include additional quality metrics to measure the solutions obtained in terms of both accuracy and fairness.

Acknowledgments. This research has been partially supported by grants PID2021-125709OA-C22 and PID2021-126605NB-I00, funded by MCIN/AEI/10.13039/501100011033 and by "ERDF A way of making Europe"; grant CIAICO/2021/224 funded by Generalitat Valenciana; grant M2988 funded by "Proyectos Impulso de la Universidad Rey Juan Carlos 2022"; and "Cátedra de Innovación y Digitalización Empresarial entre Universidad Rey Juan Carlos y Second Episode" (Ref. ID MCA06).

References

1. "Discrimination". In: Cambridge Dictionary. Cambridge Dictionary (2024). https://dictionary.cambridge.org/dictionary/english/discrimination. Accessed 29 Jan 2024
2. Becker, B., Kohavi, R.: Adult. UCI machine learning repository (1996). https://doi.org/10.24432/C5XW20
3. Bellamy, R.K., et al.: AI fairness 360: an extensible toolkit for detecting, understanding, and mitigating unwanted algorithmic bias. arXiv preprint: arXiv:1810.01943 (2018)
4. Berk, R., et al.: A convex framework for fair regression. arXiv preprint: arXiv:1706.02409 (2017)
5. Calders, T., Verwer, S.: Three Naive Bayes approaches for discrimination-free classification. Data Min. Knowl. Disc. **21**, 277–292 (2010)
6. Friedler, S.A., Scheidegger, C., Venkatasubramanian, S., Choudhary, S., Hamilton, E.P., Roth, D.: A comparative study of fairness-enhancing interventions in machine learning. In: Proceedings of the Conference on Fairness, Accountability, and Transparency, pp. 329–338 (2019)
7. for Healthcare Research, A., Quality: medical expenditure panel survey (2018). https://meps.ahrq.gov/mepsweb/. Accessed 29 Jan 2024
8. Hort, M., Zhang, J.M., Sarro, F., Harman, M.: Search-based automatic repair for fairness and accuracy in decision-making software. Empir. Softw. Eng. **29**(1), 36 (2024)
9. Kamiran, F., Calders, T.: Classifying without discriminating. In: 2009 2nd International Conference on Computer, Control and Communication, pp. 1–6. IEEE (2009)
10. Kamishima, T., Akaho, S., Asoh, H., Sakuma, J.: Fairness-aware classifier with prejudice remover regularizer. In: Flach, P.A., De Bie, T., Cristianini, N. (eds.) Machine Learning and Knowledge Discovery in Databases. Lecture Notes in Computer Science(), vol. 7524, pp. 35–50. Springer, Berlin (2012). https://doi.org/10.1007/978-3-642-33486-3_3
11. Mladenović, N., Hansen, P.: Variable neighborhood search. Comput. Oper. Res. **24**(11), 1097–1100 (1997)
12. Moro, S., Cortez, P., Rita, P.: A data-driven approach to predict the success of bank telemarketing. Decis. Support Syst. **62**, 22–31 (2014)

13. Pedreshi, D., Ruggieri, S., Turini, F.: Discrimination-aware data mining. In: Proceedings of the 14th ACM SIGKDD International Conference on Knowledge Discovery and Data Mining, pp. 560–568 (2008)
14. Propublica: Data and analysis for 'Machine Bias' (2023). https://github.com/propublica/compas-analysis/. Accessed 29 Jan 2024

A Variable Formulation Search Approach for Three Graph Layout Problems

Sergio Cavero$^{(\boxtimes)}$ (ID), J. Manuel Colmenar (ID), and Eduardo G. Pardo (ID)

Universidad Rey Juan Carlos, C/Tulipán S/N, Móstoles, 28922 Madrid, Spain
{sergio.cavero,josemanuel.colmenar,eduardo.pardo}@urjc.es

Abstract. This paper studies the relationship between three linear layout problems: minimum linear arrangement, cutwidth minimization, and bandwidth minimization. Our research suggests that, given their correlation, optimizing one problem could optimize the others. The Variable Neighborhood Search metaheuristic can take advantage of this, especially by switching problem formulations during the search process. The paper presents experiments analyzing different strategies and provides insights about their effectiveness. Our findings indicate that the proposed variant of Variable Neighborhood Search outperforms traditional single-process optimization methods in terms of both solution quality and computational efficiency.

Keywords: graph layout problems · bandwidth · cutwidth · variable neighborhood search

1 Introduction

This paper studies graph layout problems (GLP), a family of combinatorial optimization problems that aim to find an optimal arrangement of graph vertices in a metric space, typically represented by a host graph [2]. The goal of GLP is to optimize a particular function that depends on the properties of the graph, such as edge length, edge crossing, or other graph-related metrics that may reflect real-world applications. In particular, the most studied functions in the literature are the bandwidth and cutwidth functions. On one hand, the bandwidth measures the distance of an edge, i.e., the separation of two adjacent vertices in a host graph. The bandwidth is usually the maximum distance between any two vertices in a host graph. On the other hand, the cutwidth measures the maximum number of edges traversing a limited area or region of space and tries to minimize it.

In the area of GLP, regardless of the objective function being studied, researchers have traditionally focused on graph embedding problems in host

This research has been supported by the Grant Refs.: PID2021-125709OA-C22/OB-C21 and PID2021-126605NB-I00, funded by MCIN/AEI/10.13039/501100011033 and by ERDF, a way of making Europe. CIAICO/2021/224, funded by Generalitat Valenciana. M2988 and MCA06, funded by Universidad Rey Juan Carlos.

M. Sevaux et al. (Eds.): MIC 2024, LNCS 14753, pp. 390–396, 2024.
https://doi.org/10.1007/978-3-031-62912-9_38

graphs that have a known or regular structure. Among others, the most common structures are paths, cycles, grids, tori, hypercubes [1,2]. Similarly, most of the papers in the literature focus on problems with a single objective function. This specialization allows researchers to delve into individual problem characteristics.

In this paper we explore the connections between the aforementioned functions to provide a fresh perspective on graph layout issues. The research goal is to investigate how solutions to different problems are related, and whether optimizing one objective improves the quality of the others. Based on this, we propose an algorithm based on the Variable Neighborhood Search (VNS) methodology, to exploit these observations. Particularly, we focus on a subset of GLP known as Linear Layout Problems (LLP). LLP involve the embedding of graphs into a path host graph and, specifically, we concentrate on the Bandwidth Minimization Problem (BMP) [3], the Minimum Linear Arrangement Problem (MinLA) [5], which is also referred to as the Bandwidth Sum Minimization, and the Cutwidth Minimization Problem (CMP) [4].

2   Formal Description of Graph Layout Problems

Let $G = (V_G, E_G)$ represent the input graph, where V_G denotes the set of vertices and E_G denotes the set of edges. Similarly, let $H = (V_H, E_H)$ represent the host graph, where V_H denotes the set of vertices and E_H denotes the set of edges. Let (u, v) be an undirected edge that joins the vertices u and v. Finally, let $p(u, v)$ be the set of paths that connect the vertices (u, v) in E_G. s.t. $p(u, v) = (u, u_1), (u_1, u_2), \ldots, (u_{i-1}, u_i), (u_i, v)$.

Given the input graph G and the host graph H, an embedding of G in H is defined by the mathematical functions φ and ψ. The function φ assigns each vertex of the input graph to a vertex of the host graph and it is mathematically expressed in Eq. (1). The function ψ maps each edge of E_G to a set of paths in H whose endpoints are $\varphi(u)$ and $\varphi(v)$. For the considered problems, the function ψ assigns the path between $\varphi(u)$ and $\varphi(v)$ with the smallest possible cardinality, as it is presented in Eq. (2).

$$\varphi : V_G \to V_H, : \forall : u \in V_G : \exists ! : v \in V_H : | : \varphi(u) = v \tag{1}$$

$$\psi(\varphi, u, v) = \operatorname*{arg\,min}_{p(w,z) \in P} |p(\varphi(u), \varphi(v))|, \tag{2}$$

where the operator $| \cdot |$ computes the cardinality of a path.

With these definitions established, we introduce the formulation of the considered problems. The BMP and the MinLA involve the computation of the bandwidth function: $bw(\varphi, (u, v)) = |\psi(\varphi, u, v)|$, while the CMP computes the cutwidth function: $cut(\varphi, (w, z)) = |\{(u, v) \in E_G : (w, z) \in \psi(\varphi, u, v)\}|$.

Considering that Φ represents the entire set of feasible solutions to a problem. The formal definitions of the considered problems are stated as follows:

$$BMP = \arg\min_{\varphi \in \Phi} \max_{(u,v) \in E_G} bw(\varphi, (u, v)) \tag{3}$$

$$MinLA = \arg\min_{\varphi \in \Phi} \sum_{(u,v) \in E_G} bw(\varphi, (u, v)) \tag{4}$$

$$CMP = \arg\min_{\varphi \in \Phi} \max_{(w,z) \in E_H} cut(\varphi, (w, z)) \tag{5}$$

3   Algorithmic Proposal

This paper presents an algorithm based on a variant of VNS, named Variable Formulation Search (VFS). This variant exploits the similarities between the three studied problems by considering different problem formulations. It is typically used when multiple solutions of equal value exist in the solution space [4]. Initially, VFS was designed to optimize equivalent formulations, where optimal solutions are mutually optimal. However, in our methodology, the formulations may not be equivalent because they represent different but related problems.

Algorithm 1 presents the proposed VFS algorithm. It starts with an initial solution s and a set of i problem formulations $\mathcal{F}_1, \mathcal{F}_2, \ldots, \mathcal{F}_i$. The algorithm systematically explores the formulations, initializing a set of i solutions $s_1, s_2, \ldots, s_i$ with the current solution s (step 1). The outer loop continues until the predefined stopping criterion is met (step 2). Within each formulation, the inner loop (step 4) evaluates the improvement of the current solution obtained as the result of a local search procedure using the current formulation $\mathcal{F}_k$ (step 5). If a better solution s' is found, it replaces the current solution s and updates the solution s_k corresponding to that formulation (steps 6 to 8). Otherwise, the algorithm proceeds to the next formulation, incrementing k (step 9). This iterative process continues until the stopping condition is satisfied, eventually yielding a set of optimal solutions for each formulation (step 11). Using randomly generated solutions, this VFS metaheuristic adeptly navigates diverse problem formulations, dynamically adapting the optimization process and effectively traversing flat areas of the landscape.

4   Experimental Results

In this section, we examine the interconnections among the addressed problems, and subsequently compare our proposed VFS algorithm with the traditional single-process optimization. Our experiment uses a dataset of 40 representative graphs from the Harwell-Boeing collection (https://math.nist.gov/MatrixMarket/).

The first experiment used Pearson's coefficient to examine the correlation among the three problems (BMP, MinLA, and CMP) based on the quality of 100 random solutions for each instance. We found a significant positive correlation between the studied problems. Specifically, there was a correlation of 0.94

Algorithm 1: Variable Formulation Search (VFS)

Data: Initial solution s, problem formulations $\mathcal{F}_1, \mathcal{F}_2, \ldots, \mathcal{F}_i$

```
1  s₁, s₂, ..., sᵢ ← s;
2  while stopping criterion is not met do
3  │   k ← 1;
4  │   while k ≤ i do
5  │   │   s' ← LocalSearch(s, Fₖ);
6  │   │   if Fₖ(s') < Fₖ(sₖ) then
7  │   │   │   s ← s';
8  │   │   │   sₖ ← s';
9  │   │   else
10 │   │   │   k ← k + 1;

11 return s₁, s₂, ..., sₖ;
```

---

between BMP and MinLA, a correlation of 0.8 between BMP and CMP, and a correlation of 0.95 between CMP and MinLA.

In the second experiment, we implemented a local search algorithm using a swap neighborhood. Simultaneously, we evaluated the performance of the other two problems. Figure 1 shows the evolution of each problem and the quality of the others throughout the optimization process. The Y-axis represents the normalized average quality (ranging from 0 to 1) for all instances, while the X-axis represents the number of iterations, providing a perspective of the size of the solution space. The first insight from the experiment is that we can confirm the initial hypothesis: optimizing one of the three problems significantly improves the quality of the other two since the functions are all descendant. Another insight that can be gained from the figure is related to the number of moves performed in each local search. The optimization of BMP (Fig. 1a) required on average 9.15 moves to reach a local optimum. In contrast, the optimization of CMP (Fig. 1b) required an average of 1760.9 moves, and the MinLA (Fig. 1c) required 3889.47 moves. Therefore, it seems that BMP and CMP, whose objective functions are based on the maximum function, exhibit a flat landscape in the search space where many different solutions have the same objective function value. As a result, local search algorithms tend to get stuck [4]. On the other hand, MinLA's sensitivity to the objective function implies that minor variations in the solution can lead to different optimal solutions in the solution space.

Next, we ran the VFS algorithm for 99 iterations (33 optimizations per problem). Figure 2a shows the quality convergence, indicating the best solution found for each problem in each iteration, while Fig. 2b shows the solution quality across iterations for the three problems. The amalgamation of these methods is effective because it allows both diversification and, at the same time, intensification in a different, albeit related, direction. This allows for intelligent diversification since it may deteriorate momentarily in relation to a specific objective, but may lead to potential improvement in the future.

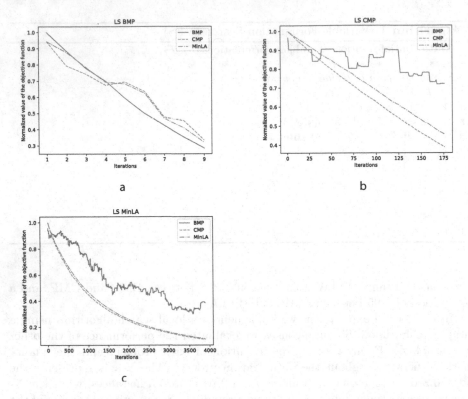

Fig. 1. Optimization of the 1a BMP, 1b CMP, and 1c MinLA, illustrating solution quality across the three considered problems

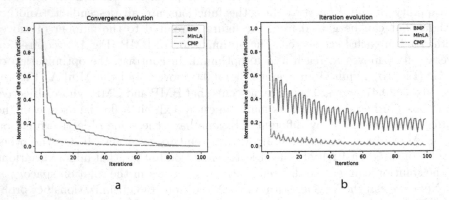

Fig. 2. 2a Objective function convergence; 2b evolution during the optimization of BMP, CMP, and MinLA through the VFS algorithm.

Table 1. Comparison of the results obtained by the different local search procedures, and the VFS, for the studied problems (BMP, MinLA, CMP).

		LS BMP	LS MinLA	LS CMP	VFS
BMP	**OF**	364.90	235.25	386.85	**190.50**
	Dev. (%)	135.33	31.46	142.00	**1304**
	#Best	0	17	0	**23**
MinLA	**OF**	256343.70	40108.75	166560.30	**38031.63**
	Dev. (%)	719.91	32.62	450.63	**16.49**
	#Best	0	**20**	1	19
CMP	**OF**	659.68	110.15	349.93	**108.90**
	Dev. (%)	651.11	26.98	319.95	**21.78**
	#Best	0	**23**	1	18
CPU Time (s)		110.71	1482.03	2210.57	**906.95**

In the last experiment, we compared each problem's optimization (LS BMP, LS MinLA, LS CMP) with VFS. The methods ran for 1 h or 99 iterations. Table 1 shows the objective function value (OF), deviation (Dev. (%)), the number of best solutions found from the experiment (#Best), and run time for each method and problem. On average, the VFS method achieves the best results in terms of solution quality. In addition, the VFS method is the fastest of the four methods, demonstrating its computational efficiency. It is worth highlighting the ability of LS MinLA to find quality solutions for other problems.

5   Conclusion

Our research focuses on three well-known linear layout problems: the bandwidth minimization, the minimum linear arrangement and the cutwidth minimization. We propose an algorithm based on the Variable Formulation Search methodology, motivated by the hypothesis that optimizing one problem can simultaneously optimize others. The algorithm alternates among different problem formulations to optimize solutions. The experimental results indicate that the Variable Formulation Search algorithm is effective in improving the quality of the solution for all three problems, including those with flat landscapes and constrained search spaces. Future research could extend these findings to other graph layout problems.

References

1. Cavero, S., Pardo, E.G., Duarte, A.: Efficient iterated greedy for the two-dimensional bandwidth minimization problem. Eur. J. Oper. Res. **306**(3), 1126–1139 (2023)
2. Díaz, J., Petit, J., Serna, M.: A survey of graph layout problems. ACM Comput. Surv. (CSUR) **34**(3), 313–356 (2002)

3. Lim, A., Lin, J., Rodrigues, B., Xiao, F.: Ant colony optimization with hill climbing for the bandwidth minimization problem. App. Soft Comp. **6**(2), 180–188 (2006)
4. Pardo, E.G., Mladenović, N., Pantrigo, J.J., Duarte, A.: Variable formulation search for the cutwidth minimization problem. App. Soft Comp. **13**(5), 2242–2252 (2013)
5. Petit, J.: Experiments on the minimum linear arrangement problem. J. Exp. Algorithmics (JEA) **8** (2003)

Author Index

Printed in the United States
by Baker & Taylor Publisher Services

Printed in the United States
by Baker & Taylor Publisher Services